WATER RESOURCE PLANNING, DEVELOPMENT AND MANAGEMENT

MODELING HYDROLOGIC EFFECTS OF MICROTOPOGRAPHIC FEATURES

XIXI WANG
EDITOR

Nova Science Publishers, Inc.
New York

LIBRARY OF CONGRESS CATALOGING-IN-PUBLICATION DATA

Modeling hydrologic effects of microtopographic features / editor, Xixi Wang.
p. cm.
Includes bibliographical references and index.
ISBN 978-1-61668-628-4 (hardcover)
1. Hydrogeology--Data processing. 2. Hydrogeology--Mathematics. 3. Groundwater flow--Computer simulation. 4. Geomorphology--Computer simulation. I. Wang, Xixi.
GB1001.72.E45M63 2010
551.48--dc22

2010044736

Published by Nova Science Publishers, Inc. † New York

CONTENTS

PREFACE

Microtopographic features, such as natural depressions, wetlands, and raised roads, are very important in regulating watershed energy and water fluxes. Taking these features into account, watershed simulation models can better mimic hydrologic processes. This new book discusses how to develop and use hydrologic models to quantify effects of microtopographic features on energy and water fluxes at watershed scale.

Chapter 1- Surface microtopography plays an important role in watershed hydrology. As one of the major factors, microtopography dominates the overland flow generation mechanism, changes the spatial and temporal variability of hydrologic processes (e.g., infiltration and surface runoff), determines the drainage pattern, and affects the fate and transport of NPS contaminants throughout surface and subsurface systems. Thus, characterization of surface microtopography is critical to watershed hydrologic and environmental modeling. Various methods and techniques have been developed for DEM-based watershed delineation, which greatly facilitate watershed modeling and management. This chapter introduces a new puddle delineation algorithm that explicitly accounts for the puddle to puddle (P2P) filling-merging-spilling overland flow process. The algorithm has been incorporated into a Windows-based P2P modeling system that can be used for characterization of surface microtopography from high-resolution DEMs, computation of depression storages, and visualization of the delineation process. This chapter also details the P2P process and presents some findings from experimental studies on the effects of surface microtopography on overland flow. The results highlight the hydrologic significance of microtopography.

Chapter 2- The construction of raised roads may modify the hydraulic connectivity of natural depression storages (e.g., wetlands) and create additional artificial storages. As a result, it would noticeably alter the natural hydrology of the inclusive watershed. Quantifying these storages is very important for watershed management that aims to increase water supply and mitigate flooding. This can be done by processing a digital elevation model (DEM) in a geographic information system (e.g., ArcGIS®) on the basis of the U.S. Geological Survey's 1.61-km by 1.61-km quadrangle sections. However, DEMs with a spatial resolution finer than 30 m are not available for most watersheds. The objective of this study was to examine whether and how the 30-m National Elevation Dataset can be adapted to estimate potential depression storage capacity (PDSC), the total volume below the lowest point (i.e., breach) of the roads surrounding a section. The study was conducted in the Forest River watershed located in northeastern North Dakota. The results indicate that the estimation of PDSC tended to have larger errors for sections with a higher topographic relief than for sections with a lower relief. In addition, the estimation was sensitive to errors in the breach elevation values. Nevertheless, the PDSC for a section within the study watershed can be estimated as the multiplication of the value computed using the NED data by an adjustment coefficient of 1.05 to 1.3. One reasonable generalization of this study is that NED can be adapted to estimate PDSCs for watersheds where high-resolution DEMs are not available.

Chapter 3- Hillslope is the basic unit of a watershed. Typical hillslopes may have a size of 1000 m long and 500 m wide. For watershed modeling, it is essential to accurately describe the hillslope-scale processes of flow, erosion and sediment transport, and solute transport. Although these processes are usually considered in experimental studies and theoretical subjects, the existing numerical models that are designed to simulate transport processes at hillslope scale rarely take microtopographic variations into account. Instead, those models assume constant slope, roughness, and infiltration rate for a given basic computational unit (i.e., hillslope). As a result, effects of microtopographic features (e.g., rills) on the aforementioned processes cannot be reflected in modeling results. However, the effects could be important because rill and sheet flows exhibit distinctly different dynamics that influence the transport processes. The objective of this chapter is to review the numerical studies for investigating the transport processes at hillslope scale. The chapter focuses particularly on the modeling efforts with the effects of microtopographic features on the dynamics of the transport processes incorporated.

Chapter 4- Specific yield and hydroperiod have proven to be useful parameters in hydrologic analysis of wetlands. Specific yield is a critical parameter to quantitatively relate hydrologic fluxes (e.g., rainfall, evapotranspiration, and runoff) and water level changes. Hydroperiod measures the temporal variability and frequency of land-surface inundation. Conventionally, hydrologic analyses used these concepts without considering the effects of land surface microtopography and assumed a smoothly-varying land surface. However, these microtopographic effects could result in small-scale variations in land surface inundation and water depth above or below the land surface, which in turn affect ecologic and hydrologic processes of wetlands. The objective of this chapter is to develop a physically-based approach for estimating specific yield and hydroperiod that enables the consideration of microtopographic features of wetlands, and to illustrate the approach at sites in the Florida Everglades. The results indicate that the physically-based approach can better capture the variations of specific yield with water level, in particular when the water level falls between the minimum and maximum land surface elevations. The suggested approach for hydroperiod computation predicted that the wetlands might be completely dry or completely wet much less frequently than suggested by the conventional approach neglecting microtopography. One reasonable generalization may be that the hydroperiod approaches presented in this chapter can be a more accurate prediction tool for water resources management to meet the specific hydroperiod threshold as required by a species of plant or animal of interest.

Chapter 5- Energy (e.g., latent and sensible heats) fluxes are important components of the land-atmosphere processes governing the hydrologic cycle. Understanding these energy fluxes as a function of topography and land management for ecosystems at different latitudes is essential to estimating water-energy exchanges between land surface and atmosphere. This chapter presents the dependence of energy fluxes on physical, geographical, and temporal factors. In addition, this chapter also discusses the estimation and evaluation of latent heat, sensible heat, evaporative energy flux, and non-evaporative energy flux for three selected sites at different latitudes. Microtopography can regulate soil moisture content and its spatial distribution, as indicated by a good correlation between topographic index and latent heat flux. The results indicate that vegetation controls the partition of energy fluxes in the three sites considered in this analysis located at different latitudes. Remote sensing can be very useful for evaluating the moisture availability, vegetative cover in wetland and grassland ecosystems, and microtopographic effects.

Chapter 6- A new site development is typically required neither to increase the peak discharge nor to degrade the water quality, of the stormwater leaving site, as compared with the corresponding values for the predevelopment conditions. In practice, stormwater detention ponds have been widely used for new site developments to satisfy these requirements. However, very limited evaluations have been conducted to document the cumulative effects of such detention ponds on hydrology and water quality at the watershed scale. The objectives of this chapter are to: 1) overview the engineering practices of stormwater detention ponds in the State of California in the United States; and 2) evaluate the cumulative effects of detention ponds within the Corte Madera Creek watershed located about 20-km north of San Francisco in California. The results from the study watershed indicate that the detention ponds designed for peak flow reduction combined with runoff volume reduction measures can be an effective means to minimize watershed environmental impacts resulting from site developments. This is consistent with the findings revealed by the practices that are overviewed in this chapter.

Chapter 7- The Red River of the North borders North Dakota and Minnesota and flows north toward Lake Winnipeg in Manitoba, Canada. The river is susceptible to flooding because of the synchrony of its discharge with spring thaw and ice jams, its shallow and sinuous channel, its low gradient, and the decrease in its gradient downstream. As a result, the property adjacent to the river is subject to frequent, damaging inundation from minor and major flood events, with a truly devastating flood about every decade. To mitigate flooding, various structural and nonstructural measures have been employed. However, the extensive flooding in 1997 necessitated reexamination of these measures and exploration of innovative concepts to augment traditional approaches. Hydrologic and hydraulic models play a key role in evaluating and identifying economical and feasible measures for flood reduction in this complex river system. In terms of complexity and modeling objective, the models developed in the past two decades can be categorized as 1) explanatory analyses, 2) floodplain and floodway management analyses, 3) land planning and management analyses, 4) flood mitigation engineering design analyses, 5) flood forecasting, and 6) miscellaneous. While a few of these models were used for some initial flood reduction analyses, they were developed mainly for other purposes. In addition, the models insufficiently address inflows from ungauged areas and overland flows, which significantly affect their calibration, verification, and application. This chapter discusses a conceptual model scheme applied to study the impacts of various storage scenarios on flood reduction in the Red River. Under this scheme, a coupled hydrologic–hydraulic model was developed by integrating two decades of modeling achievements with new algorithms specially designed for this study, employing updated modeling techniques and utilizing improved spatial and temporal data. Furthermore, this model was used to analyze storage scenarios necessary to mitigate 1997-type floods and the probable maximum flood (PMF) in the Red River.

In: Modeling Hydrologic Effects…
Editor: Xixi Wang, pp. 1-14
ISBN 978-1-61668-628-4

Chapter 1

CHARACTERIZATION OF MICROTOPOGRAPHY AND ITS HYDROLOGIC SIGNIFICANCE

Xuefeng Chu*
Department of Civil Engineering (Dept 2470), North Dakota State University
P.O. Box 6050, Fargo, North Dakota 58108-6050, USA

ABSTRACT

Surface microtopography plays an important role in watershed hydrology. As one of the major factors, microtopography dominates the overland flow generation mechanism, changes the spatial and temporal variability of hydrologic processes (e.g., infiltration and surface runoff), determines the drainage pattern, and affects the fate and transport of NPS contaminants throughout surface and subsurface systems. Thus, characterization of surface microtopography is critical to watershed hydrologic and environmental modeling. Various methods and techniques have been developed for DEM-based watershed delineation, which greatly facilitate watershed modeling and management. This chapter introduces a new puddle delineation algorithm that explicitly accounts for the puddle to puddle (P2P) filling-merging-spilling overland flow process. The algorithm has been incorporated into a Windows-based P2P modeling system that can be used for characterization of surface microtopography from high-resolution DEMs, computation of depression storages, and visualization of the delineation process. This chapter also details the P2P process and presents some findings from experimental studies on the effects of surface microtopography on overland flow. The results highlight the hydrologic significance of microtopography.

Keywords: DEM, depression storage, microtopography, overland flow, watershed delineation

* Tel. (701) 231-9758, Email: xuefeng.chu@ndsu.edu.

INTRODUCTION

Topography plays an important role in water flow and distribution over natural landscape (ASCE 1999), and influences insolation, water flow, and organism movement as an essential controlling variable in many ecological processes (Johnston 1998). Not only does topography control local flow direction and accumulation, but it also dominates the development and evolution of the entire drainage system. Surface microtopography is often quantified by using digital elevation models (DEMs), which may have different spatial scales, resolutions, and accuracy. New technologies have allowed us to acquire small-scale, high-resolution DEMs (e.g., Huang et al. 1988; Huang and Bradford 1990b, 1992; Darboux and Huang 2003). A variety of approaches and techniques have been developed for delineating topographic surfaces using DEMs. Some delineation techniques also have been incorporated into comprehensive hydrologic and environmental modeling software packages. These computer software packages particularly facilitate the characterization of surface microtopography and automate watershed delineation. Some software packages have been widely used for watershed modeling and management.

Surface microtopography can be characterized as topographic features such as depressions/pits/puddles, mounts/peaks, ridges, and channels. These features are topographically significant and hydrologically sensitive. Puddles may have different sizes, shapes, spatial distributions, and relationships with others (i.e., topologies). Puddles are linked to each other in a hierarchical fashion across a surface area. During a rainfall event, excess rainwater fills individual puddles; two puddles can be merged to form a larger, high-level puddle; and a fully-filled puddle spills and links to a downstream puddle. Herein, this puddle-to-puddle filling-merging-spilling dynamic process is referred to as the P2P process. The hydrologic role of puddles can be far beyond storing excess rainwater and can be complicated. Puddles can change the overland flow generation mechanism and hydrologic processes (e.g., infiltration and runoff partition). In accumulation, puddles can change the drainage pattern of the entire hydrologic system and alter the overall water mass balance for the surface and/or subsurface subsystems.

ACQUISITION OF HIGH-RESOLUTION DEM AND DATA PROCESSING

DEMs are commonly used for terrain analysis and watershed modeling and management. The resolution and accuracy of a DEM depend primarily on its grid spacing. Although DEMs with 30- and 90-m grid spacings had been utilized in a wide range of applications that have local, regional, or global spatial scales, DEMs with smaller (1- and 10-m) grid spacings are now available for many geographic areas. Further, new technologies of laser scanner have been developed for acquiring DEM data with a grid spacing of less than 1 mm, which is particularly useful for small-scale hydrologic studies. Huang et al. (1988) developed a laser scanner for measuring laboratory-scale soil surface microtopography. The laser scanner was further improved in early 1990s (Huang and Bradford 1990b, 1992). Subsequently, Darboux and Huang (2003) developed a second-generation, instantaneous-profile, high-resolution laser scanner, which has a horizontal resolution of 0.98 mm and a vertical resolution of 0.5 mm, and used this laser scanner for a runoff and soil erosion research project. This laser scanner used two laser diodes and an 8-bit monochrome CCD camera. The camera was connected to a computer that controls scanning, processes the scanned data, and stores the processed data. The point coordinates acquired by the camera were rectified in accordance with the benchmarks and in turn, the rectified points were interpolated to generate the high-resolution DEM. Chu et al. (2009) used such a scanner (Figure 1-1) to quantify microtopography of a series of smooth and rough soil surfaces in an experimental study of infiltration and overland runoff.

Figure 1-1. The instantaneous-profile laser scanner (top) and two samples of scanned soil surfaces (bottom).

DEM-based Watershed Delineation

Techniques and Software Packages for Automated Watershed Delineation

Watershed-scale hydrologic and environmental modeling often involves dealing with large sizes of spatially-distributed data, which can be greatly facilitated by using a geographic information system (GIS). In the past decades, significant progress has been made in GIS-based watershed modeling (DeVantier and Feldman 1993; Ross and Tara 1993; Olivera and

Maidment 2000; Olivera 2001; Vieux 2001). Ogden et al. (2001) summarized GIS-based hydrologic modeling and applications, and discussed a series of key implementation issues associated with the use of GIS in watershed hydrologic modeling. Because watershed delineation from DEMs is the key to GIS-based watershed modeling and management, researchers (e.g., Marks et al. 1984; O'Callaghan and Mark 1984; Jenson and Domingue 1988; Jenson 1991; Martz and Garbrecht 1993; Garbrecht and Martz 1997; Martz and Garbrecht 1999) have developed a number of delineation approaches and techniques. Some of those delineation techniques have been incorporated into widely-used watershed modeling software packages, including Arc Hydro (Maidment 2002), HEC-GeoHMS (USACE 2003; 2009), and WMS (2008), to provide powerful tools that can facilitate automated processing and visualization of spatially-distributed data, characterization of surface microtopography, hydrologic analysis, environmental modeling, and watershed management.

Arc Hydro is an ArcGIS data model for building hydrologic information systems that synthesize geospatial and temporal water resources data for supporting hydrologic analysis and modeling (Maidment 2002). It serves as a tool for using and processing water resources data in the ESRI ArcGIS environment, and takes full advantage of the powerful analysis capability of ArcGIS.

HEC-GeoHMS was originally designed as an ArcView extension (USACE 2003). An upgraded version is currently available for the latest ArcGIS (version 9.2) environment (USACE 2009). HEC-GeoHMS is a geospatial, hydrologic tool for watershed delineation using DEM. It is capable of visualizing spatial data, performing spatial analysis, quantifying watershed characteristics, delineating stream channels and subbasins, and preparing inputs for commonly-used hydrologic models (e.g., HEC-HMS).

WMS is a comprehensive modeling environment for watershed-scale hydrologic and hydraulic analysis. It incorporates a number of existing models, including HEC-HMS, HEC-RAS, TR-55, HSPF, and others (WMS 2008). In WMS, a modified TOPAZ (TOpographic PArameteriZation) model (Garbrecht and Martz 1995; Garbrecht and Martz 2000; Garbrecht et al. 2004) is used for watershed delineation using DEM. TOPAZ was developed primarily for topographic evaluation and watershed parameterization in support of hydrologic modeling and analysis (Garbrecht et al. 2004). WMS provides automated digital tools for processing raster DEM data, computing overland flow directions and accumulations, identifying the drainage network and stream channels, and determining subbasin boundaries.

Besides, PCRaster (Van Deursen and Wesseling 1992; Van Deursen 1995; Wesseling et al. 1996) has also been widely used for topographic analysis and hydrologic and environmental modeling. PCRaster seamlessly integrates environmental modeling with GIS because it uses a powerful dynamic modeling language. The language provides functionality for data storage, modeling, and visualization through a set of spatial and temporal operators (PCRaster 2009).

Conventional DEM-Based Procedure for Watershed Delineation

Most existing watershed delineation and terrain processing software packages implement a five-step procedure to extract from a raw DEM the information needed for hydrologic modeling. Hereinafter, the Mona Lake watershed located in west Michigan is selected to explain this procedure (Figure 1-2).

Step 1: *Creation of depressionless DEM*

Raw DEMs (e.g., Figure 1-2a) usually contain various depressions or sinks. As a critical step of terrain preprocessing, a depressionless DEM is first created by using a set of automated algorithms. This step involves filling all sinks or depressions in the original DEM by raising their elevations to the surrounding levels, which ensures to produce a continuously-flowing drainage network across the entire watershed, allowing the surface runoff to concentrate at the outlet of the watershed. The details of this step can be found in Jenson and Domingue (1988).

Step 2: *Determination of flow direction*

Based on the depressionless DEM (e.g., Figure 1-2b), this procedure determines flow directions using either a single-direction method or a multiple-direction method. The commonly-used single-direction methods include deterministic eight-node (D8; O'Callaghan and Mark 1984) and random eight-node (Rho8; Fairfield and Leymarie 1991). The widely-used multiple-direction methods are FD8 (Quinn et al. 1991), FRho8 (Moore et al. 1993), and D∞ (Tarboton 1997). The single-direction methods assume that the runoff in a given cell flows toward only one of its eight neighboring cells along the steepest gradient. In contrast, the multiple-direction methods allow the runoff in a given cell to flow toward one of its eight neighboring cells or more. The runoff is partitioned using a weighting system among the receiving cells.

Step 3: *Determination of flow accumulation*

Based on the flow direction result, this procedure determines flow accumulations by tracing all cells that contribute runoff to a given cell. That is, this procedure counts the total number of cells that are hydraulically connected with this given cell.

Step 4: *Determination of stream network*

A series of cells that have flow accumulations greater than an empirical threshold are determined as channel cells, forming the stream network (e.g., Figure 1-2c). A smaller threshold results in a denser drainage network (i.e., a network with more streams), whereas, a larger threshold results in fewer streams to be delineated. In practice, an appropriate threshold should be determined using a trial-and-error approach to make the delineated streams closely match the field surveyed data.

Step 5: *Determination of subbasin boundary*

Based on the stream network, the procedure delineates the watershed and its subbasin boundaries (e.g., Figure 1-2d).

Step 1 of the conventional delineation procedure can be problematic for some practical applications. All depressions in the Mona Lake watershed, including the Carr Lake, are filled when the raw DEM is processed using either HEC-GeoHMS (Figure 1-3b versus Figure 1-3a) or WMS (Figure 1-4 versus Figure 1-3a). The flow directions and stream channels delineated by both HEC-GeoHMS and WMS clearly indicate that the lake is completely filled, which can result in an erroneous representation of the actual drainage pattern. Thus, this step of the conventional procedure for watershed delineation needs to be modified to account for detailed topographic conditions such as the Carr Lake. In this regard, Chu and Zhang (2010) developed a new puddle delineation algorithm, described in detail in the next section. Chu et al. (2010) evaluated this algorithm in terms of using DEMs with distinctly different resolutions (i.e., grid sizes), and compared the delineation results from this algorithm with those from Arc Hydro, HEC-GeoHMS, and PCRaster.

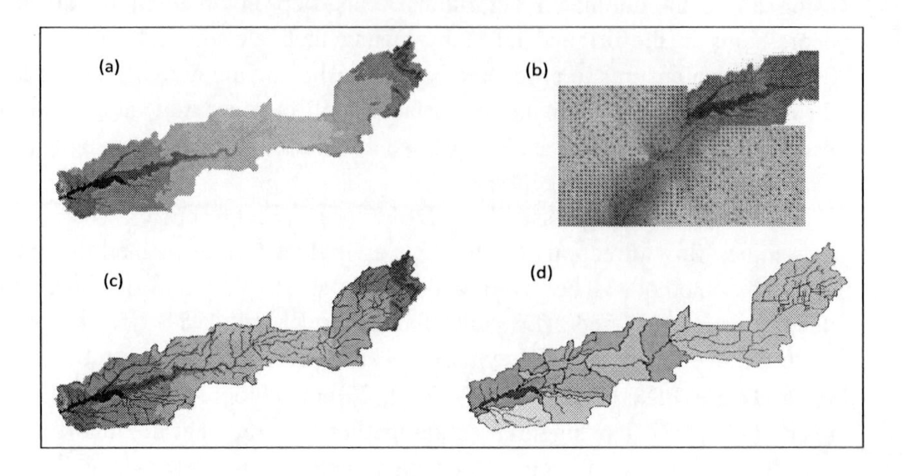

Figure 1-2. Illustration of the conventional DEM-based watershed delineation procedure showing the (a) raw DEM, (b) determined flow directions, (c) delineated stream network, and (d) delineated boundaries, of the Mona Lake watershed located in west Michigan.

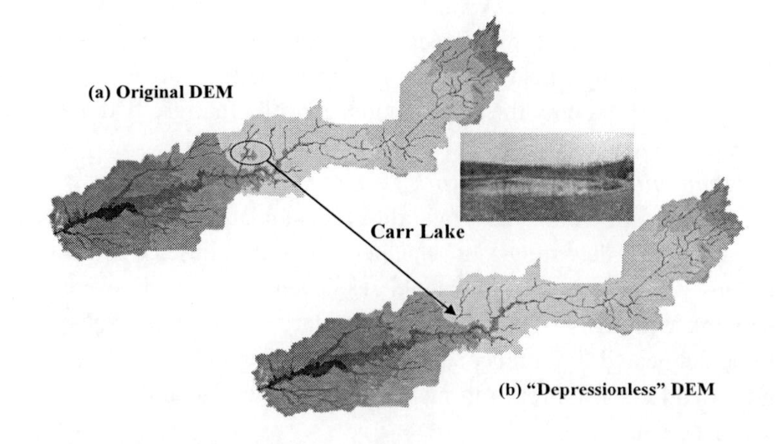

Figure 1-3. The (a) raw and (b) depressionless DEMs of the Mona Lake watershed located in west Michigan. The depressionless DEM is the output of HEC-GeoHMS.

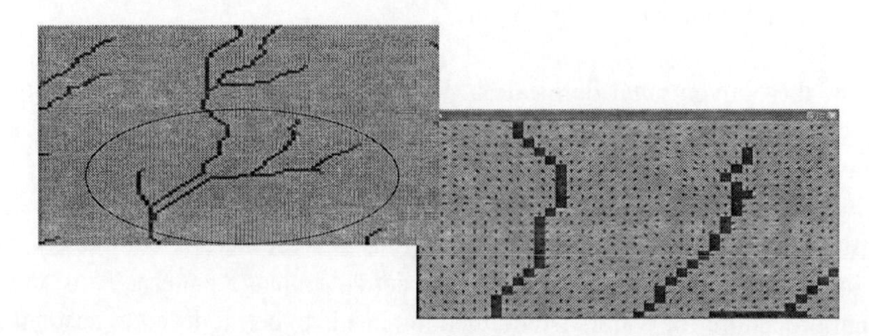

Figure 1-4. Flow directions and channels delineated from the depressionless DEM by WMS.

A New Algorithm for Watershed Delineation

Chu and Zhang (2010) developed a new algorithm for delineating surface depressions/puddles and calculating depression storage based on high-resolution DEMs from the instantaneous-profile laser scanner (Darboux and Huang 2003). The algorithm identifies individual puddles, computes their storages, determines their topologies, and defines their hydraulic relations. Using a DEM, the algorithm is implemented in series to: 1) identify the centers and flats; 2) search from low to high levels all puddles based on the threshold-controlled P2P process and a set of other criteria (Chu and Zhang 2010); and 3) define the spatial and hydraulic relations of the puddles. The algorithm can cope with special schemes related to flats and boundary conditions for surfaces with complex microtopographic characteristics. When the puddle delineation is completed, the algorithm computes the storage of each individual puddle and the maximum depression storage over the entire area.

This new algorithm has been programmed into a Windows-based P2P modeling system (Chu and Zhang 2010). The system can greatly facilitate puddle delineation from high-resolution DEMs, computation of depression storages, and visualization of the puddle delineation process. The delineation results can be imported into hydrologic models. Figure 1-5 shows the main interface of the Windows-based P2P modeling system and the delineation program. The interface consists of a map area and a control panel. The control panel is designed for loading DEM data, implementing the delineation program, and visualizing the delineation results.

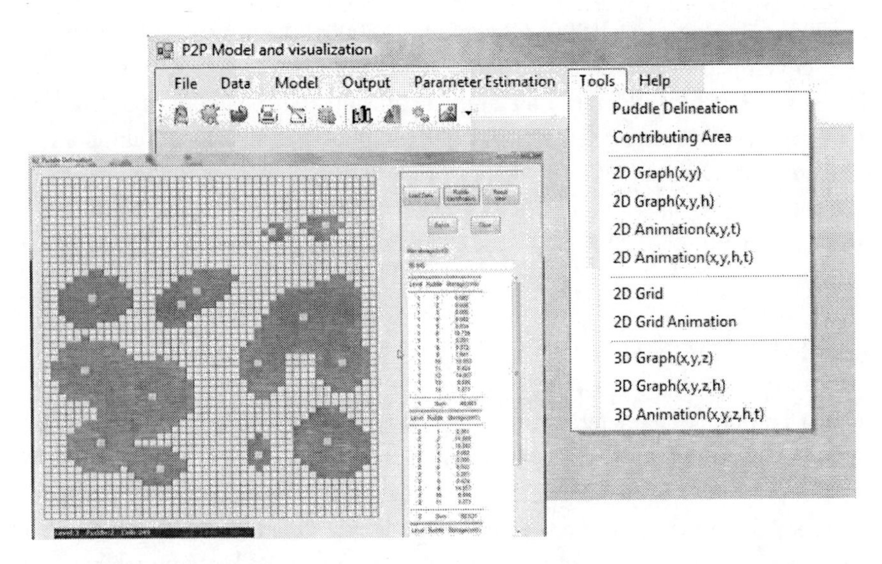

Figure 1-5. The main interface of the Windows-based P2P modeling system and the delineation program.

Effects of Microtopography on Hydrology

Microtopography is one of the primary variables that control hydrologic processes. It may have significant effects on overland flow generation, surface runoff, infiltration, soil erosion and sediment transport, nonpoint source (NPS) pollutant transport, and surface and subsurface hydrologic interaction. For this reason, considerable efforts (e.g., Huang and Bradford 1990a; Kamphorst et al. 2000; Hansen 2000; Kamphorst and Duval 2001; Darboux and Huang 2003, 2005; Abedini et al. 2006) have been made to quantify these effects. Based on laboratory-scale experimental studies, Huang and Bradford (1990a) found that surface topography controls the magnitude and spatial/temporal distribution of surface runoff depth. Increasing soil surface roughness tends to increase the storage of surface depressions, resulting in the decrease of surface runoff quantity (Johnson et al. 1979; Steichen 1984; Cogo et al. 1984; Kamphorst et al. 2000). Also, the increase in surface roughness will reduce flow velocity (Cogo et al. 1984) and alter overland flow direction (Darboux et al. 2001). The process of storing water in surface depressions can delay the initiation of surface runoff and increase infiltration from the ponded water into soil (Darboux et al. 2001; Darboux et al. 2004; Darboux and Huang 2005). The watershed microrelief can affect runoff retention (Abedini et al. 2006).

Surface Roughness and Depression Storage

A raster DEM can be used to present the information on the surface microtopography that is characterized by depressions/puddles/pits, mounts/peaks, ridges, and channels. Depression storage is one of the major variables in hydrologic analysis because it represents the overall effect of surface roughness. The relationship between surface roughness and depression storage has been well studied and understood (Onstad 1984; Linden et al. 1988; Huang and Bradford 1990a; Hansen et al. 1999; Govers et al. 2000). As a result, a number of methods have been developed to estimate surface depression storage. Some methods calculate depression storage using DEM (Ullah and Dickinson 1979; Huang and Bradford 1990a; Martz and Garbrecht 1993; Hansen et al. 1999; Kamphorst et al. 2000; Kamphorst and Duval 2001; Planchon and Darboux 2002), whereas, the other methods estimate depression storage in terms of surface roughness indices (e.g., Onstad 1984; Mwendera and Feyen 1992; Hansen et al. 1999; Kamphorst et al. 2000).

Puddle and the P2P Process

A typical puddle is characterized by a center (i.e., the point with a lowest elevation), a threshold for overflowing to occur, and a number of body cells. The puddle may have multiple thresholds, through which water pours out. A flat, which may be fully open, partially open, or closed, can be viewed as a special puddle that is able to transfer runoff water but has zero storage. A closed flat actually forms a puddle that has a flat bottom (center).

The puddles within an area or a watershed can be categorized in terms of their "levels." A higher-level puddle can be formed by combining two or more lower-level puddles. For example, all four puddles A to D in Figure 1-6a are first-level because they are isolated from each other. The two or more puddles that share an identical threshold can be combined together to form a higher-level puddle. In Figure 1-6b, puddles B and C are combined to form a second-level puddle BC. Puddle BC is further combined with puddle A to form the third-level puddle ABC (Figure 1-6c). By the end of this combination process, all four first-level puddles (Figure 1-6a) are merged into the fourth-level (highest-level for this example) puddle

ABCD (Figure 1-6d). Once the water surface level in puddle ABCD reaches its threshold, water will start to pour out of the highest-level puddle through its outlet (Figure 1-6e).

Depending on its microtopographic characteristics, a puddle may have different relationships with its adjacent puddles. The typical relationships can be upstream-downstream, outflow-inflow, isolated from each other, or combined to form another higher-level puddle. During rainfall/snowmelt event, the excess precipitation will fill the puddles. A fully-filled puddle will be merged with the others or spill water out into one downstream puddle or more. As illustrated in Figure 1-6, this is a puddle-to-puddle (P2P) filling-merging-spilling dynamic overland flow process. This process has a cascade flow pattern and varies spatially and temporally.

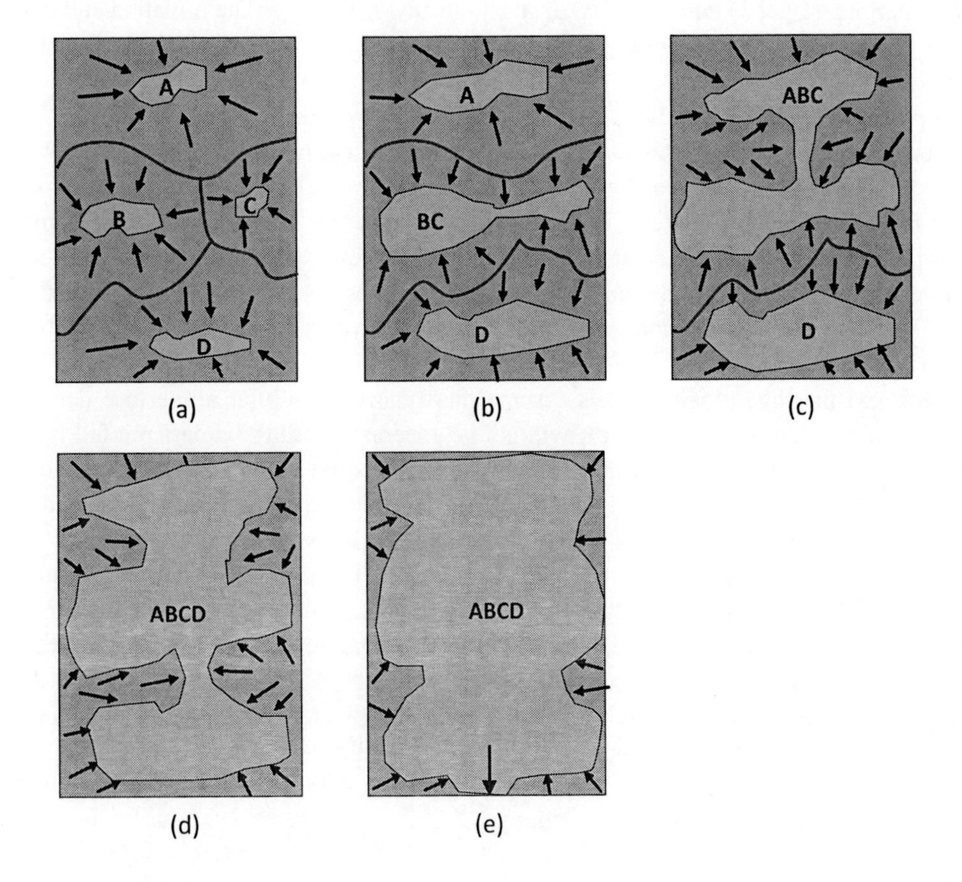

Figure 1-6. Illustration of (a) the four first-level puddles, and (b), (c), (d), and (e) the puddle-to-puddle (P2P) filling-merging-spilling process.

Findings from Experimental Studies on Overland Flow

Chu et al. (2009) evaluated the effects of soil surface microtopography on overland flow generation and rainfall-runoff processes by conducting a set of overland flow experiments for both rough and smooth soil surfaces. A 1.0 m wide by 1.2 m long soil box (Figure 1-1) was used for the experimental studies. The high-resolution DEM data of the soil surfaces were obtained using the instantaneous-profile laser scanner (Darboux and Huang 2003). A 4-head

Norton-style multiple-intensity rainfall simulator (Meyer and McCune 1958; Meyer and Harmon 1979; Meyer 1994) was used to generate steady rainfall events. The data collected during the experiments included the P2P filling-merging process, movement of the wetting front along the soil profile, percolating water through the bottom of the soil box, and discharge at the outlet of the soil box. Subsequently, the experimental data were processed and analyzed in terms of the mass conservation principle. The results for these two soil surfaces were compared to quantify any discrepancies for the P2P staring time, the P2P ending time, the outlet flow initiation time, and the steady flow occurrence time.

Based on the analysis results, Chu et al. (2009) found that the rainfall-runoff process can be characterized into four unique successive stages: 1) infiltration-dominated stage; 2) P2P filling-merging stage; 3) transition stage; and 4) steady-state stage. The initiation and duration of the P2P process vary depending on the surface microtopography, soil, rainfall, and other conditions.

Figure 1-7 compares the infiltration rates, outlet flow rates, and surface storage change rates between the rough and smooth surfaces for an identical rainfall intensity of 3.39 cm/hr. For the rough soil surface, the flow hydrograph at the outlet exhibited an uneven, stepwise increasing pattern. This pattern is primarily attributed to the discontinuous, threshold-controlled overland flow associated with the P2P process. Such influence of the surface microtopography on the runoff and infiltration processes was propagated beyond the P2P stage. In contract, for the smooth soil surface, the flow hydrograph at the outlet exhibited an even, rapid increasing pattern.

Most existing hydrologic models assume that runoff starts after all surface depressions are fully filled. That is, surface runoff will not be generated until the excess rainfall is greater than a threshold depth at which the total depression storage volume is satisfied. However, Chu et al. (2009) demonstrated that the P2P filling-merging process continued after the commencement of surface runoff at the outlet and that the runoff started prior to that all surface depressions were completely filled (Figure 1-7). This finding is consistent with those of Onstad (1984) and Hansen (2000). The experimental studies conducted by Chu et al. (2009) also indicated that there was a time lag between the initiations of the P2P process and outlet discharge. The initiation time and the duration of the P2P process varied with the soil surface microtopography and other conditions (e.g., soil water content). The rougher a soil surface is, the longer the P2P process will last. The studies also revealed that using a point infiltration rate to represent the areal infiltration rate for rough surface is likely to overestimate surface runoff (Chu et al. 2009).

Conclusions

Presently, various techniques and tools are available for watershed delineation. However, few of these techniques and tools have the capability for analyzing detailed surface microtopographic features and the related puddle to puddle filling-merging-spilling process. In this regard, this chapter introduced the P2P algorithm and a Windows-based P2P modeling system. The experimental studies indicated that depressions/puddles are of special significance in hydrology and can control overland flow direction, flow accumulation, and drainage pattern. Surface roughness tends to increase the retention and detention of surface runoff and to strengthen the infiltration process. Surface microtopography determines the initiation timing and duration of the P2P process as well as the quantity and distribution of surface depression storage. This in turn affects the physical mechanisms of hydrologic processes, such as infiltration and overland runoff. Further, surface microtopography affects these hydrologic processes both during and after the P2P stage (i.e., all puddles are completely filled).

ACKNOWLEDGMENTS

This material is partially based upon the work supported by the National Science Foundation under Grant No. EAR-0907588 and ND NASA EPSCoR through NASA grant #NNXO7AK91A. I would like to thank Jianli Zhang (Postdoc) and Jessica Higgins (graduate student) for their contribution to the research project.

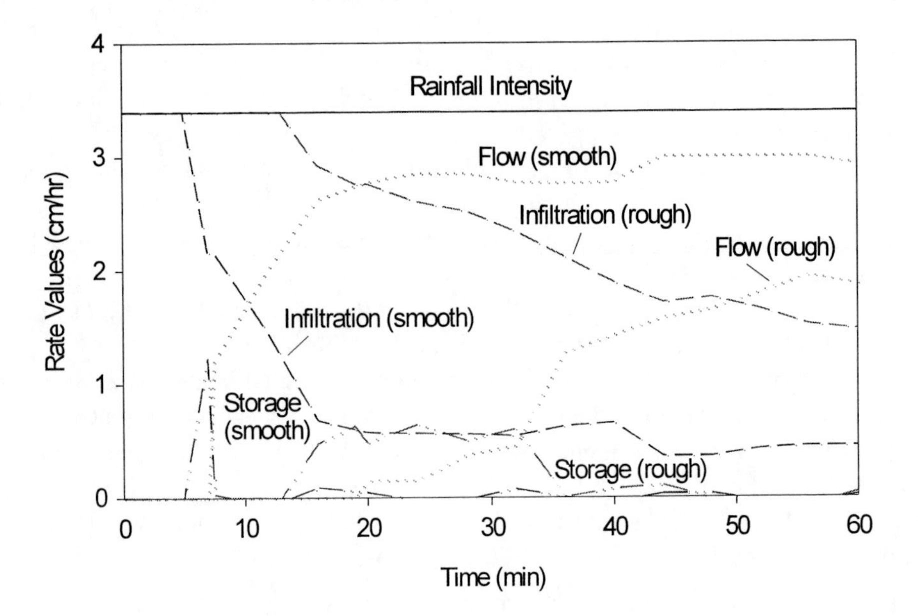

Figure 1-7. The infiltration rates, outlet discharge rates, and surface storage changes for the rough soil surface versus the smooth soil surface.

REFERENCES

Abedini, M.J., Dickinson, W.T., Rudra, R.P., 2006. On depressional storages: The effect of DEM spatial resolution. *Journal of Hydrology* 318, 138-150.

ASCE (American Society of Civil Engineers), 1999. *GIS Modules and Distributed Models of the Watershed, ASCE Task Committee on GIS Modules and Distributed Models of the Watershed*. American Society of Civil Engineers, Reston, Virginia.

Chu, X., Higgins, J., Zhang, J., Huang, C., 2009. Effects of soil surface microtopography on overland flow and infiltration processes. *Journal of Hydrology*, (in review).

Chu, X., Zhang, J., 2010. Development of a new algorithm for puddle delineation and determination of maximum depression storage. *Journal of Hydrology*, (in submission).

Chu, X., Zhang, J., Yang, J., Chi, Y., 2010. Quantitative evaluation of the relationship between grid spacing of DEMs and surface depression storage, p4447-4457. In: *Challenges of Change, Proceedings of the 2010 World Environmental and Water Resources Congress*, edited by R. N. Palmer. American Society of Civil Engineers.

Cogo, N.P., Moldenhauer, W.C., Foster, G.R., 1984. Soil loss reductions from conservation tillage practices. *Soil Science Society of America Journal* 48, 368-373.

Darboux, F., Davy, P., Gascuel-Odoux, C., Huang, C., 2001. Evolution of soil surface roughness and flowpath connectivity in overland flow experiments. *Catena* 46, 125-139.

Darboux, F., Huang, C., 2003. An instantaneous-profile laser scanner to measure soil surface microtopography. *Soil Science Society of America Journal* 67, 92-99.

Darboux, F., Huang, C., 2005. Does soil surface roughness increase or decrease water and particle transfers? *Soil Science Society of America Journal* 69, 748-756.

Darboux, F., Reichert, J.M., Huang, C., 2004. Soil roughness effects on runoff and sediment production. In: *Conserving Soil and Water for Society: Sharing Solutions, Proceedings of 13th International Soil Conservation Organization Conference*, edited by S.R. Raine, A.J.W. Biggs, N.W. Menzies, D.M. Freebairn, and P.E. Tolmie. 4-9th July, Brisbane. ASSSI/ IECA. Paper No. 116. p1-6.

DeVantier, B.A., Feldman, A.D., 1993. Review of GIS applications in hydrologic modeling. *Journal of Water Resources Planning and Management* 119(2), 246-261.

Fairfield, J., Leymarie, P., 1991. Drainage networks from grid digital elevation models. *Water Resources Research* 27, 709–717.

Garbrecht, J.D., Campbell, J., Martz, L.W., 2004. *TOPAZ User Manual - Updated Manual.* Grazinglands Research Laboratory Miscellaneous Publication.

Garbrecht, J., Martz, L.W., 1995. TOPAZ: An automated digital landscape analysis tool for topographic evaluation, drainage identification, watershed segmentation and subcatchment parameterization: Overview. *ARS-NAWQL 95-1*, 17pp., USDA-ARS, Durant, Oklahoma.

Garbrecht, J., Martz, L.W., 1997. The assignment of drainage direction over flat surfaces in raster digital elevation models. *Journal of Hydrology* 193, 204-213.

Garbrecht, J., Martz, L.W., 2000. *TOPAZ: An Automated Digital Landscape Analysis Tool for Topographic Evaluation, Drainage Identification, Watershed Segmentation and Subcatchment Parameterization: TOPAZ User Manual.* Grazinglands Research Laboratory, USDA Agricultural Research Services, El Reno, OK. ARS Pub No. GRL 2-00, 144pp.

Govers, G., Takken, I., Helming, K., 2000. Soil roughness and overland flow. *Agronomie* 20, 131-146.

Hansen, B., 2000. Estimation of surface runoff and water-covered area during filling of surface microrelief depressions. *Hydrological Processes* 14, 1235-1243.

Hansen, B., Schjonning, P., Sibbesen, E., 1999. Roughness indices for estimation of depression storage capacity of tilled soil surfaces. *Soil & Tillage Research* 52, 103-111.

Huang, C., Bradford, J.M., 1990a. Depressional storage for Markov-Gaussian surfaces. *Water Resources Research* 26(9), 2235-2242.

Huang, C., Bradford, J.M., 1990b. Portable laser scanner for measuring soil surface roughness. *Soil Science Society of America Journal* 54, 1402-1406.

Huang, C., Bradford, J.M., 1992. Applications of laser scanner to quantify soil microtopography. *Soil Science Society of America Journal* 56, 14-21.

Huang, C., White, I., Thwaite, E.G., Bendeli, A., 1988. A noncontact laser system for measuring soil surface topography. *Soil Science Society of America Journal* 52, 350-355.

Jenson, S.K., 1991. Applications of hydrologic information automatically extracted from digital elevation models. *Hydrological Processes* 5(1), 31-44.

Jenson, S.K., Domingue, J.O., 1988. Extracting topographic structure from digital elevation data for geographic information system analysis. *Photogrammetric Engineering and Remote Sensing* 54(11), 1593-1600.

Johnson, C.B., Mannering, J.V., Moldenhauer, W.C., 1979. Influence of surface roughness and clod size and stability on soil and water losses. *Soil Science Society of America Journal* 43, 772-777.

Johnston, C.A., 1998. *Geographic Information Systems in Ecology, Methods in Ecology.* Blackwell Science, Inc. Malden, MA.

Kamphorst, E.C., Duval, Y., 2001. Validation of a numerical method to quantify depression storage by direct measurements on moulded surfaces. *Catena* 43, 1-14.

Kamphorst, E.C., Jetten, V., Guerif, J., PJitkanen, J., Iversen, B.V., Douglas, J.T., Paz, A., 2000. Predicting depressional storage from soil surface roughness. *Soil Science Society of America Journal* 64, 1749-1758.

Linden, D.R., Van Doren, Jr., D.M., Allmaras, R.R., 1988. A model of the effects of tillage-induced soil surface roughness on erosion. In: *Proceedings of the 11th Conference of the International Soil Tillage Research Organization*. Edinburgh, July 11-15, 1988. pp. 373-378.

Maidment, D.R, 2002. *Arc Hydro GIS for Water Resources*. ESRI Press, California.

Marks, D., Dozier, J., Frew, J., 1984. Automated basin delineation from digital elevation data. *Geo-Processing* 2, 299-311.

Martz, L.M., Garbrecht, J., 1993. Automated extraction of drainage network and watershed data from digital elevation models. *Water Resources Bulletin* 29(6), 901-908.

Martz, L.M., Garbrecht, J., 1999. An outlet breaching algorithm for the treatment of closed depressions in a raster DEM. *Computers & Geosciences* 25, 835-844.

Meyer, L.D., 1994. Rainfall simulators for soil erosion research. In: *Proceedings of Soil Erosion Research Methods*, 2nd Edition, edited by R. Lal. CRC Press.

Meyer, L.D., Harmon, W.C., 1979. Multiple-intensity rainfall simulator for erosion research on row sideslopes. *Transactions of the American Society of Agricultural Engineers* 22, 100-103.

Meyer, L.D., McCune, D.L, 1958. Rainfall simulator for run-off plots. *Agricultural Engineering* 39(10), 644-648.

Moore, I.D., Turner, A.K., Wilson, J.P., Jenson, S.K., Band, L.E., 1993. GIS and land surface-subsurface modeling. In: *Environmental Modeling with GIS*, Goodchild, M.F., Parks, B.O., Steyaert, L.T. (eds). Oxford University Press: New York, 196–230.

Mwendera, E.J., Feyen, J., 1992. Estimation of depression storage and Manning's resistance coefficient from random roughness measurements. *Geoderma* 52, 235-250.

O'Callaghan, J.F., Mark, D.M., 1984. The extraction of drainage networks from digital elevation data. *Computer Vision, Graphics and Image Processing* 28, 323-344.

Ogden, F.L., Garbrecht, J., DeBarry, P.A., Johnson, L.E., 2001. GIS and distributed watershed models II: Modules, interfaces, and models. *Journal of Hydrologic Engineering* 6(6), 515-523.

Olivera, F., 2001. Extracting Hydrologic Information from Spatial Data for HMS Modeling. *Journal of Hydrologic Engineering* 6(6), 524-530.

Olivera, F., Maidment, D.R., 2000. *GIS Tools for HMS Modeling Support. In Chapter 5, Hydrologic and Hydraulic Modeling Support with Geographic Information Systems.* Maidment, D.R., Djokic, D. (eds). ESRI Press: Redlands, California.

Onstad, C.A., 1984. Depressional storage in tilled soil surfaces. *Transactions of the American Society of Agricultural Engineers* 27, 729-732.

PCRaster, 2009. *Department of Physical Geography, Utrecht University, The Netherlands.* Available at http://pcraster.geo.uu.nl/index.html, Accessed on October 25, 2009.

Planchon, O., Darboux, F., 2002. A fast, simple and versatile algorithm to fill the depressions of digital elevation models. *Catena* 46, 159-176.

Quinn, P.F., Beven, K.J., Chevallier, P., Planchon, O., 1991. The prediction of hillslope flow paths for distributed hydrological modeling using digital terrain models. *Hydrological Processes* 5, 59–79.

Ross, M.A., Tara, P.D., 1993. Integrated hydrologic modeling with geographic information systems. *Journal of Water Resources Planning and Management* 119(2), 129-140.

Steichen, J.M., 1984. Infiltration and random roughness of a tilled and untilled claypan soil. *Soil Tillage Research* 4, 251-262.

Tarboton, D.G., 1997. A new method for the determination of flow directions and upslope areas in grid digital elevation models. *Water Resources Research* 33, 309-319.

Ullah, W., Dickinson, W.T., 1979. Quantitative description of depression storage using a digital surface model: I. determination of depression storage. *Journal of Hydrology* 42, 63-75.

USACE (U.S. Army Corps of Engineers), 2003. *HEC-GeoHMS Geospatial Hydrologic Modeling Extension, User's Manual, Version 1.1, CPD-77.* USACE-HEC, Davis, CA.

USACE (U.S. Army Corps of Engineers), 2009. *HEC-GeoHMS Geospatial Hydrologic Modeling Extension, User's Manual, Version 4.2, CPD-77.* USACE-HEC, Davis, CA.

Van Deursen, W.P.A., 1995. *Geographical Information Systems and Dynamic Models: Development and Application of a Prototype Spatial Modelling Language.* Ph.D. Thesis, Utrecht University, NGS Publication 190, 198 pp.

Van Deursen, W.P.A., Wesseling, C.G., 1992. *The PC Raster Package.* Department of Physical Geography, Utrecht University, Netherlands.

Vieux, B.E., 2001. Distribute Hydrologic Modeling Using GIS. In: *Water Science and Technology Library* V. 38. Kluwer Academic Publishers: Boston.

Wesseling, C.G., Karssenberg, D., van Deursen, W.P.A., Burrough, P.A., 1996. Integrating dynamic environmental models in GIS: the development of a dynamic modelling language. *Transactions in GIS* 1, 40-48.

WMS, 2008. *Watershed Modeling System, WMS v8.1 Tutorials.* Aquaveo, LLC, Provo, Utah.

In: Modeling Hydrologic Effects...
Editor: Xixi Wang, pp. 15-31
ISBN 978-1-61668-628-4

Chapter 2

ADAPTATION OF NATIONAL ELEVATION DATATSET TO ESTIMATE DEPRESSION STORAGE IN A GLACIATED WATERSHED

Sarita Pachhai Karki[1,*] and Xixi Wang[2,†]

[1]Senior GIS Analyst I, Post, Buckley, Schuh & Jernigan, Inc (PBS&J), 7406 Fullerton Street, Suite 350, Jacksonville, Florida 32256, USA

[2]Assistant Professor and Coordinator, Hydrology and Watershed Management Program, Department of Engineering and Physics, Tarleton State University, BOX T-0390, Stephenville, Texas 76402, USA

ABSTRACT

The construction of raised roads may modify the hydraulic connectivity of natural depression storages (e.g., wetlands) and create additional artificial storages. As a result, it would noticeably alter the natural hydrology of the inclusive watershed. Quantifying these storages is very important for watershed management that aims to increase water supply and mitigate flooding. This can be done by processing a digital elevation model (DEM) in a geographic information system (e.g., ArcGIS®) on the basis of the U.S. Geological Survey's 1.61-km by 1.61-km quadrangle sections. However, DEMs with a spatial resolution finer than 30 m are not available for most watersheds. The objective of this study was to examine whether and how the 30-m National Elevation Dataset can be adapted to estimate potential depression storage capacity (PDSC), the total volume below the lowest point (i.e., breach) of the roads surrounding a section. The study was conducted in the Forest River watershed located in northeastern North Dakota. The results indicate that the estimation of PDSC tended to have larger errors for sections with a higher topographic relief than for sections with a lower relief. In addition, the estimation was sensitive to errors in the breach elevation values. Nevertheless, the PDSC

* E-mail: spachhai@gnail.com
† Tel. (254) 968-9164, E-mail: xxqqwang@gmail.com

for a section within the study watershed can be estimated as the multiplication of the value computed using the NED data by an adjustment coefficient of 1.05 to 1.3. One reasonable generalization of this study is that NED can be adapted to estimate PDSCs for watersheds where high-resolution DEMs are not available.

Keywords: DEM, GIS, depression, hydrology, PDSC, relief, roads

INTRODUCTION

The construction of roads and their associated drainage ditches can noticeably alter the natural hydrology of the inclusive watershed (Jones et al., 2000; Wigmosta and Perkins, 2001). This is particularly true for watersheds with gentle terrain, such as the Forest River watershed located in northeastern North Dakota (Figure 2-1). On the one hand, ditches may hydraulically connect natural depression areas (e.g., wetlands) with streams, increasing the contribution drainage area but shortening the distance required for a drop of water to travel through the overland into a stream (Magner et al., 2004). As a result, concentrated flow processes would become more prevalent and thus the time of concentration, the time required for a drop of water to travel from the most hydrologically remote point within a watershed to its outlet (Viessman and Lewis, 2003), would be reduced, leading to increased peak discharge (LaMarche and Lettenmaier, 2001). On the other hand, raised roads with a top above the adjacent ground levels could create artificial depression storages to store and attenuate the concentrated flows when conveyance structures (e.g., culverts and bridges) have insufficient capacities or are nonexistent (Duke et al., 2003). The artificial storages and the modified natural depression storages play a detention role, reducing the runoff contribution area, streamflow volume, and peak discharge. These two opposite effects of roads on watershed hydrology have been documented by a number of researchers (e.g., Harr et al., 1975; Ziemer, 1981; King and Tennyson, 1984; Wright et al., 1990; Lewis et al., 2001).

Harr et al. (1975) conducted a paired watershed study in the Orgeon Coast Range and found that peak discharges for the combined fall and winter periods noticeably increased when roads occupied more than 12% of the watershed area. They also found that the flow hydrograph was shifted to have a delayed rising limb as a result of the detention function effects of artificial storages formed by the raised roads. In contrast, King and Tennyson (1984) showed a decrease in peak discharge through a paired watershed study conducted in north central Idaho. The authors stated that the noted decrease was the result of insufficient conveyance structures. Similarly, Ziemer et al. (1981) and Wright et al. (1990) conducted paired watershed studies in northwestern California and indicated no increase in peak discharge due to the construction of roads. The authors speculated that because the roads occupied only 5% of the watershed in size, which is much smaller than the 12% threshold value presented by Harr et al. (1975), the effects might not be detectable. Using the streamflow data collected at 14 gauging stations within the same watershed, Lewis et al. (2001) conducted a multiple regression analysis to study how the peak discharges and streamflow volumes depended on the areal extent of, and the number of stream crossings beneath, the roads. The results indicate an insignificant effect at a significance level of 0.05.

The inconsistent findings from the aforementioned studies as well as the others are unsurprising because effects of roads on the natural hydrology of a watershed depend on multiple factors, including: 1) the topology of road and stream networks (Wemple et al., 1996); 2) the road class (e.g., raised versus cut roads) and drainage type (e.g., relief culvert, mitre drain, and pushout drain; Croke and Mockler, 2001); 3) the road density, defined as the ratio of the total length of roads to the total drainage area of the watershed (Jones and Grant, 1996); 4) the topography (Veldhuisen and Russell, 1999); and 5) the weather (i.e.., wet versus dry) condition (Wemple and Jones, 2003). For a given weather condition, the first four factors affect how the natural storages would be modified and how many artificial storages would be created as a result of the construction of roads. When the density of raised roads is high, which is the case for the Forest River watershed, the artificial depression storages can greatly augment the natural depression storages, modifying the natural hydrology. Therefore, quantifying both the artificial and natural storages is very important for understanding consequences of such modifications as well as for making efforts to utilize the storages to increase water supply and mitigate flooding. FIPR (1998) explored the feasibility to use depressions from mining activities to recharge aquifers and increase water supply, whereas, Bolles and Wang (2003) studied how storages formed by raised roads could be utilized to temporarily store spring snowmelt runoff to reduce flooding.

The quantification can be implemented using a digital elevation model (DEM) and a geographic information system (e.g., ArcGIS®; Ormsby et al., 2001) on the basis of the U.S. Geological Survey (USGS) 1.61-km by 1.61-km quadrangle sections within a watershed. Herein, the key point is the availability and the required accuracy of DEM. The 30-m NED (National Elevation Dataset) is available for the conterminous United States and its territories. However, the question arised is whether the NED can be adapted to estimate the total (i.e., artificial plus natural depression) storages with sufficient accuracy. The answer to this question is seldom available in literature, though numerous studies (e.g., Garbrecht and Martz, 1997; Noman et al., 2003; Cochrane and Flanagan, 2005; Ziadat, 2007) have indicated that the spatial resolution of DEM affects the delineated watershed boundary and stream network as well as the estimated topographic parameters (e.g., slope, slope length, and relief). The objective of this study was to answer this question by comparing the storages estimated using the 30-m NED with those estimated using a 1-m LiDAR (Light Detection and Ranging) DEM for the Forest River watershed. To simplify the analysis, the total storage for a USGS quadrangle section, which is simply designated "section" for descriptive purposes, was defined as the total volume below the lowest top elevation of the roads surrounding the section.

Materials and Methods

The Forest River Watershed

The 2131 km^2 Forest River watershed, which is delineated as the USGS eight-digit hydrologic unit code of 09020308, covers portions of three North Dakota counties: Walsh, Nelson, and Grand Forks (Figure 2-1). The watershed comprises 823 USGS 1.61-km by 1.61-km sections, which form 22 townships that belong to the three counties. With the recently glaciated terrain, the watershed has a physiography of the Drift Prairie in the upper (i.e., western) portion and the Red River Valley Lake Plain in the lower (i.e., eastern) portion (Stoner et al., 1993; LeFever et al., 1999).

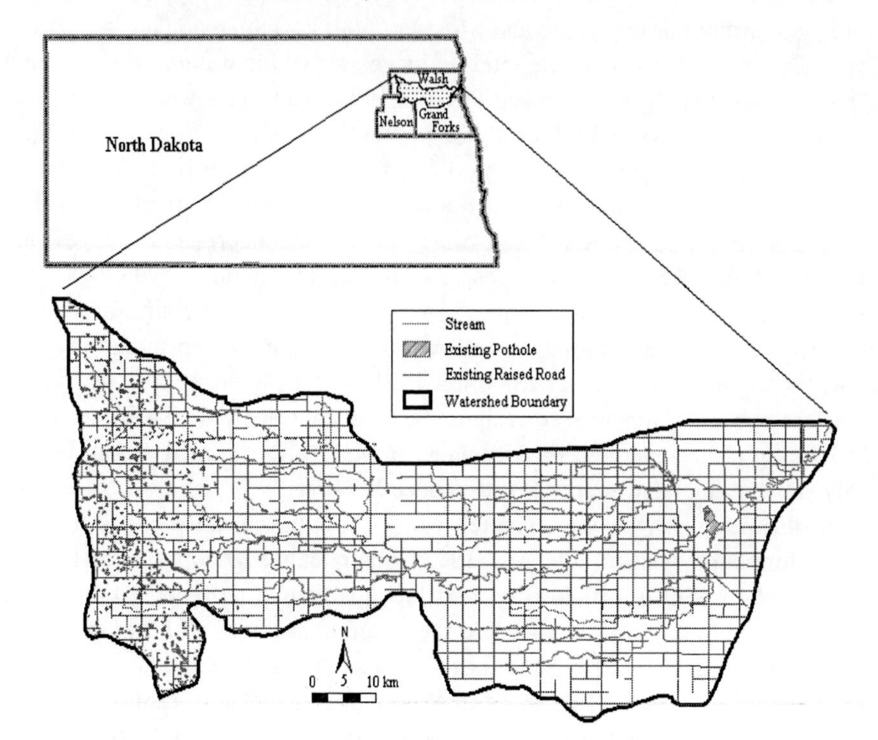

Figure 2-1. Map showing the location and boundary of the Forest River watershed, superimposed by the stream network and existing potholes presented by the U.S. Geological Survey's National Hydrography Dataset (NHD).

The Drift Prairie physiographic area is characterized by rolling hills and numerous prairie potholes (Mark, 1988; Dredge, 2000). Prairie potholes are small ponds and wetlands surrounded by marshy borders that occupy closed basins and do not contribute runoff to streams. Based on the National Hydrography Dataset (NHD; USGS, 2001), the prairie potholes account for approximately 0.9% (19.4 km^2) of the Forest River watershed in size. NHD is a comprehensive set of digital spatial data that contain the information about surface water features such as lakes, ponds, streams, rivers, springs, and wells. Along with the potholes that have been drained, dredged, filled, and leveled since colonization (USGS, 2001), which are not included in NHD, the watershed used to have about 80 km^2 prairie potholes as estimated by Gleason and Euliss (1998). In contrast, the Red River Valley Lake Plain physiographic area has a very low topographic relief and thus few ponds and lakes, as indicated by the information presented by the NHD. The potholes are natural depression storages for runoff. On the other hand, the existing raised roads, with a total length of approximately 2070 km and constructed along the section and quarter-section lines in the Forest River watershed (Figure 2-1; USGS, 1996), created a large amount of artificial storages (Duke et al., 2003). Because of the combined effects of the natural and artificial storages, the runoff contribution area accounts for only 84% (i.e., 1785 out of 2131 km^2) of the watershed in size.

Roads have been cited as one of the major factors inducing the altered hydrology in watersheds (e.g., USEPA, 2006). A raised road has a top elevation higher than that of its adjacent ditches and/or fields, modifying the direction, velocity, and conveyance process of overland flow (Wemple et al., 1996; Jones et al., 2000; Tague and Band, 2001). The installation of culverts at locations where the overland flow tends to accumulate can facilitate the conveyance process, whereas, the absence of culverts would cause the flow to be ponded at these locations. The ponded water is ultimately lost to evaporation and infiltration. Hereafter, these locations are designated "breaches" for descriptive purposes. For a given section, the volume below the breach elevation is designated the "potential depression storage capacity (PDSC)," the total volume of water that can be stored by the natural and artificial storages assuming that no culvert is installed. That is, for the section, its PDSC is greater than its actual storage capacity when one culvert or more exists. Also, the PDSC excludes the volumes below the annual average water surface elevations of the potholes. These volumes are perennially occupied by water and not available for storing additional overland flow.

Datasets Used

Three national datasets, namely NHD, DLG (Digital Line Graph), and NED, were downloaded from the USGS website http://edcftp. cr.usgs.gov/pub/data. As mentioned above, NHD is a comprehensive set of digital data that contain the information about surface water features such as lakes, ponds, streams, rivers, springs, and wells (USGS, 2001). This study used the NHD data layer to define the boundary of, and to identify the stream network and potholes in, the Forest River watershed.

DLG, with a spatial scale of 1:24,000, is the digital representation of cartographic information (USGS, 1996). DLG consists of nine data layers, namely: 1) Public Land Survey System (PLSS) that provides the information on township, range, and section line; 2) boundaries of state, county, city, and other national and state lands (e.g., forests and parks); 3) transportation (e.g., roads, trails, and railroads) and utility supply network (e.g., pipelines and transmission lines); 4) hydrology, such as flowing water, standing water, and wetlands; 5) hypsography, such as contours and supplementary spot elevations; 6) non-vegetative features of lava, sand, and gravel; 7) survey control and bench markers defined by horizontal and vertical positions; 8) manmade features (e.g., building); and 9) vegetative surface cover, such as woods, scrub, orchards, vineyards, and vegetative features associated with wetlands. In this study, the PLSS data layer was used to define the section lines within the Forest River watershed, whereas, the transportation layer was used to identify types of the roads existed along the section lines. For a given section, the elevations at the four corners and four mid points of the section lines were extracted from the hypsography data layer, giving eight known road surface elevations. Because the elevations at other locations along the roads are not available, this study assumed that the road surfaces have linear slopes between any two of the adjacent known-elevation points. Also, this study assumed that the minimal value of these eight known elevations be the breach elevation of the section.

In order to verify these assumptions, three sections were randomly selected for the Forest River watershed. A differential global positioning system (DGPS), which was calibrated to have a horizontal accuracy of less than 1 m and a vertical accuracy of less than 10 cm (Wikimedia Foundation Inc., 2006), was used to survey more than one million points along the four roads surrounding each of the selected sections (Pachhai, 2005; K. William, Research Scientist in the Energy & Environmental Research Center at University of North Dakota, personal communication, 2006). The profiles (not shown), plotted using these surveyed elevations, indicate that for each selected section, the lowest road surface location coincides with the breach that was assumed to be the lowest one of the eight known-elevation points.

NED was designed to provide National elevation data in a seamless form with a consistent datum, elevation unit, and projection (USGS, 2006). NED has a resolution of one arc-second (approximately 30 m) for the conterminous United States, Hawaii, Puerto Rico and the island territories and a resolution of two arc-second for Alaska. The data sources used to develop NED have a variety of elevation units, horizontal units, and map projections. As a result, NED has an accuracy that varies from one area to another. The NED data for the Forest River watershed has a vertical accuracy of between 1 and 2 m (Holmes et al., 2000; Gesch et al., 2002).

In the early spring of 2004, when the vegetations (in particular, trees and tall grasses) did not germinate and thus had a minimal coverage of the ground, Sanborn (http://www.sanborn.com) collected the evaluation data for the Forest River watershed using a LiDAR system that utilizes an airborne laser beam as the sensing carrier (Wehr and Lohr, 1999) and can achieve a very high accuracy when the laser pulses are least prevented from reaching the ground surface (Lillesand et al., 2004). Subsequently, Sanborn conducted a post processing to remove buildings, automobiles, and vegetation stubs from the collected data. The processed data were further analyzed and resampled to develop a 1-m elevation dataset, which was designated the "LiDAR DEM" for descriptive purposes. To assess the vertical accuracy of the LiDAR DEM, the ground elevations for 106 locations that were randomly selected across the watershed were surveyed using the aforementioned DGPS (K. William, Research Scientist in the Energy & Environmental Research Center at University of North Dakota, personal communication, 2006) and compared with the corresponding values presented by the LiDAR DEM. The assessment indicated that the LiDAR DEM has a vertical accuracy of between 0.03 and 0.46 m (Pachhai, 2005). For a given location within the study watershed, the absolute difference between the elevation presented by the LiDAR DEM and that presented by the NED data can be very large (Figure 2-2). The difference tends to be smaller for the Red River Valley Lake Plain than for the Drift Prairie, with the largest differences for the areas adjacent to the streams.

Study Methods

The values for PDSC estimated using the LiDAR DEM were presumed to be accurate and thus taken as the references for assessing the values for PDSC estimated using the NED data. Pachhai (2005) indicated that for a section, the difference of the estimated PDSC values using these two datasets depends on the topographic relief of the section. For the purpose of fair comparison, the 823 sections within the Forest River watershed were classified into three categories. The Category–I sections are characterized by having a low topographic relief, whereas, the Category–III sections are characterized by having a high topographic relief. The Category–II sections have a moderate topographic relief. The index that is commonly used to measure topographic relief is defined as the difference of the highest elevation than the lowest one within the area of interest (herein, section). Obviously, this index is a function of the spatial resolution (i.e., cell size) of DEM (Gao, 1997; Kienzle, 2004): a DEM with a higher resolution tends to give a greater value of the index than the one with a lower resolution, and vice versa. For this reason, the index was not used to classify the sections in this study because the LiDAR DEM and NED have distinctly different spatial resolutions, probably giving incompatible classifications.

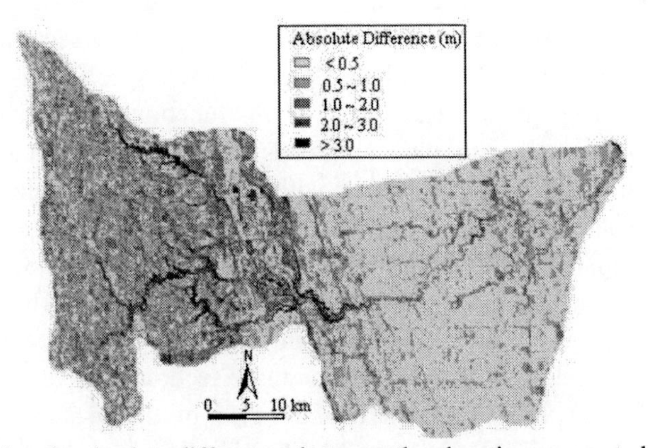

Figure 2-2. Map showing the absolute differences between the elevations presented by the LiDAR DEM and those presented by the U.S. Geological Survey's National Elevation Dataset (NED).

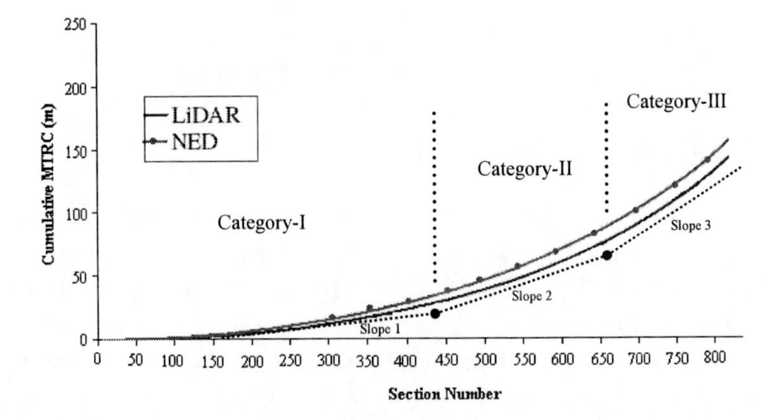

Figure 2-3. Plot showing the cumulative modified topographic relief classifier (MTRC) computed using the 1-m LiDAR DEM and 30-m National Elevation Dataset (NED). Based on the MTRC, the sections in the Forest River watershed were classified into Category-I, -II, and –III.

Instead, this study developed and used a new index, which is defined as the difference of the average elevation of a section minus the median elevation of the same section. To be differentiated from the aforementioned conventional index, this newly developed index was designated MTRC (modified topographic relief classifier) for descriptive purposes. The MTRC is almost independent of the resolution of DEM, as indicated by that the values of MTRC computed using the LiDAR DEM and those computed using the NED data for the sections within the study watershed are visually same (Figure 2-3). This was further verified by a paired t-test, which indicated that the two estimated values for a given section are statistically identical at a significance level of 0.05 (p-value = 0.786). Thus, this study used the MTRC to classify the 823 sections by taking the points (Figure 2-3), at which the slope of the curves has a noticeable change, as the breakpoints for the three categories. A Category–I section was defined to have a MTRC value less than 0.35 m, whereas, a Category–III section was defined to have a MTRC value greater than 0.80 m. A Category–II section was defined to have a MTRC value between 0.35 and 0.80 m.

Along each of the 17 south-north transects that are about 3 km apart, three to six sections were randomly selected to compute the values of PDSC using the LiDAR DEM and the NED data, respectively. The computation was implemented section by section using the ArcGIS® "Surface Volume" function, with the "Plane Height Below" specified as the breach elevation of a selected section. The detailed description of this function can be found in Ormsby et al. (2001). Subsequently, visualization plots and maps were generated to compare the values for PDSC computed using the LiDAR DEM data with those computed using the NED data.

The breach elevation value presented by the hypsography data layer of DLG has an accuracy of ±0.45 m (M. Bearden, Research Geographer, the U.S. Geological Survey, personal communication, 2005). To assess the sensitivity of PDSC to errors included in a breach elevation value, one Category–I section, one Category–II section and one Category–III section were randomly selected. For each of the three sections, the values for PDSC with the "Plane Height Below" specified as ±0.15 and ±0.45 of the breach elevation value were computed using both the LiDAR DEM and NED data. The computed values for PDSC for each section were plotted separately to examine the sensitivity.

RESULTS AND DISCUSSION

Classification of the Sections

Base on the MTRC, 85% (i.e., 703 out of 823) of the sections had a consistent classification using the LiDAR DEM and the NED data. However, each of the remaining 120 sections was assigned into different categories (Table 2-1 and Figure 2-4). This indicated that the MTRC is a powerful classifier irrespective of the spatial resolution of DEM. A close examination revealed that most of the misclassified sections are adjacent to the streams (Figure 2-4 versus Figure 2-1), portions of which are located in the floodplains and could be inundated during flooding season (i.e., spring) but may become dry for the other seasons. The elevations presented both by the LiDAR DEM and the NED data are likely to be the water surfaces at the time when the data were collected (Lillesand et al., 2004; USGS, 2006). Thus, the misclassification of these sections might be because the water surfaces at the time when the LiDAR data were collected were very different from those at the time when the data used to develop NED were collected. Another possible reason is that the dense tall vegetations (e.g., trees) along, and in the vicinity of, the streams caused large errors in the collected data (Hopkinson et al., 2004), as indicated by the larger differences between the elevations presented by the LiDAR DEM data and those presented by the NED data (Figure 2-2). For these reasons, the misclassified sections were excluded from further analysis.

Computed PDSC

In this study, 44 Category–I, 30 Category–II, and 24 Category–III sections were randomly selected to compute the values for PDSC using the LiDAR DEM and NED data. Averaged across the 98 sections, the value for PDSC was estimated to be 2.7×10^5 m^3 per section using the NED data and 3.3×10^5 m^3 per section using the LiDAR DEM data. For all three categories, the values for PDSC computed using the NED data were smaller than the corresponding values computed using the LiDAR DEM data, with a greatest difference of 37.5% for the Category–III sections and a smallest difference of 7.7% for the Category–II sections (Figure 2-5). The difference for the Category–I sections was 19.4%.

Table 2-1. The classes of the 823 sections in the Forest River watershed

Class	Definition[1]	Number of Sections	Consistent for the LiDAR DEM and the NED data? [2]
Category–I	MTRC < 0.35 m	504	Yes
Category–II	0.35 m ≤ MTRC ≤ 0.80 m	99	Yes
Category–III	MTRC > 0.80 m	100	Yes
Misclassified		120	No

[1] MTRC is the modified topographic relief classifier, defined as the average elevation of a section minus the median elevation of the section.

[2] The LiDAR DEM is the 1-m Light Detection and Ranging Digital Elevation Model, whereas, the NED data are the U.S. Geological Survey's 30-m National Elevation Dataset.

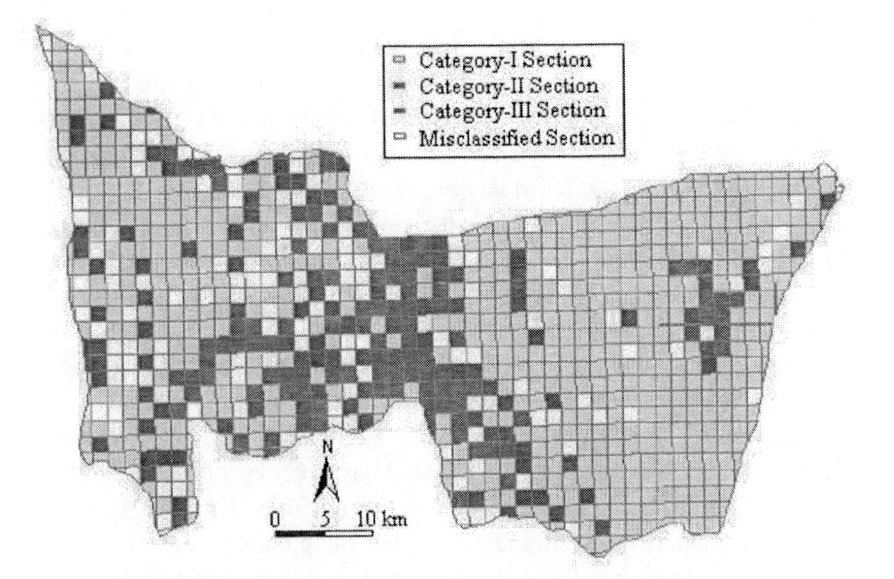

Figure 2-4. Map showing the classification consistency using the 1-m LiDAR DEM and the U.S. Geological Survey's 30-m National Elevation Dataset (NED).

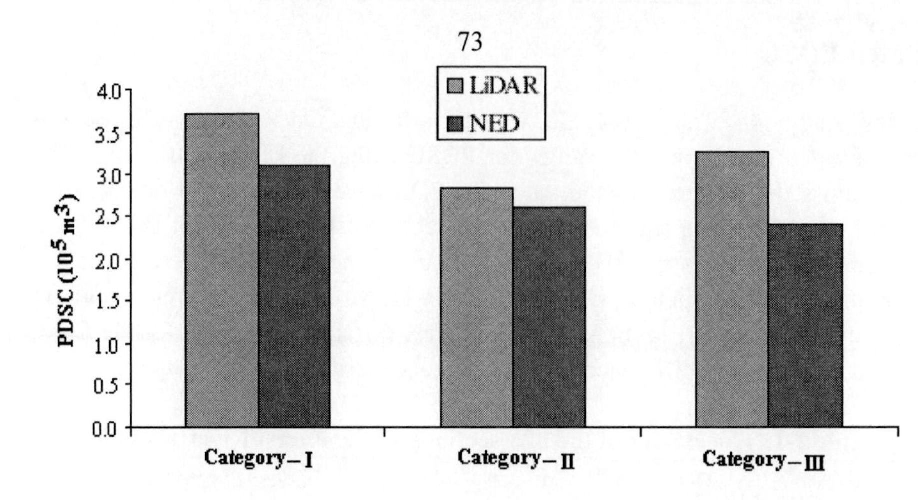

Figure 2-5. The section-averaged values for potential depression storage capacity (PDSC) computed using the 1-m LiDAR DEM versus those computed using the 30-m National Elevation Dataset (NED) for selected Category-I, -II, and –III sections.

With a higher spatial resolution, the LiDAR DEM presents the information that can be used to accurately characterize the minor terrain features such as draws and spurs. A draw is a less developed stream course, whereas, a spur is a short, continuous sloping line of higher ground, and is often formed by two roughly parallel streams (DOD, 1999). However, these minor terrain features are smoothed out in the 30-m NED data. As a result, the storages created by these minor features could not be quantified using the NED data, leading to the underestimation of PDSC. The Category–III sections have more minor features than the Category–I and Category–II sections (Stoner et al., 1993). Thus, the difference of the values for PDSC computed using these two DEM datasets was largest for the Category–III sections (Figure 2-6). In addition, because the Category–I sections have a low topographic relief, a small discrepancy of the elevations provided by the LiDAR DEM and the NED data (Figure 2-2) can result in a large difference of the computed storage areas (Figure 2-7) and thus in a large difference of the computed values for PDSC. In contrast, the Category–II sections have fewer minor terrain features and the PDSC is least sensitive to the small discrepancy of the elevations presented by these two DEM datasets (Figure 2-8).

However, for the Category–III sections, the values for PDSC computed using the LiDAR DEM were well correlated with the corresponding values computed using the NED data, as indicated by a high coefficient of determination (R^2) of 0.95 (Figure 2-9c). In contrast, the computed values for PDSC using these two DEM datasets exhibited a moderate inconsistency for the Category–I and Category–II sections (Figures 9a and 9b), indicating a relatively greater sensitivity to data resolution. This is consistent with the findings of Noman et al. (2003), Hill and Neary (2005), and Martin et al. (2008). The results showed that for watersheds where a high-resolution LiDAR DEM is not available, NED can be used to estimate the PDSCs with an adjustment coefficient of 1.05 to 1.3. That is, the PDSC for a section can be estimated as the multiplication of the value computed using the NED data by the adjustment coefficient. A coefficient approaching to the upper bound should be applied to Category–III sections, whereas, a coefficient close to the lower bound is more appropriate for

Category–II sections. A coefficient around the midpoint of the range (i.e., 1.15) is probably a best choice for Category–I sections.

(a) (b)

Figure 2-6. The potential depression storage area for a typical Category–III section determined using the (a) 1-m LiDAR DEM and (b) 30-m National Elevation Dataset (NED). The areas in blue color are water.

(a) (b)

Figure 2-7. The potential depression storage area for a typical Category–I section determined using the (a) 1-m LiDAR DEM and (b) 30-m National Elevation Dataset (NED). The areas in blue color are water.

(a) (b)

Figure 2-8. The potential depression storage area for a typical Category–II section determined using the (a) 1-m LiDAR DEM and (b) 30-m National Elevation Dataset (NED). The areas in blue color are water.

Sensitivity of PDSC Estimation to Breach Elevation Value

For a given section, errors in the breach elevation value affect the estimation of PSDC (Figure 2-10). As expected, a positive error would result in the overestimation of PSDC, whereas, a negative error would result in the underestimation of PSDC. In addition, the estimation of PSDC is more sensitive to negative than positive errors. Nevertheless, the sensitivity is independent of data resolution, as indicated by that for a given variation of the breach elevation value (e.g., +0.15 m), the corresponding variation of the values for PSDC computed using the LiDAR DEM is almost identical to that of the values for PSDC computed using the NED data (Figure 2-10). This is probably because the water at 0.45 m below the breach elevation may be deep enough to fill the draws, minimizing influences of the minor terrain features on the estimation of PSDC. Compared with the Category–II and –III sections, the Category–I section was determined to be most sensitive to errors in the breach elevation value. Again, because the Category–I section has a low topographic relief, a small discrepancy of the elevation value can result in a large difference of the estimated storage area.

SUMMARY AND CONCLUSIONS

This study developed MTRC, which is computed as the difference of the average elevation of a section from the median elevation of this same section, and showed that MTRC is independent of the spatial resolution of DEM. Using this newly developed index, this study classified the 823 sections of the Forest River watershed, which is located in northeastern North Dakota, into three categories. The Category–I sections have a MTRC value less than 0.35 m, whereas, the Category–III sections have a MTRC value greater than 0.80 m. The Category–II sections have a MTRC value between 0.35 and 0.80. A section with a larger MTRC value tends to have a greater topographic relief than the one with a smaller MTRC value. Subsequently, this study evaluated the feasibility to estimate PDSCs using the available NED data. The evaluation was implemented by comparing the values for PDSC computed using a 1-m LiDAR DEM with those computed using the 30-m NED data for 98 sections that were randomly selected from these three categories.

For a section, compared with the NED data, the LiDAR DEM tended to give a greater value for PDSC. Across the study watershed, the average value for PDSC was estimated to be 2.7×10^5 m^3 per section using the NED data and 3.3×10^5 m^3 per section using the LiDAR DEM. The difference between the values for PDSC computed using the LiDAR DEM and those computed using the NED data was largest for the Category–III sections, followed by the Category–I and then Category–II sections. The estimation of PDSC is more sensitive to errors in the breach elevation value for a Category–I section than for Category-II and –III sections. Nevertheless, for all three category sections, the values of PDSC computed using the LiDAR DEM were well correlated with the values computed using the NED data ($R^2 \geq 0.83$). The results indicate that the PDSCs for the study watershed can be estimated as the multiplications of the values for PDSC computed using the NED data by an adjustment coefficient of 1.05 to 1.3. A coefficient approaching to the upper bound should be applied to the Category–III sections, whereas, a coefficient close to the lower bound is appropriate for the Category–II sections. A coefficient around the midpoint of the range (i.e., 1.15) is probably a best choice for the Category–I sections. One reasonable generalization is that NED can be adapted to estimate PDSCs for watersheds where a high-resolution LiDAR DEM is not available.

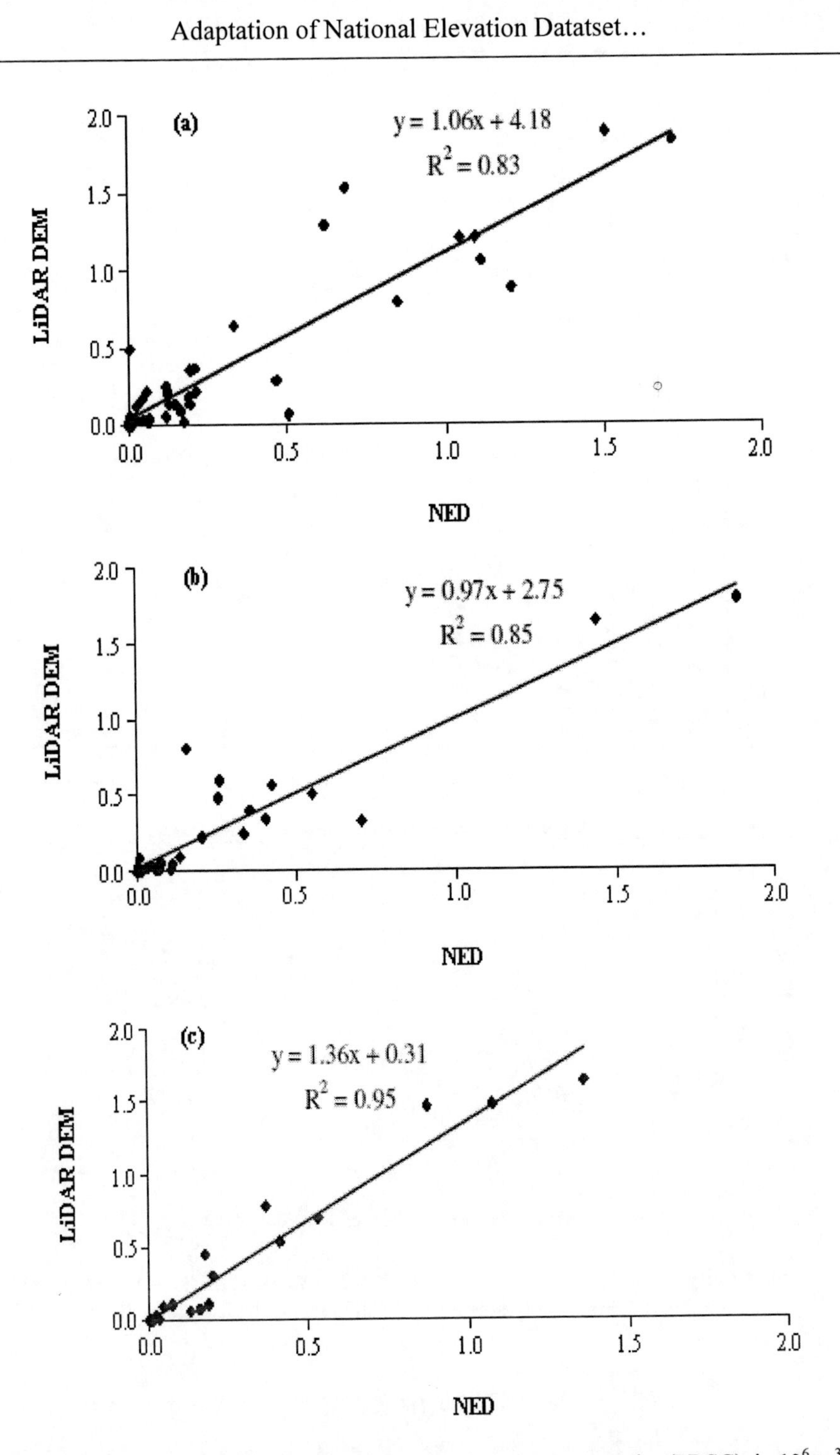

Figure 2-9. Plot showing the values for potential depression storage capacity (PDSC), in 10^6 m^3, computed using the 1-m LiDAR DEM versus those computed using the 30-m National Elevation Dataset (NED) for randomly selected (a) Category-I, (b) Category-II, and (c) category-III, sections.

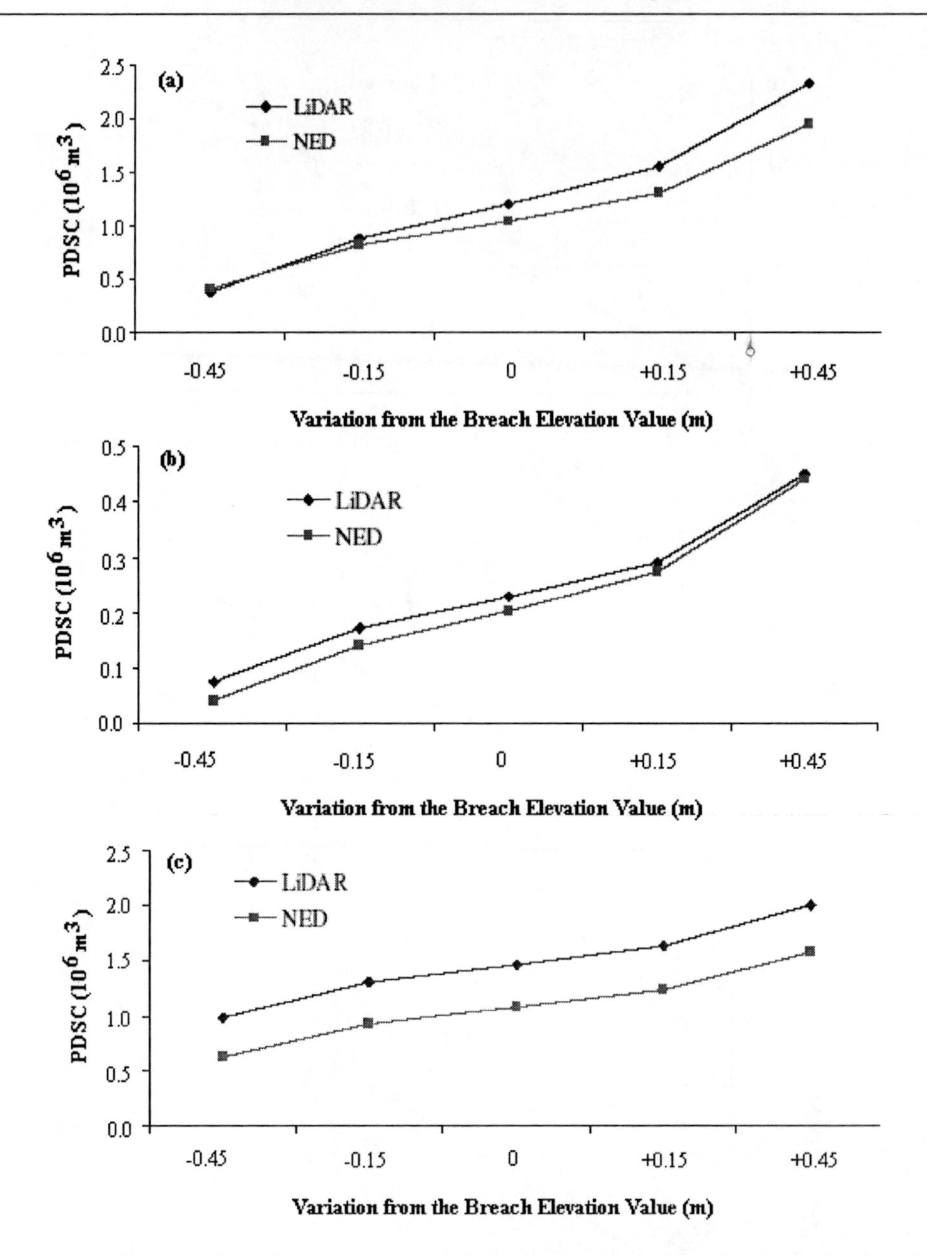

Figure 2-10. Sensitivity of the estimation of potential depression storage capacity (PDSC) to breach elevation value as evaluated for randomly selected (a) Category–I, (b) Category–II, and (c) Category–III, sections.

REFERENCES

Bolles, B.A., and Wang, X. (2003). Distributed basinwide storage for flood mitigation. In *Proceedings of the 20th Annual Red River Basin Land and Water International Summit Conference*. Winnipeg, Manitoba, Canada.

Cochrane, T.A., and Flanagan, D.C. (2005). Effect of DEM resolutions in the runoff and soil loss predictions of the WEPP watershed model. *Transactions of the ASAE* 48(1): 109 – 120.

Croke, J., and Mockler, S. (2001). Gully initiation and road-to-stream linkage in a forested catchment, southeastern Australia. *Earth Surface Processes and Landforms* 26: 205 – 217.

DOD (Department of Defense). (1999). Map Reading and Land Navigation. Washington D.C., Apple Pie Publishers.

Dredge, L. A. (2000). Age and origin of upland block fields on Melville Peninsula, eastern Canadian Arctic. *Geografiska Annaler* 82A(4): 443 – 454.

Duke, G. D., Kienzle, S.W., Johnson, D. L., and Byrne, J. M. (2003). Improving overland flow routing by incorporating ancillary road data into digital elevation models. *Journal of Spatial Hydrology* 3(2): 1 – 27.

FIRP (Florida Institute of Phosphate Research). (1998). Evaluation of the feasibility of water storage reservoir on mined lands to meet future agricultural, potable, industrial, and public water supply demands. WWW document, http://www.fipr.state. fl.us/ar97/ ar97wate.htm.

Gao, J. (1997). Resolution and accuracy of terrain representation by grid DEMs at a micro-scale. *International Journal of Geographical Information Science* 11(2): 199 – 212.

Garbrecht, J., and Martz, L. W. (1997). TOPAZ: An Automated Digital Landscape Analysis Tool for Topographic Evaluation, Drainage Identification, Watershed Segmentation, and Subcatchment Parameterization: Overview. Durant, Oklahoma, the U.S. Department of Agriculture Agricultural Research Service ARS-NAWQL 95 – 1.

Gesch, D., Oimoen, M., Greenlee, S., Nelson, S., Steuck, C., and Tyler, D. (2002). The national elevation dataset. *Photogrammetric Engineering & Remote Sensing* 68(1): 5 – 15.

Gleason, R.A., and Euliss, N.H. Jr. (1998). Sedimentation of prairie wetlands. *Great Plains Research* 8(1): 97 – 112.

Harr, D. H., Harper, W.C., and Krieger, J.T. (1975). Changes in storm hydrographs after road building and clear-cutting in the Oregon Coast Range. *Water Resources Research* 11(3): 436 – 444.

Hill, A. J., and Neary, V.S. (2005). Factors affecting estimates of average watershed slope. *Journal of Hydrologic Engineering* 10(2): 133 – 140.

Holmes, K.W., Chadwick, O. A., and Kyriakidis, P. C. (2000). Error in a USGS 30-meter digital elevation model and its impact on terrain modeling. *Journal of Hydrology* 233: 154 – 173.

Hopkinson, C., Chasmer, L.E., Zsigovics, G., Greed, I.F., Sitar, M., Treitz, P., and Maher, R. V. (2004). Errors in LiDAR ground elevation and wetland vegetation height estimates. *International Archives of Photogrammetry, Remote Sensing and Spatial Information Services* XXXVI–8/W2.

Jones, J. A., and Grant, G. E. (1996). Peak flow responses to clear-cutting and roads in small and large basins, Western Cascades, Oregon. *Water Resources Research* 32(4): 959 – 974.

Jones, J. A., Swanson, F. J., Wemple, B. C., and Snyder, K. U. (2000). Effects of roads on hydrology, geomorphology, and disturbance patches in stream networks. *Conservation Biology* 14(1): 76 – 85.

Kienzle, S. (2004). The effect of DEM raster resolution on first order, second order and compound terrain derivatives. *Transactions in GIS* 8(1): 83 – 111.

King, J. G., and Tennyson, L. C. (1984). Alteration of streamflow characteristics following road construction in north central Idaho. *Water Resources Research* 20(8): 1159 – 1163.

LaMarche, J. L., and Lettenmaier, D. P. (2001). Effects of forest roads on flood flows in the Deschutes River, Washington. *Earth Surface Processes and Landforms* 26: 115 – 134.

LeFever, J. A., Bluemle, J. P., and Waldkirch, R. P. (1999). Flooding in the Grand Forks–East Grand Forks, North Dakota and Minnesota Area. Bismarck, North Dakota, North Dakota Geological Survey Educational series No 25.

Lewis, J., Mori, S.R., Keppeler, E.T., and Ziemer, R.R. (2001). Impacts of logging on storm peak flows, flow volumes and suspended sediment loads in Caspar Creek, California. In *Proceedings of Land Use and Watersheds: Human Influence on Hydrology and Geomorphology in Urban and Forest Areas* (ed. M S Wigmosta and S J Burges). Washington D.C., American Geophysical Union: 127 – 144.

Lillesand, T.M., Kiefer, R.W., and Chipman, J.W. (2004). Remote Sensing and Image Interpretation (5th ed.). New York, John Wiley & Sons Ltd.

Magner, J.A., Payne, G.A., and Steffen, L.J. (2004). Drainage effects on stream nitrate-N and hydrology in south-central Minnesota (USA). *Environmental Monitoring and Assessment* 91(1 – 3): 183 – 198.

Mark, D.M. (1988). Network models in geomorphology. In Proceedings of Modeling Geomorphological Systems (ed. Anderson M G), New York, John Wiley & Sons Ltd. 73 – 97.

Martin, Y.E., Valeo, C., Tait, M., and Johnson, E. (2008). Impacts of centimetre-scale landscape roughness on depression storage, runoff and sediment transport. In *Proceedings of American Geophysical Union Fall Meeting 2008*. Washington D.C., American Geophysical Union, Abstract #H51D-0848.

Noman, N.S., Nelson, E.J., and Zundel, A.K. (2003). Improved process for floodplain delineation from digital terrain models. *Journal of Water Resources Planning and Management* 129(5): 427 – 436.

Ormsby, T., Napoleon, E., Burke, R., Groessl, C., and Feaster, L. (2001). *Getting to Know ArcGIS Desktop*. Redlands, Calofornia, ESRI Press.

Pachhai, S. (2005). Comparison Between 1-m LiDAR and 30-m NED Digital Elevation Models for Temporary Floodwater Storage Estimations in the Red River of the North Basin. Grand Forks, North Dakota, University of North Dakota, Geography Department, Master Thesis.

Stoner, J.D., Lorenz, D.L., Wiche, G.J., and Goldstein, R.M. (1993). Red River of the North Basin, Minnesota, North Dakota, and South Dakota. *Water Resources Bulletin* 29(4): 575 – 615.

Tague, C., and Band, L. (2001). Simulating the impact of road construction and forest harvesting on hydrologic response. *Earth Surface Processes and Landforms* 26(2): 135 – 152.

USEPA (U.S. Environmental Protection Agency). (2006). Hydrology. WWW document, http://www.epa.gov/owow/watershed/wacademy/wam/hydrology.html.

USGS (U.S. Geological Survey). (1996). US GeoData: Digital line graphs. WWW document, http://erg.usgs.gov/isb/pubs/factsheets/fs07896t.pdf.

USGS (U.S. Geological Survey). (2001). National Hydrography Dataset. WWW document, http://nhd.usgs.gov.

USGS (U.S. Geological Survey). (2006). National Elevation Dataset. WWW document, http://ned.usgs.gov/Ned/about.asp.

Veldhuisen, C., and Russell, P. (1999). Forest Road Drainage and Erosion Initiation in Four West-Cascade Watersheds. TFW Effectiveness Monitoring Report TFW-MAG1-99-001.

Viessman, W. Jr., and Lewis, G.L. (2003). *Introduction to Hydrology* (5th ed.). Prentice Hall, Upper Saddle River, NJ.

Wehr, A., and Lohr, U. (1999). Airborne laser scanning: An introduction and overview. *ISPRS Journal of Photogrammetry and Remote Sensing* 54(2): 68 – 82.

Wemple, B.C., and Jones, J.A. (2003). Runoff production on forest roads in a steep, mountain catchment. *Water Resources Research* 39(8): 1220 – 1236.

Wemple, B.C., Jones, J.A., and Grant, G.E. (1996). Channel network extension by logging roads in two basins, Western Cascades, Oregon. *Water resources Bulletin* 32(6): 1195 – 1207.

Wigmosta, M. S., and Perkins, W.A. (2001). Simulating the effects of forest roads on watershed hydrology. In *Proceedings of Land Use and Watersheds: Human Influence on Hydrology and Geomorphology in Urban and Forest Areas* (ed. M S Wigmosta and S J Burges). Washington D.C., American Geophysical Union: 127 – 144.

Wikimedia Foundation Inc. (2006). Differential GPS. WWW document, http://en.wikipedia.org/wiki/Differential_GPS.

Wright, K.A., Sendek, K.H., Rice, R.M., and Thomas, R.B. (1990). Logging effects on streamflow: Storm runoff at Caspar Creek in northwestern California. *Water Resources Research* 26(7): 1657 – 1667.

Ziadat, F.M. (2007). Effect of contour intervals and grid cell size on the accuracy of DEMs and slope derivatives. *Transactions in GIS* 11(1): 67 – 81.

Ziemer, R.R. (1981). Storm flow response to road building and partial cutting in small streams of northern California. *Water Resources Research* 17(4): 907 – 917.

In: Modeling Hydrologic Effect...
Editor: Xixi Wang, pp. 33-58

ISBN 978-1-61668-628-4

Chapter 3

NUMERICAL MODELING OF TRANSPORT PROCESSES AT HILLSLOPE SCALE ACCOUNTING FOR LOCAL PHYSICAL FEATURES

Gokmen Tayfur*
[1]Department of Civil Engineering, Izmir Institute of Technology,
Gulbahce Kampus, Urla, Izmir 35430, Turkey

ABSTRACT

Hillslope is the basic unit of a watershed. Typical hillslopes may have a size of 1000 m long and 500 m wide. For watershed modeling, it is essential to accurately describe the hillslope-scale processes of flow, erosion and sediment transport, and solute transport. Although these processes are usually considered in experimental studies and theoretical subjects, the existing numerical models that are designed to simulate transport processes at hillslope scale rarely take microtopographic variations into account. Instead, those models assume constant slope, roughness, and infiltration rate for a given basic computational unit (i.e., hillslope). As a result, effects of microtopographic features (e.g., rills) on the aforementioned processes cannot be reflected in modeling results. However, the effects could be important because rill and sheet flows exhibit distinctly different dynamics that influence the transport processes. The objective of this chapter is to review the numerical studies for investigating the transport processes at hillslope scale. The chapter focuses particularly on the modeling efforts with the effects of microtopographic features on the dynamics of the transport processes incorporated.

Keywords: Hillslope, flow, sediment, simulation, microtopography, rill, interrill, roughness, infiltration

---

* E-mail: gokmentayfur@iyte.edu.tr

INTRODUCTION

In the past decades, variety of models have been developed to simulate hydrologic and hydraulic processes in watersheds with a size of thousands of square kilometers (e.g.,the Euphrates river basin in Turkey). These models are usually used for: 1) planning, design, and operation of projects to conserve water and soil resources and to protect their quality (e.g., planning and designing soil conservation practices, irrigation water management, wetland restoration, stream restoration, and water-table management, flood protection projects, rehabilitation of aging dams, floodplain management, water-quality evaluation, and water supply forecasting); 2) water resources assessment, development, and management (e.g., analyzing the quantity and quality of streamflow, reservoir system operations, groundwater development and protection, surface water and groundwater conjunctive use management, water distribution systems, and water use); 3) assessing impacts of climate change on national water resources and agricultural productivity; and 4) quantifying impacts of watershed management strategies on environmental and water resources protection (Wurbs 1998; Mankin et al. 1999; Rudra et al. 1999; Singh and Woolhiser 2002).

The use of these watershed models can be greatly facilitated with the available data on topography, soil, land use, and hydrography (Engman and Gurney 1991). For example, digital imagery provides mapping of spatially varying landscape attributes, while radar is being employed for rainfall measurements. Digital elevation models (DEMs), with a typical resolution of 30 m by 30 m, can be used to derive basin geometry, stream networks, slope, aspect, flow direction (Singh and Woolhiser 2002). The use of a geographic information system (GIS) facilitates the: 1) subdivision of a watershed into hydrologically homogeneous subareas in both horizontal and vertical domains; 2) determination of soil loss rates; 3) identification of potential areas of nonpoint source contaminants; 4) maping of groundwater contamination susceptibility; and 5) incorporation of spatial details beyond the existing capability of watershed hydrologic models (Singh and Woolhiser 2002).

Hillsopes form subsections within a watershed. The sizes of the subsections can range from 100 to 500,000 m^2, and a hillslope usually includes one computational cell or more, depending on the cell size. For example, a 900 m by 90 m hillslope consists of 90 numerical computational cells with a size of 30 m by 30 m. As a result, a small size hillslope may form one single computational cell of a watershed, but a large size hillslope would constitute hundreds computational cells. Thus, the dynamics of hydrologic processes at the hillslope scale can greatly influence the ones of hydrologic processes at the watershed scale. The accurate description of the hillslope-scale dynamics is very important for watershed modeling and analysis.

Conventionally, numerical models for transport processes over hillslope assume a smooth surface and do not consider the microtopographic variations on the surface. One justification for this simplification is that considering microtopography could noticeably increase the complexity of the numerical procedure and mandate extra efforts to obtain high-resolution microtopographic data. Zhang and Cundy (1989) and Tayfur et al. (1993) qualitatively investigated the flow over varying microtopographic surfaces at the hillslope scale. In a separate study, Tayfur and Singh (2004) modeled sediment transport over microtopographic surfaces. These studies revealed the importance of varying the infiltration rate, roughness, and local slope, for the analysis of hillslope hydrologic dynamics.

Land surfaces, on which transport processes occur, contain irregular microtopography and/or rills. Transport over such surfaces occurs in both rill and interrill areas (Figure 3-1). Runoff over hillslopes or agricultural watersheds initially starts as sheet flow, and then it concentrates into a series of small channels. The flow concentrations depend on either topographic irregularities or differences in soil erodibility or both. As runoff continues and the erosion progresses, these channels are deepen and widen as a function of slope steepness, runoff characteristics, and soil erodibility. Such erosion-formed microchannels are defined as rills (Emmett 1978; Li et al. 1980). The importance of rills on flow dynamics and sediment transport has been well observed in field and laboratory experimental studies. For example, Meyer et al. (1975) studied the influence of rilling in determining the source of eroded soil in agricultural plots, and observed that there was a significant increase in sediment loss due to the presence of rills. They found that the transport capacity of the rill flow is much greater than that of sheet flow over interrill areas; soil loss increases three to five times when rill develops on a surface. These results were verified in subsequent independent experiemental studies (e.g., Moss and Walker 1978; Abrahams et al. 1989; Abrahams and Parsons 1990; Govindaraju and Kavvas 1992). In addition, Kavvas and Govindaraju (1992) and Tayfur and Kavvas (1994; and 1998) investigated numerical modeling flow over rilled-surfaces employing physically-based flow equations. Tayfur (2007) investigated erosion and sediment transport from rilled surfaces using physically-based erosion and sediment transport equations. These three studies indicate that rills have a significant effect on the transport processes.

These experimental results provide a good opportunity to improve the existing numerical models to reflect the differences in transport capacities of rills and their adjunct interrills. The purpose of this chapter is to review the numerical modeling efforts for simulating transport of flow and sediment through hillslopes, which take into account effects of local features (e.g., microtopographic rills).

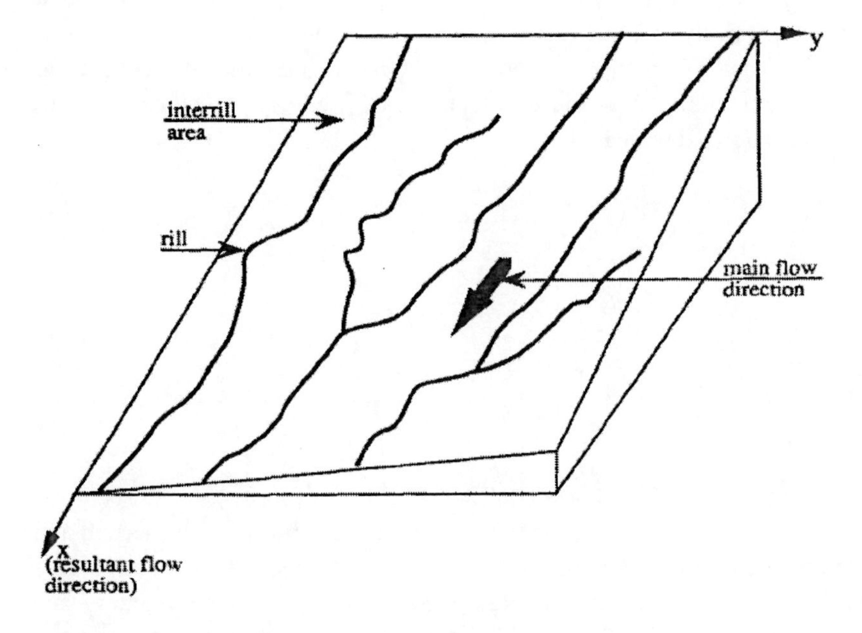

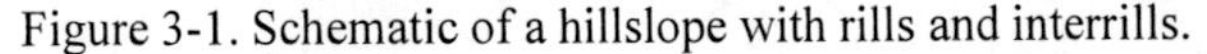
Figure 3-1. Schematic of a hillslope with rills and interrills.

MODELING OVERLAND FLOW OVER MICROTOPOGRAPHIC SURFACE

Overland flows may cause surface erosion and result in flash response in the stream hydrograph. The hydraulics of overland flows is very important in determining flow depth, velocity, and its transport capacity of sediment and chemicals (Moore and Foster 1989). In addition, overland flows can carry nonpoint source pollutants from agricultural lands into receiving water bodies (e.g., channels, natural lakes, and reservoirs). Although numerical techniques (e.g., finite difference, finite volume, or finite element) are effective to solve the governing equations, the direction solution to the St. Venant equations can incur problems of instability and divergence because of the highly nonlinear nature of these equations. Alternatively, researchers simplify the St. Venant equations by only considering important hydraulic processes for the purpose of solving practical problems, resulting in the commonly used kinematic and diffusion wave models.

The two-dimensional St.Venant equations can be expressed as (Tayfur et al. 1993):

$$\frac{\partial h}{\partial t}+\frac{\partial (hu)}{\partial x}+\frac{\partial (hv)}{\partial y}=[r-i] \tag{1}$$

$$\frac{\partial u}{\partial t}+u\frac{\partial u}{\partial x}+v\frac{\partial u}{\partial y}+g\frac{\partial h}{\partial x}=g\left[S_{ox}-S_{fx}\right] \tag{2}$$

$$\frac{\partial v}{\partial t}+u\frac{\partial v}{\partial x}+v\frac{\partial v}{\partial y}+g\frac{\partial h}{\partial y}=g\left[S_{oy}-S_{fy}\right] \tag{3}$$

where h is the flow depth; u and v are the depth-averaged flow velocities in x- and y-directions, respectively; r is the rainfall intensity; i is the infiltration rate; S_{ox} and S_{oy} are the bed slopes in x- and y- directions, respectively; g is the gravitational acceleration; and S_{fx} and S_{fy} are the friction slopes in x- and y- directions, respectively, which can be computed using Manning's equation expressed as:

$$S_{fx}=\frac{n^2u\sqrt{u^2+v^2}}{h^{4/3}} \tag{4a}$$

$$S_{fy}=\frac{n^2v\sqrt{u^2+v^2}}{h^{4/3}} \tag{4b}$$

where n is the Manning's roughness coefficient.

The diffusion wave model neglects the local inertia terms (i.e., the second and third terms in Eqs. 2 and 3), whereas, the kinematic model neglects the first three terms in Eqs. (2) and (3). Zhang and Cundy (1989) investigated effects of varying slope, roughness, and infiltration rate on predicting flows over an artificial domain. Tayfur et al. (1993) used the St. Venant equations as well as the kinematic and diffusion wave models to route flow over an experimental plot (Figure 3-2), and compared the numerical results with the observed hydrographs (Figs. 3, 4, 5, and 6). The plot has average slopes in y- and x-directions of 8.6 and 0.86%, respectively, but the local slopes were computed to be as steep as 15% for the 0.6-m grid resolution. The depressions and crests on the surface may form nodal locations with steep negative slopes, function as storages, and have backwater effects. These physical situations invalidate the kinematic wave model because it assumes that the characteristics move in the forward direction only. The numerical solutions of the St. Venant equations and the diffusion wave model failed to converge even with very small time steps because of the rapidly changing flow regime. The regime of flow with highly variable microtopography is very different from that of sheet flow as assumed by the St. Venant equations. In order to stabilize the solutions, the authors smoothed the surface to obtain a more gradually varying topographic profile. In the flow direction, the local slopes were averaged/smoothed within a 0.6-m window. In contrast, because the local slopes have abrupt changes in the transverse direction, the slopes were averaged within a 1.2-m window to get consistent numerical results. Thus, while the numerical procedure requires a very fine grid resolution to achieve sufficient computational accuracy, the tiny topographic variations have to be smoothed out to satisfy the gradually varying assumption of the St. Venant flow equations.

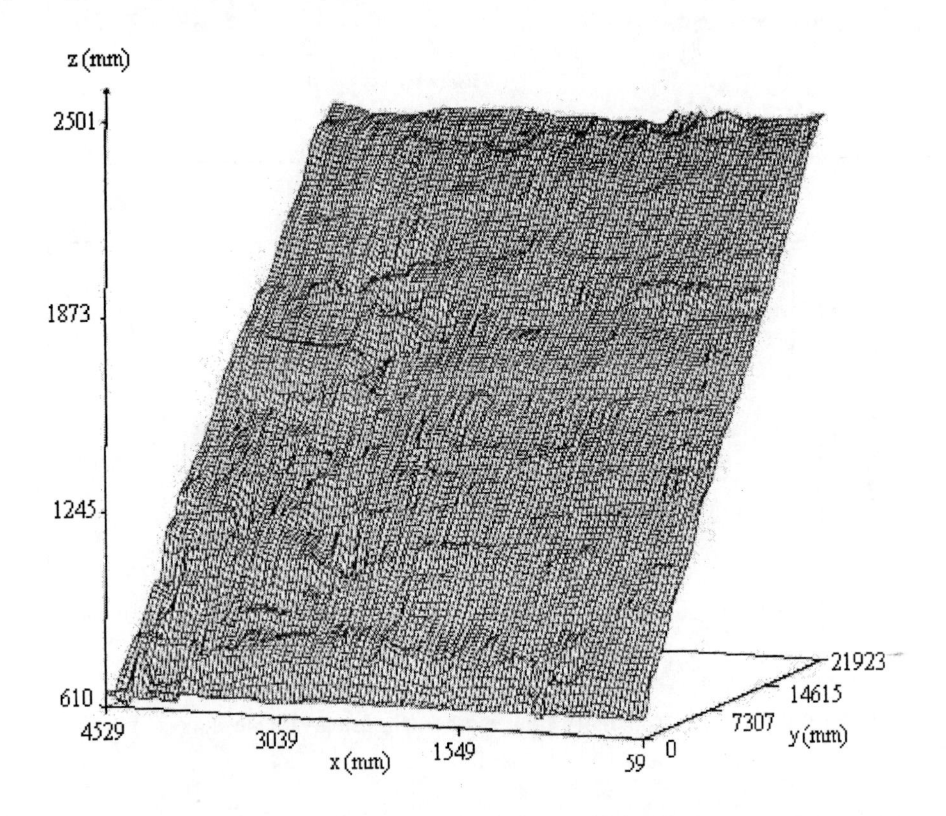

Figure 3-2. The mircotopography of the study plot S3R2A. (Barfield, B.J. and Storm, D.E., Department of Agricultural Engineering at University of Kentucky, Lexington, Kentucky, USA, personal communications, 1989).

The results indicate that there are negligible differences between flow hydrographs predicted using the average slope (Figure 3-3) and those predicted using varying slopes (Figure 3-4). This is probably because the smoothing process removed the partial storage and backwater effects of the microtopographic features, as indicated by the near-identical rising limbs of the hydrographs shown in Figs. 3-3 and 3-4.

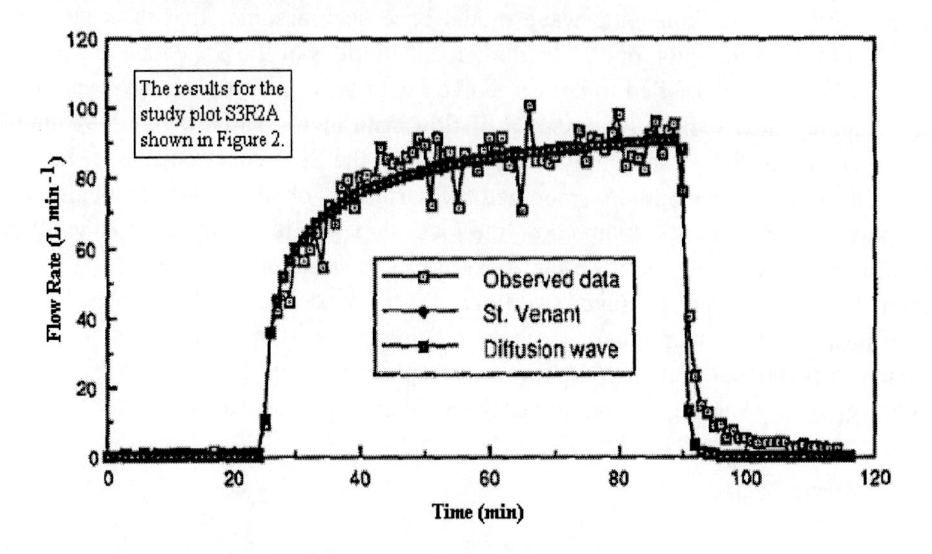

Figure 3-3.The observed versus simulated flow hydrographs using the average slope. (After Tayfur et al. 1993).

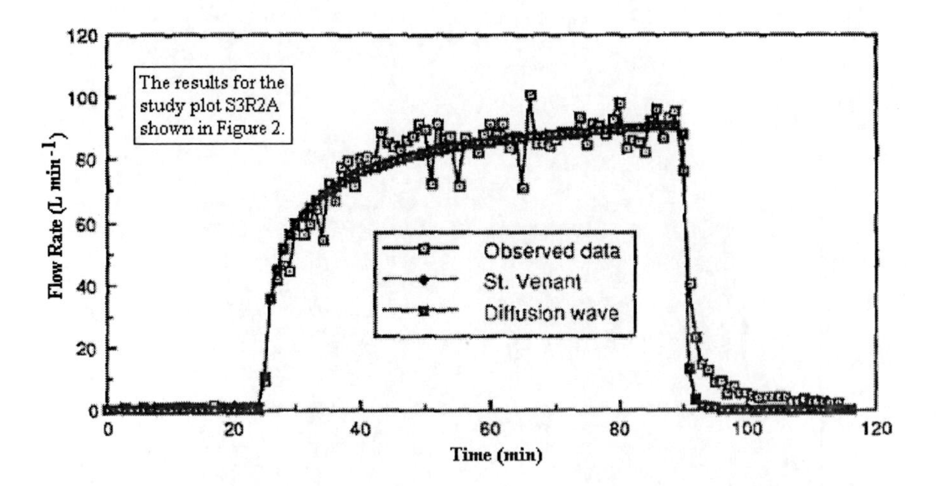

Figure 3-4.The observed versus simulated flow hydrographs using variable slopes. (After Tayfur et al. 1993).

However, the smoothing process did not remove influences of the microtopography on predicting the spatial variations of the local flow depths (Figure 3-5a versus Figure 3-5b) and velocities (Figure 3-6a versus Figure 3-6b), as indicated by the distinctly different predicted spatial patterns. These differences reflect the effects of the microtopographic features. While there is a gradual increase in the flow depth with increasing distance downstream in the x- and y-directions, the increase pattern is not smooth. The velocities predicted using the average slope exhibit a gradual variation both in magnitude and direction (Figure 3-6a), which is different from the pattern of the velocities predicted using variable slopes (Figure 3-6b). When the microtopography is considered, the predicted velocity magnitude gradually increases with slope length, but the velocity direction show great deviations from this pattern at the upstream end of the hillslope, where the water depth is shallow. The variations tend to become smaller as the increase of water depth towards the downstream.

MODELING OVERLAND FLOW OVER RILLED SURFACE

In order to study dynamics of flow over rilled surface, Tayfur and Kavvas (1994; 1998) developed a physically-based model. The model simulates overland flows by combining dynamics of rill flow with those of interrill sheet flow at hillslope scale. The model assumes the interrill sheet flow to be two-dimensional and considers the natural variability of microtopography. On the other hand, the model treats the rill flow to be one-dimensional. A rill receives lateral flows from its adjunct interrill areas, with no reverse flows (Figure 3-7).

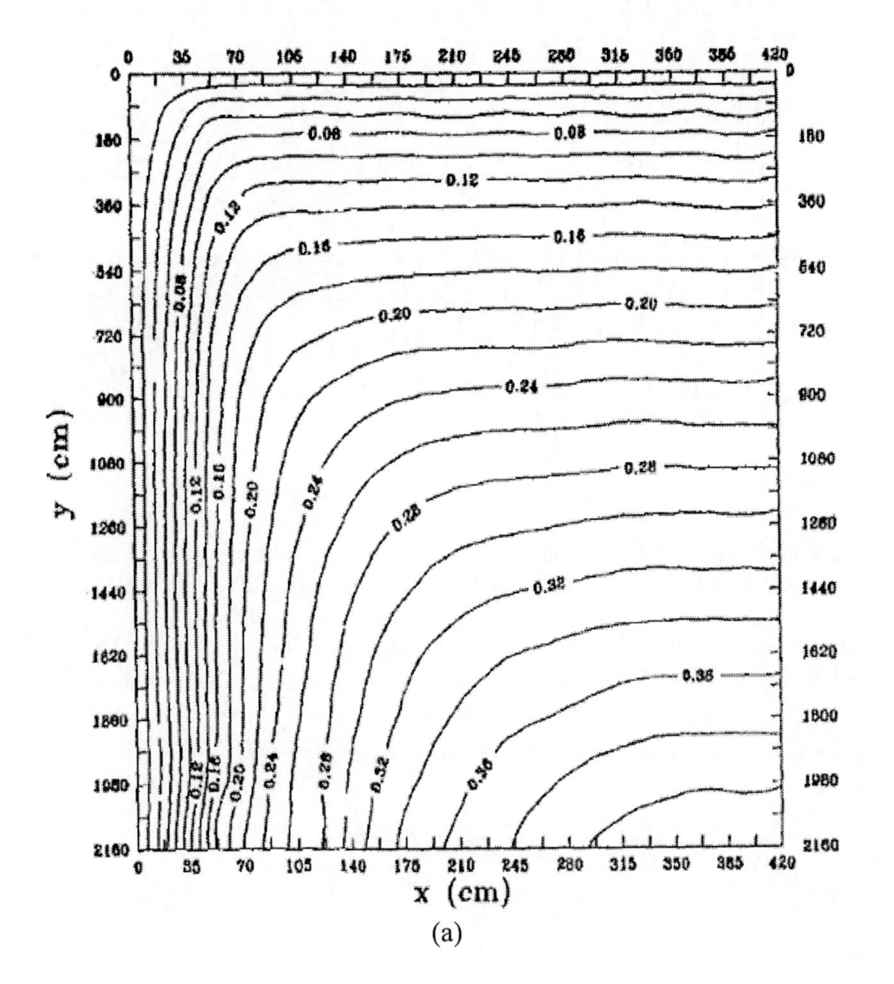

(a)

Figure 3-5. (Continued)

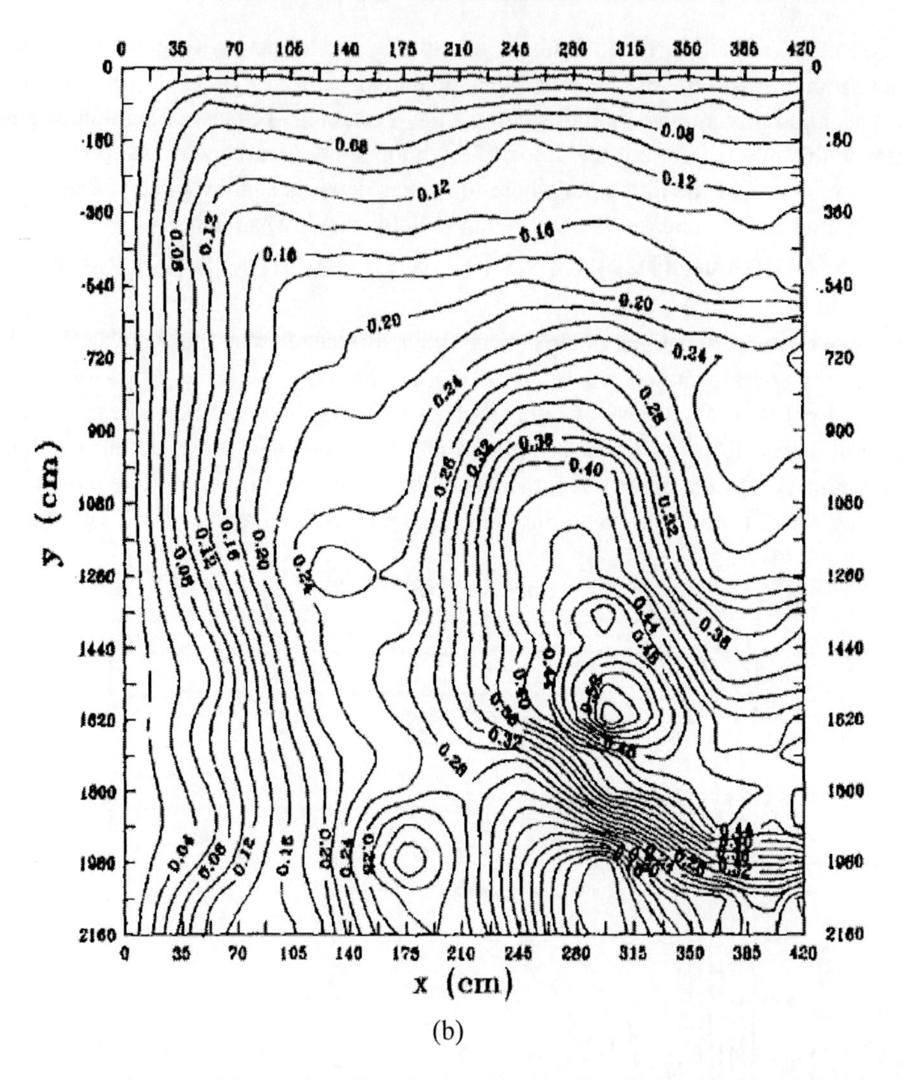

(b)

Figure 3-5. Contour maps pf the overland flow depth, in cm, predicted using (a) the average slope and (b) variable slopes smoothed within a 1.2-m by 0.6-m window. (After Tayfur et al 1993).

For an infinitesimal interrill area, the two-dimensional sheet flow can be approximated by the kinematic wave locally and described by a one-dimensional interrill sheet flow equation (Tayfur and Kavvas 1994). The equation uses the flow averaged over the width of the interrill area (Figure 3-8; Tayfur and Kavvas 1994), and is expressed as:

$$\frac{\partial \bar{h}_o}{\partial t} + \frac{\partial}{\partial x}\left(K'_x \bar{h}_o^{\,3/2}\right) = \bar{q}_l - 1.97\frac{K_y}{l}\bar{h}_o^{\,3/2} \qquad (5)$$

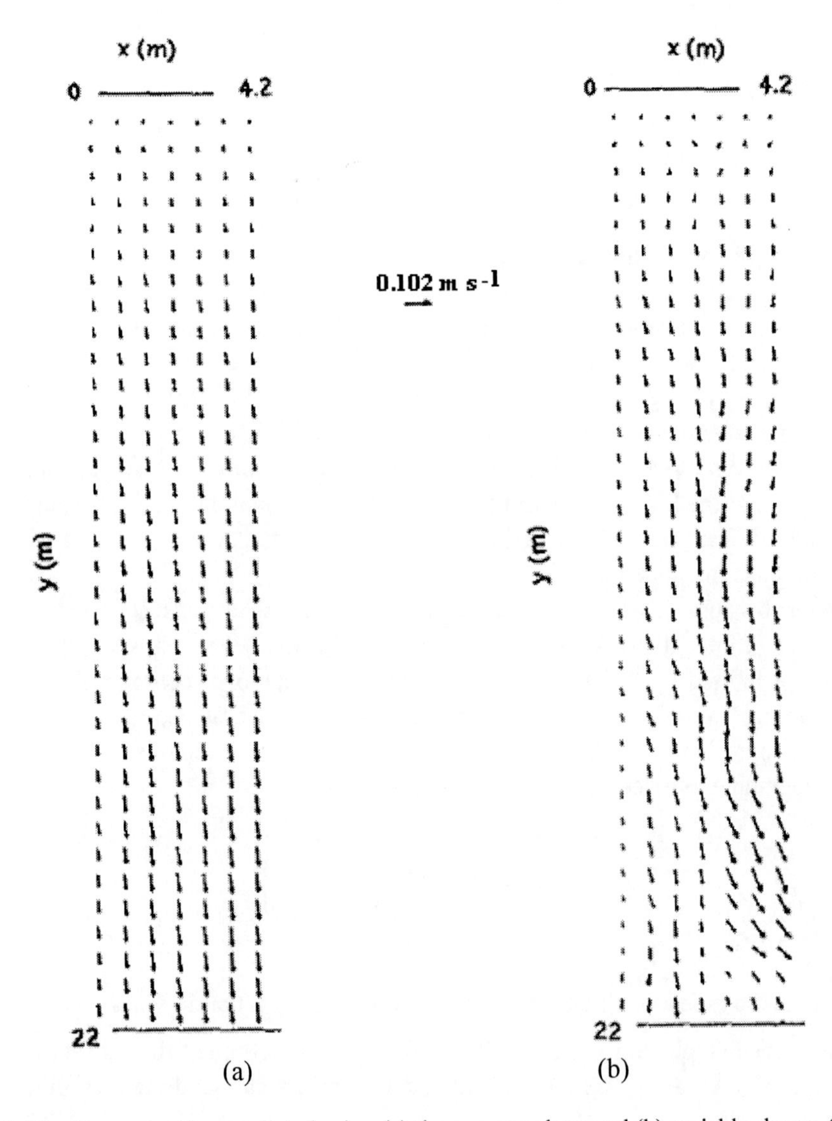

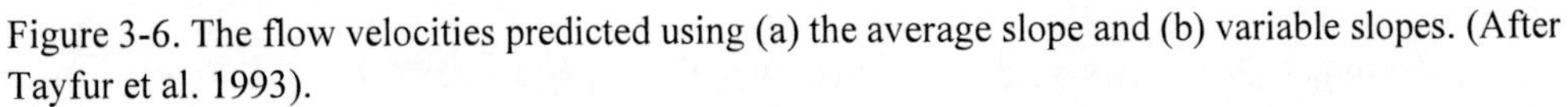

Figure 3-6. The flow velocities predicted using (a) the average slope and (b) variable slopes. (After Tayfur et al. 1993).

where $\overline{h}_o$ is the averaged interrill sheet flow depth; $\overline{q}_l$ is the averaged net lateral flow (i.e., the difference between rainfall and infiltration); K'_x is the expected value of K_x over the interrill area (Figure 3-8).

K_x and K_y are defined as:

$$K_x = \frac{C_z S_{ox}^{0.5}}{\left[1+\left(\frac{S_{oy}}{S_{ox}}\right)^2\right]^{1/4}} \tag{6a}$$

$$K_y = \frac{C_z S_{oy}^{0.5}}{\left[1+\left(\frac{S_{ox}}{S_{oy}}\right)^2\right]^{1/4}} \tag{6b}$$

where C_z is the Chezy roughness coefficient.

The advantage gained by this averaging is that it is not necessary to solve the flow in two dimensions so that there is no need for a finite-difference mesh. This local averaging explicitly yieldes the term representing the water flux from interrill areas into rills, which is represented by the last term on the right hand side of Eq (5). This averaging approach is extended to transects of a hillslope along the orthogonal direction to the main resultant flow (Tayfur and Kavvas 1994) to derive an equation that combines rill flow and interrill sheet flow at the scale of the hillsope transect (Figure 3-8). This equation does not separately model flows in each rill and over each interrill area because the model parameters are individually computed for each hillslope transect.

However, because this model routes flows from transect to transect, it requires detailed geometric data of interrill areas and rills (Tayfur and Kavvas 1994). Thus, applications of this model may be limited by the data availability and computational complexity. To overcome this shortcoming, Tayfur and Kavvas (1998) suggested to integrate Eq. (5) over the hillslope length and to approximate rill flows using an one-dimensional rectangular channel flow (Figs. 9 and 10) equation expressed as:

$$\frac{\partial h_r}{\partial t} + \frac{\partial}{\partial x}\left(K_r \frac{w_r^{0.5} h_r^{1.5}}{\left[w_r + 2h_r\right]^{0.5}}\right) = q_l + 1.97\bar{h}_o^{3/2}\left(\frac{K_{y1}}{w_r} + \frac{K_{y2}}{w_r}\right) \tag{7}$$

where h_r is the cross-sectionally averaged rill flow depth; w_r is the rill width; $K_r = C_{zr}\sqrt{S_{rx}}$; S_{rx} is the rill bed slope; and C_{zr} is the Chezy roughness coefficient for rill sections; and K_{y1} and K_{y2} are computed using Eq. (6b) for the local interrill area 1 and the local interrill area 2 illustrated in Figure 3-7.

Eq. (7) assumes that flows in the rill and over the adjunct interrill areas follow a sine profile and that the rill width at a given location does not change with time. It also assumes that there is no overflow from the rill to the interrill areas. The locally areal-averaged interrill sheet flow and rill flow equations are expressed as (Tayfur and Kavvas 1998):

$$\frac{\partial \hat{h}_o}{\partial t} + \frac{1.97}{L_x}\left(\hat{K}_{xL_x}\hat{h}_o^{3/2}\right) = \hat{q}_l - 3.88\hat{K}_{yl}\hat{h}_o^{3/2} \tag{8}$$

$$\frac{\partial \hat{h}_r}{\partial t} + \left(\frac{1.97 K_R \hat{h}_r^{1.5}}{\left[w_{rL_x} + \pi\hat{h}_r\right]^{0.5}}\right) = \hat{q}_l + 3.88\hat{h}_o^{3/2}\left(\hat{K}_{Y_1} + \hat{K}_{Y_2}\right) \tag{9}$$

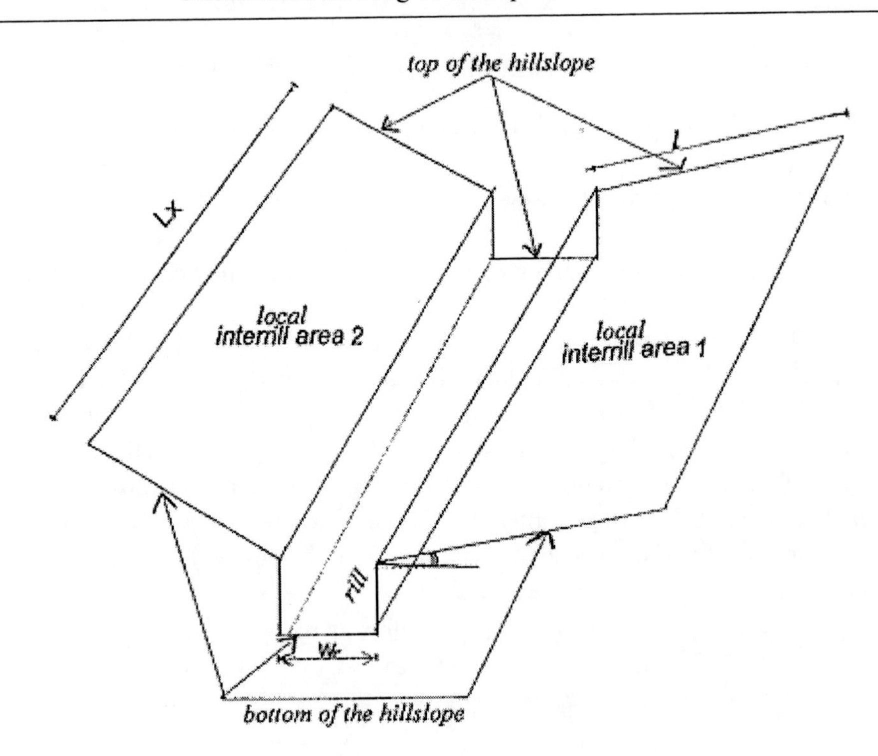

Figure 3-7. The conceptualization of a rill and its adjunct interrill areas.

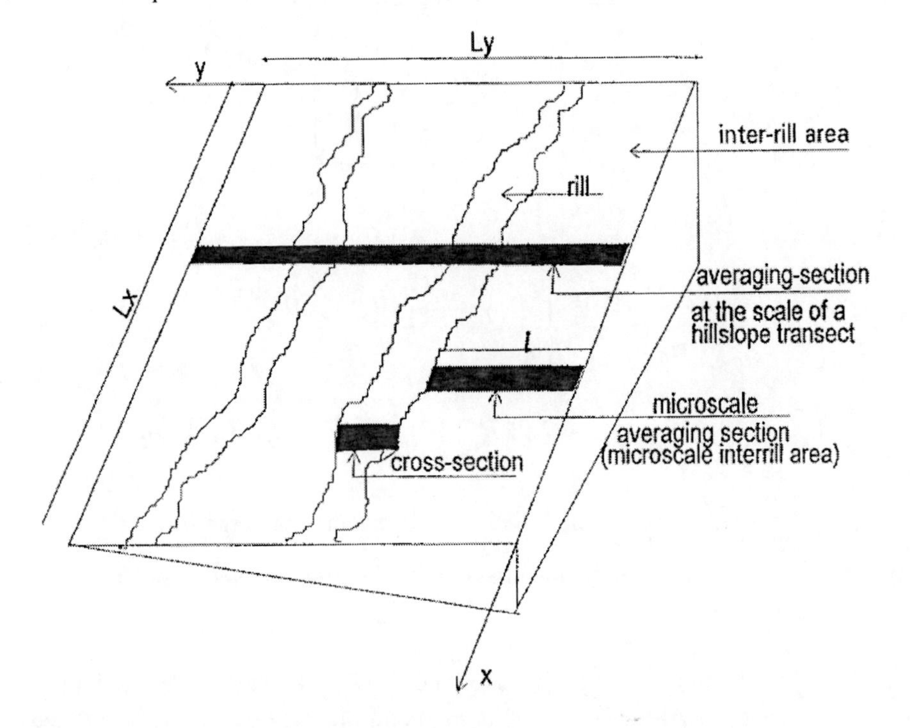

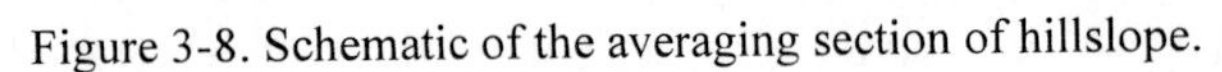
Figure 3-8. Schematic of the averaging section of hillslope.

where $\hat{h}_o$ and $\hat{h}_r$ are the areal-averaged flow depths over the interrill area and in the rill, respectively; $\hat{q}_l$ is the areal-averaged net lateral flow; K_{yl} = K_y/l; and $\hat{K}_{XL_x}$ is the interrill areal-averaged K_x at the hillslope bottom; $K_R = \frac{K_{rLx} w_{rLx}}{Lx}$ and $K_{Yi} = \frac{K_{yi}}{w_r}$, $i=1,2$; K_{rLx} = K_r as determined for the bottom of the hillslope; and w_{rLx} is the rill width at the bottom of the hillslope.

Eq. (8) is for modeling flow of an individual interrill area section (Fig.3-9), whereas, Eq. (9) is for modeling flows of an individual rill section (Fig.3-9). However, given the large number of rill and interrill sections for a hilllsope, it may be ineffective to solve the dynamics section by section. Instead, Tayfur and Kavvas (1998) suggested to statistically aggregate Eqs. (8) and (9) over a whole hilllsope section (Fig.3-10) using the regular perturbation method that considers the first two terms of a Taylor series expansion. The method assumes that the randomness of variables is solely inherited from the physical model. The aggregated equation for an interrill section is expressed as:

$$\frac{\partial h_o'(\bar{r}')}{\partial t}+0.985\sum_{i=1}^{n}\sum_{j=1}^{n}Cov(r_i,r_j)\left\{\frac{\partial^2\left[K_X'(\bar{r}')h_o'^{1.5}(\bar{r}')\right]}{\partial r_i'\partial r_j'}+1.97\frac{\partial^2\left[K_{yl}'(\bar{r}')h_o'^{1.5}(\bar{r}')\right]}{\partial r_i'\partial r_j'}\right\}$$
$$+1.97\left\{K_X'(\bar{r}')h_o'^{1.5}(\bar{r}')+1.97K_{yl}'(\bar{r}')h_o'^{1.5}(\bar{r}')\right\}=\langle q_l'\rangle \tag{10}$$

The aggregated equation for a rill section is expressed as:

$$\frac{\partial h_r'(\bar{r}')}{\partial t}+0.985\sum_{i=1}^{n}\sum_{j=1}^{n}Cov(r_i,r_j)\left\{\frac{\partial^2\left[\frac{K_R(\bar{r}')\,h_r'^{1.5}(\bar{r}')}{\left(w_{r_{Lx}}+\pi h_r'(\bar{r}')\right)^{0.5}}\right]}{\partial r_i'\,\partial r_j'}\right. -$$
$$\left.-1.97\frac{\partial^2\left[K_{Y_1}'(\bar{r}')\,h_o'^{1.5}(\bar{r}')\right]}{\partial r_i'\partial r_j'}-1.97\frac{\partial^2\left[K_{Y_2}'(\bar{r}')\,h_o'^{1.5}(\bar{r}')\right]}{\partial r_i'\partial r_j'}\right\} \tag{11}$$
$$1.97\left\{\frac{K_R(\bar{r}')\,h_r'^{1.5}(\bar{r}')}{\left[w_{r_{Lx}}+\pi h_r'(\bar{r}')\right]^{0.5}}-1.97\,h_o'^{1.5}(\bar{r}')\left[K_{Y_1}'(\bar{r}')+K_{Y_2}'(\bar{r}')\right]\right\}=\langle q'\rangle$$

where $h_o'(\bar{r}')$ and $h_r'(\bar{r}')$ are the hillslope areal-averaged interrill sheet flow and rill flow depths, respectively; and $\bar{r}'$ is the mean vector of the hillslope vector random variable $\bar{r}$ = $(C_z, S_{ox}, S_{oy}, S_r, w_r, l, L_x)$.

To obtain the complete solution to overland flow at the scale of a hillslope, the large-scale areal-averaged interrill sheet flow Eq. (10) and the large-scale areal-averaged rill flow Eq. (11) are solved conjunctively. Eq.(10) is solved first to obtain the areal-averaged discharge flowing into a rill from its adjunct interrill areas as well as the areal-averaged discharge into the stream at the hillslope bottom. Sequentially, Eq. (11) is solved to calculate the areal-averaged discharge from a rill to the stream. The total discharge into the stream is the summation of the discharges from all rills of the hillslope after corrected by considering a probability of rill occurrence λ (Govindaraju and Kavvas 1992; Kavvas and Govindaraju 1992). The large-scale areal-averaged rill flow discharge to the stream is multiplied by λ, while the large-scale areal-averaged interrill sheet flow discharge to the stream is multiplied by (1-λ). These productions are summed up to determine the total discharge into the stream.

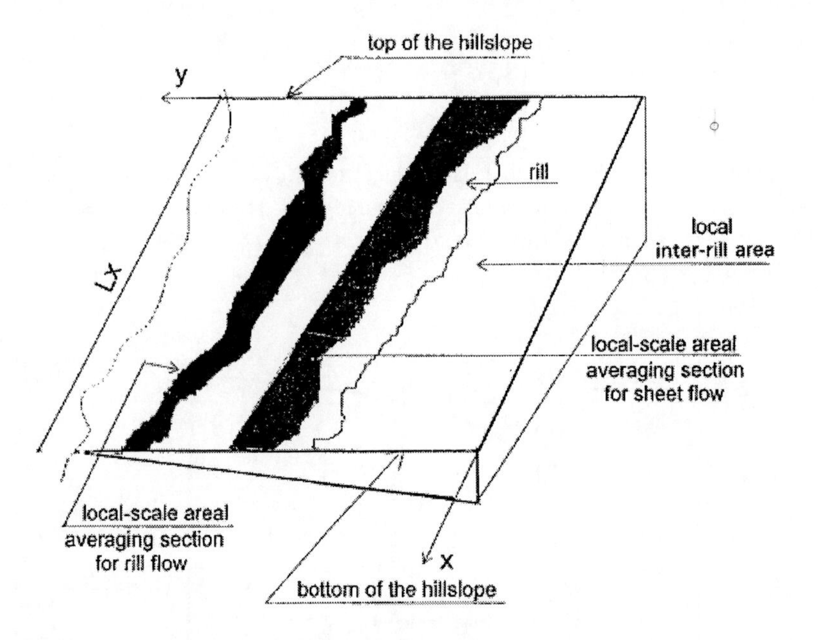

Figure 3-9. Schematic of a local-scale areal averaging section.

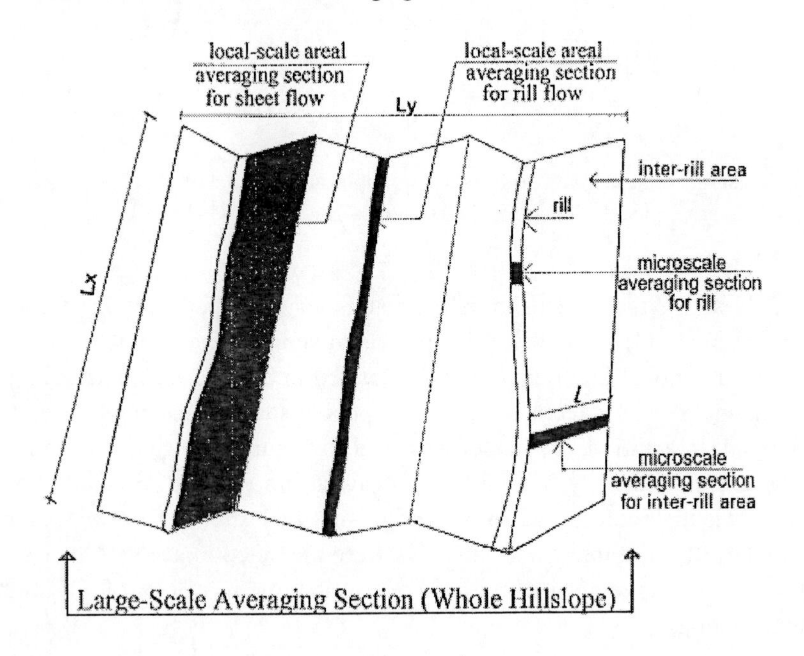

Figure 3-10. Schematic of a hillslope-scale averaging section.

Table 3-1. Summary of the inputs into the model as defined by Eqs. (10) and (11)[1]

Variable	Value	Variable	Value	Variable	Value
L_x (m)	22	S_{oy} (%)	5.2	Cov (S_{ox}, S_{oy})[2]	1.02×10^{-3}
L_y (m)	4.2	l (m)	0.31	Cov (S_{ox}, S_{oy})[2]	0.0041
Rill Number	6	w_r (m)	0.10	C_h (cm hr^{-1})	0.65
R (mm hr^{-1})	97	λ (%)	13.8	p	0.42
t_r (min)	90	Var (S_{ox})[2]	3.24×10^{-4}	Ψ (cm)	18
t_p (min)	20	Var (S_{oy})[2]	0.00112		
S_{ox} (%)	9.1	Var (L_y)[2]	0.095		

r is rainfall intensity; t_r is rainfall duration; t_p is ponding time; C_h is saturated hydraulic conductivity; p is available porosity; Ψ is wetting front capillary pressured head. The other variables are defined in Eqs. (10) and (11). Var ( ) is variance and Cov ( ) is covariance.

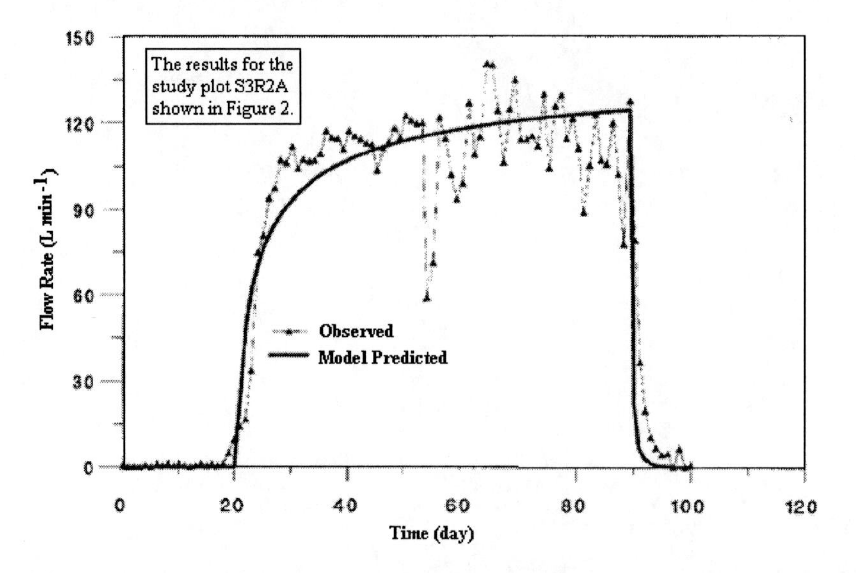

Figure 3-11. Observed versus simulated flow hydrographs at the outlet of the experimental plot S3R2A using the hillslope-scale model. The model is defined by Eqs. (10) and (11). (After Tayfur and Kavvas 1998).

Figure 3-11 shows the runoff simulation results for the experimental plot S3R2A (Figure 3-2; Barfield et al. 1983) using the hillslope-scale averaged flow model as defined by Eqs. (10) and (11). The model input data are summarized in Table 3-1. The results indicate that although the model does not require intensive inputs, it successfully reproduced the observed flow hydrograph. In addition, the model was used to simulate the runoff from a hypothetical hillslope that has a rill density of about 11% (Tayfur and Kavvas, 1994) and the simulation results are shown in Figure 3-12. As expected, the total stream flow was predicted to be mainly from the rills. The interrill sheet flows were predicted to account for less than 5% of the stream flow. This reveals the importance of considering the effects of hills and interills on hydrologic modeling.

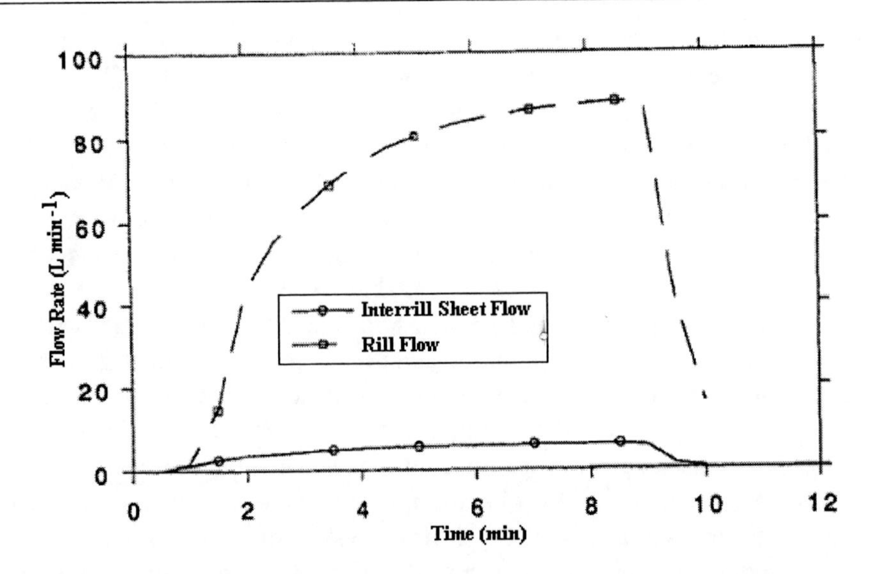

Figure 3-12. Predicted rill flow and interrill sheet flow. (After Tayfur and Kavvas 1994).

Modeling Sedimentation over Microtopographic Surface

Watershed sediment yield is a direct indication of overland erosion rates and quantities of suspended solids that are transported through aquatic systems. Water erosion may incept three sequential processes of detachment, transport, and deposition of soil particles. These processes are controlled by raindrop energy and runoff transport capacity (Foster 1982). Detachment would occur when the erosive force of raindrop or overland flow exceeds the resistance of soil to erosion. Detached particles may then be carried downstream by overland flow. On the other hand, deposition would occur when the sediment load exceeds the runoff transport capacity. Water erosion reduces productivity of cropland and its subsequent sedimentation may degrade water quality because of the agricultural chemicals associated with the fine sediment particles. Further, deposition in water conveyance structures, such as irrigation canals, stream channels, reservoir, estuaries, and harbors, could adversely impact their functionalities (Foster 1982).

Water erosion has been widely studied using laboratory and field experiments (e.g., Kilinc and Richardson 1973; Abrahams et al. 1989; and Govindaraju and Kavvas 1992). Kilinc and Richardson (1973) did extensive rainfall-runoff simulations over slopes of 5.7 to 40% to study the mechanics of soil erosion from overland flow. In addition, Mosley (1974) examined effects of slope and catchment size and slope on rill morphology, discharge, and sediment transport from interrill areas and rills. The author used eight different slopes ranging from 3 to 12%. Further, Moss and Walker (1978), Moss (1979) and Moss et al. (1980; 1982) conducted rainfall-runoff simulations over slopes of 0.1 to 4.2%, measured total sediment concentrations in sheet flow, and examined the formation of rills. Loch and Donnollan (1983a, b) and Loch (1984) measured sediment loadings associated with artificial steady-state runoff over a tilted slope of 4%. Govindaraju et al. (1992) did rainfall-runoff simulations over steep slopes of weathered granite to assess the erosion from cuttings and/or fillings.

Rainfall-induced overland erosion has also been widely studied using physically-based mathematical models by many researchers (e.g., Negev 1967; Foster and Meyer 1972; and Govindaraju and Kavvas 1991). These studies did not consider the microtopography of overland surfaces, which might oversimplify the aforementioned erosion processes. As advancement, Tayfur (2001) developed a two-dimensional erosion and sediment transport equation and examined typical values of the variables. The equation still approximates the highly irregular microtopography using a smooth surface to avoid complications arising in the numerical solution and extra efforts in obtaining the grid-scale microtopographic data required by the solution. Also, the equation assumes homogeneous soil properties and thus does not allow the roughness and infiltration rate to be varied spatially. Moreover, the equation uses the kinematic wave approximation, which would become invalid when backwater effects are important. As with the previous modeling studies cited above, the equation tends to oversimplify the physical processes of soil erosion. As a further improvement, Tayfur and Singh (2004) incorporated effects of mircotopography on erosion and sediment transport into the equation.

The improved two-dimensional erosion and sediment transport equation (Tayfur, 2001; Tayfur and Singh 2004) can be expressed as:

$$\frac{\partial(hc)}{\partial t} + \frac{\partial}{\partial x}(q_x c) + \frac{\partial}{\partial y}(q_y c) = \frac{1}{\rho_s}\left[\alpha r^{\beta} + \sigma(T_c - q_s)\right] \qquad (12)$$

where q_x and q_y are the flow fluxes in the x and y directions, respectively (L^2 T^{-1}); $q_s = \rho_s c(q_x^2 + q_y^2)^{0.5}$ is the sediment flux (M L^{-1} T^{-1}); $T_c = \eta\left[\gamma h(S_x^2 + S_y^2)^{0.5} - \delta_s(\gamma_s - \gamma)d\right]^{k_1}$ is the flow transport capacity (M L^{-1} T^{-1}); c is the sediment concentration by volume (L^3 L^{-3}); ρ_s is the sediment particle density (M L^{-3}); α is the soil detachability coefficient ranging from 0.00012 to 0.0086 kg m^{-2} mm^{-1} (Sharma et al. 1993); β is a constant ranging from 1 to 2; σ is the transfer rate coefficient ranging from 3 to 33 m^{-1} (Foster 1982); η is the soil erodibility coefficient ranging from is 0 to 1.0 (Foster 1982); γ_s is the specific weight of sediment (M L^{-2} T^{-2}); γ is the specific weight of water (M L^{-2} T^{-2}); δ_s is a constant of 0.047 (Gessler 1965); d is the particle diameter (L); and k_1 is an exponent ranging from 1.0 to 2.5 (Foster 1982).

In Eq. (12), αr^{β} describes the soil detachment by raindrop, while $\sigma(T_c - q_s)$ represents the soil detachment and deposition by sheet flow. When $T_c > q_s$, soil particles would be detached. Otherwise, particles will be deposited. This equation is solved conjunctively with Eqs. (1) to (3) to determine the parameters of flow and sediment transport. Eqs (1) to (3) are first solved for the flow variables (i.e., depth, velocity, and flux), which in turn are used in Eq. (12) to determine the parameters for sediment transport. When the water depth and sediment concentration at the upper and lower boundaries are near-zero, they are assumed very small values of 0.00001 m and 0.0001 t m^{-3}. This will eliminate the singularity problem in the numerical solution (Tayfur 2001). The inputs into these equations are summarized in Table 3-2, and the simulation results for the experimental plot S3R2A (Figure 3-2) are shown in Figure 3-13.

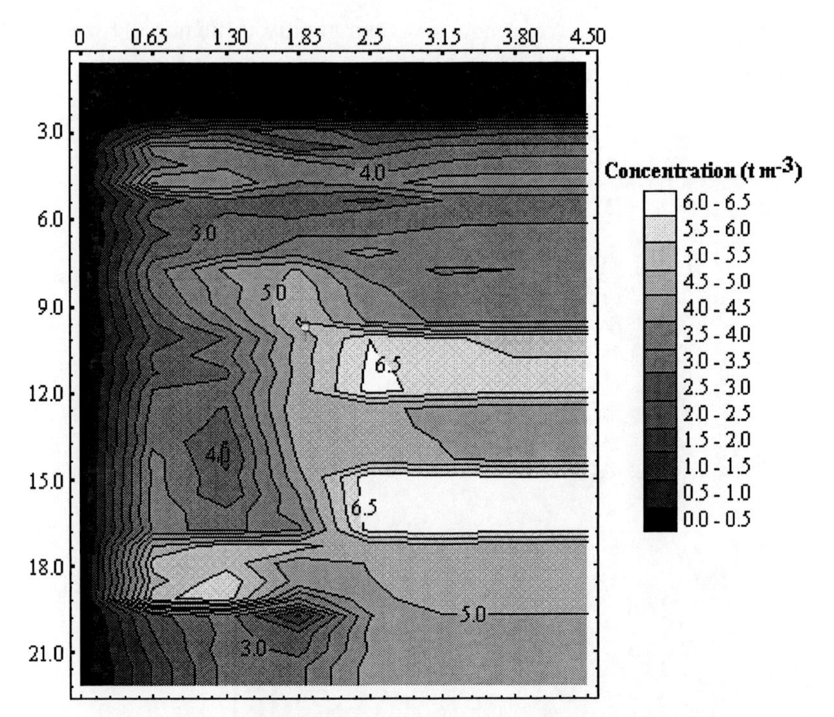

Figure 3-13. Contour map showing the sediment concentration predicted using variable slopes over the experimental plot S3R2A illustrated in Figure 3-2. (After Tayfur and Singh 2004).

Table 3-2. Inputs into Eqs. (1) to (3) and Eq. (12)

Variable	Description	Value
n	Manning's roughness coefficient	0.012
α (kg m^{-2} mm^{-1})	Soil detachability coefficient	0.0022
B	Exponent	1.80
σ (m^{-1})	Transfer rate coefficient	24.0
k_1	Exponent	1.5
H	Soil erodibility coefficient	0.12
r (mm h^{-1})	Rainfall intensity	117
Δt (min)	Rainfall duration	20
i (mm h^{-1})	Constant infiltration rate	7
d (mm)	Particle diameter	1
ρ (kg m^{-3})	Soil bulk density	1500

Although the plot surface was smoothed using a 1.2-m by 0.6-m window (Tayfur and Singh 2004) to stabilize the numerical solution, the predicted sediment concentrations exhibit noticeable spatial variations (Figure 3-13), which reflect the effects of microtopographic slopes. The similar effects can be resulted from heterogeneities of roughness (Figure 3-14) and infiltration rate (Figure 3-15). In this study, the roughness has a mean of 0.0187 and a standard deviation of 0.0066, while the infiltration rate has a mean of 13.89 mm h^{-1} and a standard deviation of 7.95 mm h^{-1}. Compared with those of roughness and local slope, the effects of infiltration rate may be smaller. Nevertheless, the temporal variation of infiltration rate, as described by the Green-Ampt equation, can greatly influence the predicted sediment loadings (Figure 3-16).

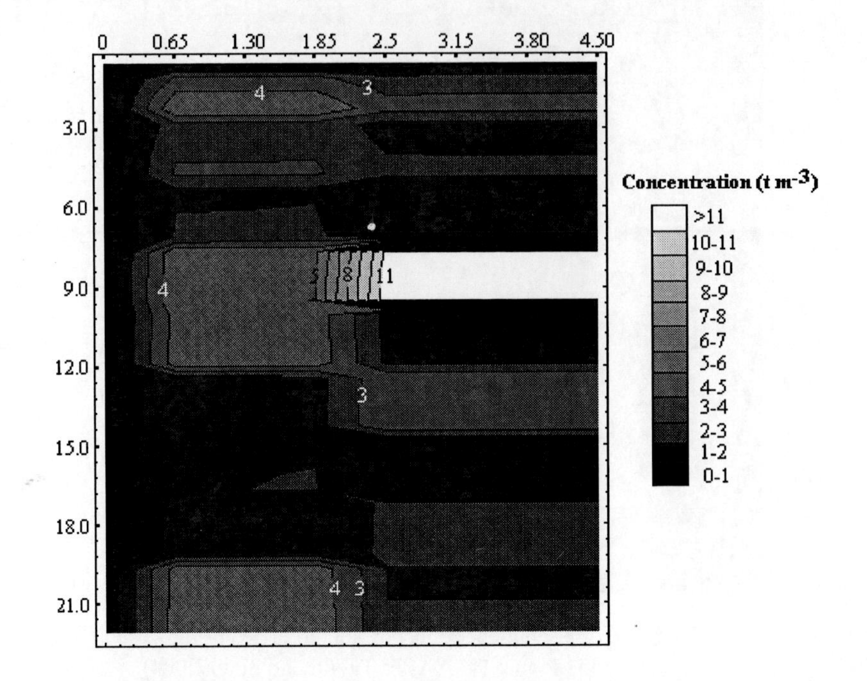

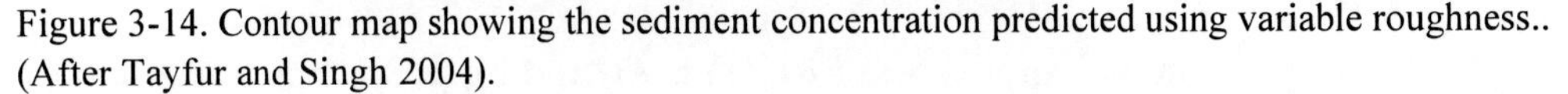

Figure 3-14. Contour map showing the sediment concentration predicted using variable roughness.. (After Tayfur and Singh 2004).

MODELING SEDIMENTATION OVER RILLED SURFACE

As discussed above, microtopographic features such as rills have large effects on hydrologic and erosion processes (Govindaraju and Kavvas 1991; Hairsane and Ross 1992a; Sander et al. 1996; Lisle et al. 1998; Parlange et al. 1999; Hairsane et al. 1999; and Tayfur 2001, 2002). Hairsane and Ross (1992b) developed a theoretical steady-state model for one-dimensional sediment transport from rilled surface. This model assumes that: 1) rills are parallel to each other; 2) rills receive sediment and water fluxes at the transverse direction; 3) rills occur at a fixed frequency of N rills per unit width measured transverse the slope; 4) runoff from an interrill area is directly captured by its adjunct rill and thus the downslope delivery of water solely occurs in the rill; 5) rills have an identical volumetric flow rate; and 6) soils are homogenous. The Water Erosion Prediction Project (WEPP) model also uses a one-dimensional steady-state sediment continuity equation to describe the movement of sediment on rilled surface (Bulygin et al. 2002). In the WEPP model, the interrill sediment delivery is considered to be location independent, and the sediment is conceptualized either to be carried off the hillslope by rill flows or deposited in the rill. These two models are based on equilibrium sediment transport in a rill section and do not consider the transport processes over interrill areas.

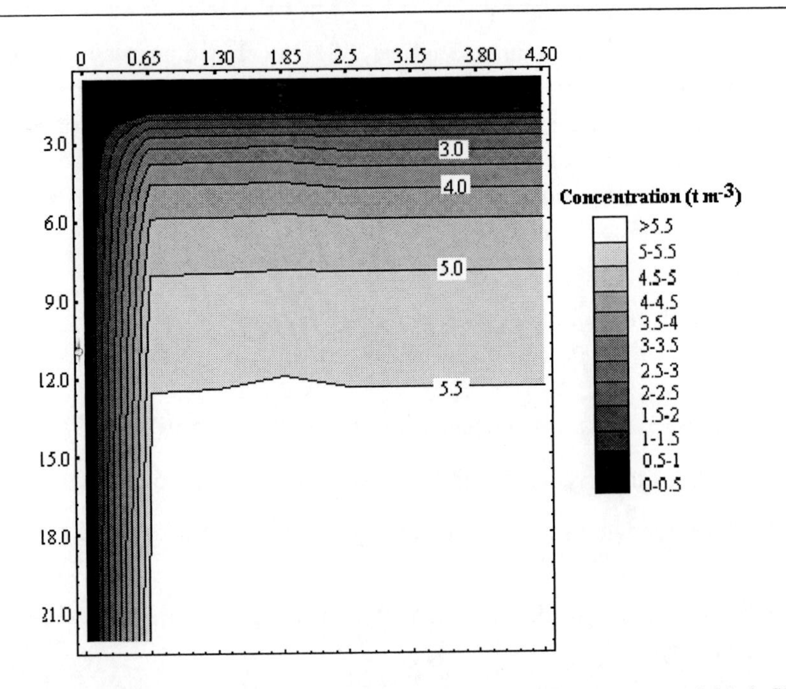

Figure 3-15. Contour map showing the sediment concentration predicted using variable infiltration rate. (After Tayfur and Singh 2004).

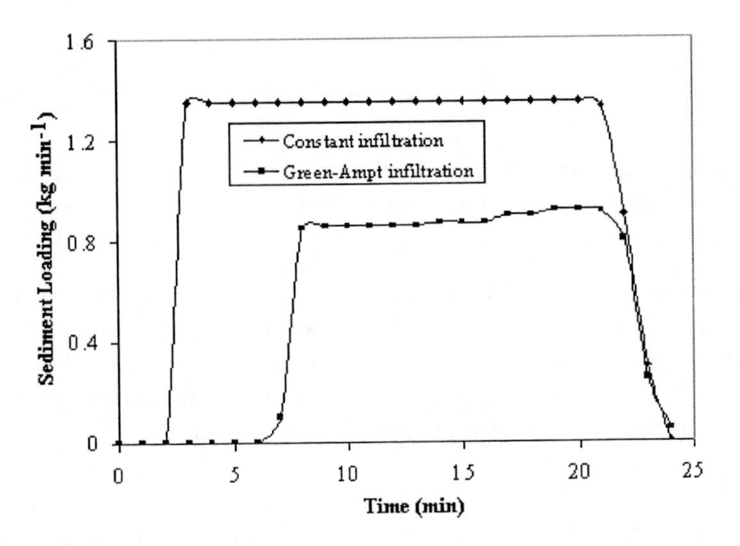

Figure 3-16. The predicted sediment loadings by assuming constant versus temporally variable infiltration rate. (After Tayfur and Singh 2004).

In contrast, Kavvas and Govindaraju (1992) developed an unsteady-state one-dimensional model that considers sheet sediment transport processes over rilled surface. This model takes into account the flow and sediment processes over interrill areas and in rills, but assumes that there is no interaction between these two types of processes. Also, this model does not consider the variability in local microtopography and the lateral sediment inputs from an interrill area into its adjunct rill. To eliminate these limitations and by integrating Eq.

(12) over the length of an interrill area, Tayfur (2007) developed an areal-averaged equation for unsteady-state, non-uniform sheet sediment transport. This two-dimensional equation considers the interactions between rills and interrill areas and the natural variability of surface microtopography. The values for the variables in this equation can be estimated at hillslope scale using digital elevation models. This equation is based on the mass and momentum conservations at hillslope scale and can be expressed as:

$$\frac{\partial(\bar{h}_o\bar{c}_o)}{\partial t}+\frac{\partial}{\partial x}\left(K'_x\bar{h}_o^{1.5}\bar{c}_o\right)=\frac{1}{\rho_s}\left(\bar{D}_{do}+\bar{D}_{fo}\right)-2.95\frac{K_{yl}}{l}\bar{h}_o^{1.5}\bar{c}_o \tag{13}$$

where $\bar{c}_o$ is the averaged sediment concentration ($L^3\ L^{-3}$) over interrill area; $\bar{h}_o$ is the averaged flow depth (L); $\bar{D}_{do}$ is the averaged soil detachment rate of raindrop on interrill area ($M\ L^{-2}\ T^{-1}$); and $\bar{D}_{fo}$ is the averaged soil detachment/deposition rate of sheet flow over interrill area ($M\ L^{-2}\ T^{-1}$).

The cross-sectional averaged, one-dimensional rill sediment transport equation (Tayfur 2007) can be expressed as:

$$\frac{\partial(h_rc_r)}{\partial t}+\frac{\partial}{\partial x}\left[K_rR^{0.5}h_rc_r\right]=\frac{1}{\rho_s}D_{fr}+\left(2.95\bar{h}_o^{1.5}\bar{c}_o\right)\left(\frac{K_{yl_i}}{w_r}\right) \tag{14}$$

$$D_{fr} = \varphi\left[\eta(\tau_r-\tau_c)^k-\rho_sc_rK_rR^{0.5}h_r\right] \tag{15}$$

where c_r is the cross-sectional averaged sediment concentration from the rill ($L^3\ L^{-3}$); D_{fr} is the cross-sectional averaged soil detachment/deposition rate by rill flow ($M\ L^{-2}\ T^{-1}$); and $\tau_r=\gamma RS_r$ is the cross-sectional averaged rill shear stress ($M\ L^{-2}$).

Eq (14) neglects the soil detachment due to raindrop from rill section because the raindrop impact is a dominant factor for the detachment of soil particles on interrill areas but in rills detachment and transport by flow are dominant (Foster 1982). The last term on the right hand side of this equation represents the local-scale lateral sediment flux into the rill from its two adjunct interrill areas illustrated in Figure 3-7.

In practice, the numerical solution needs to be simplified and extra efforts to collect very high-resolution data should be minimized. Eqs. (13) and (14) are integrated over the hillslope length (Figs. 3-8 and 3-9) using a local-scale averaging procedure (Tayfur and Kavvas 1998). The areal-averaged equations (Tayfur 2007) are expressed as:

$$\frac{\partial(h'_oc'_o)}{\partial t}+\frac{2.95}{L_x}\left(K'_{x_{Lx}}h_o'^{1.5}c'_o\right)=\frac{1}{\rho_s}\left(D'_{do}+D'_{fo}\right)-2.95K'_{yl}h_o'^{1.5}c'_o \tag{16}$$

$$\frac{\partial(h'_rc'_r)}{\partial t}+\left[\frac{2.95K_Rh_r'^{1.5}c'_r}{\left(w_{r_{Lx}}+\pi h'_r\right)^{0.5}}\right]=\frac{1}{\rho_s}D'_{fr}+2.95h_o'^{1.5}c'_oK'_{Y_i} \tag{17}$$

where c'_o and c'_r are the local-scale areal averaged interrill area and rill sediment concentrations ($L^3\ L^{-3}$), respectively; h'_o and h'_r are the local-scale areal averaged interrill area and rill flow depths (L), respectively; D'_{do} is the local-scale areal averaged soil detachment rate due to raindrop over interrill area ($M\ L^{-2}\ T^{-1}$); and D'_{fo} and D'_{fr} are the local-scale areal averaged soil detachment/deposition rate ($M\ L^{-2}\ T^{-1}$) by sheet and rill flows, respectively; R_{Lx} is the hydraulic radius of the section at the downstream end of a rill; and $K_R = \frac{K_{r_{Lx}} w_{r_{Lx}}^{0.5}}{L_x}$.

Eqs. (16) and (17) are for modelling the sheet sediment transport over an individual interrill area and the sediment transport in an individual rill. These two equations are statistically averaged over the whole hillslope (Figure 3-10) using a procedure developed by Tayfur and Kavvas (1998). The resulted hillslope-scale sediment transport equations (Tayfur 2007) are expressed as:

$$\frac{\partial\left(h'_o(\bar{r}')c'_o(\bar{r}')\right)}{\partial t} + 1.48\sum_{i=1}^{n}\sum_{j=1}^{n} Cov(r_i, r_j)\left\{\frac{\partial^2\left[K'_X(\bar{r}')h_o'^{1.5}(\bar{r}')c'_o(\bar{r}')\right]}{\partial r'_i \partial r'_j} + \frac{\partial\left[K'_{yl}(\bar{r}')h_o'^{1.5}(\bar{r}')c'_c(\bar{r}')\right]}{\partial r'_i \partial r'_j}\right\}$$
$$+2.95\left\{K'_X(\bar{r}')h_o'^{1.5}(\bar{r}')c'_o(\bar{r}') + K'_{yl}(\bar{r}')h_o'^{1.5}(\bar{r}')c'_o(\bar{r}')\right\} = \frac{1}{\rho_s}\left[D'_{do}(\bar{r}') + D'_{fo}(\bar{r}')\right] \tag{18}$$

$$\frac{\partial\left(h'_r(\bar{r}')c'_r(\bar{r}')\right)}{\partial t} + 1.48\sum_{i=1}^{n}\sum_{j=1}^{n} Cov(r_i, r_j)\left\{\frac{\partial^2\left[\frac{K_R(\bar{r}')h_r'^{1.5}(\bar{r}')c'_r(\bar{r}')}{\left(w_{r_{Lx}} + \pi h'_r(\bar{r}')\right)^{0.5}}\right]}{\partial r'_i\ \partial r'_j} - \frac{\partial^2\left[K'_{Y_1}(\bar{r}')h_o'^{1.5}(\bar{r}')c'_r(\bar{r}')\right]}{\partial r'_i \partial r'_j} - \frac{\partial^2\left[K'_{Y_2}(\bar{r}')h_o'^{1.5}(\bar{r}')c'_r(\bar{r}')\right]}{\partial r'_i \partial r'_j}\right\}$$
$$2.95\left\{\frac{K_R(\bar{r}')h_r'^{1.5}(\bar{r}')c'_r(\bar{r}')}{\left[w_{r_{Lx}} + \pi h'_r(\bar{r}')\right]^{0.5}} - h_o'^{1.5}(\bar{r}')c'_r(\bar{r}')\left[K'_{Y_1}(\bar{r}') + K'_{Y_2}(\bar{r}')\right]\right\} = \frac{1}{\rho_s}D'_{fr}(\bar{r}') \tag{19}$$

where $\bar{r}\,[\,C_z, S_{ox}, S_{oy}, L_x\,]$ is the vector random variable and $\bar{r}'$ is its hillslope-scale mean vector; $h'_o(\bar{r}')$ is the hillslope-scale interrill area sheet flow depth computed by Eq. (10); $c'_o(\bar{r}')$ is the hillslope-scale interrill area sediment concentration; $D'_{do}(\bar{r}')$ is the hillslope-scale soil detachment rate by raindrop over interrill area; $D'_{fo}(\bar{r}')$ is the hillslope-scale soil detachment/deposition rate by sheet flow over interrill area; $h'_r(\bar{r}')$ is the hillslope-scale rill flow depth computed by Eq. (11); $c'_r(\bar{r}')$ is the hillslope-scale rill sediment concentration; $D'_{fr}(\bar{r}')$ is the hillslope-scale rill soil detachment/deposition rate; and $K'_X = \frac{K'_{x_{Lx}}}{L_x}$.

Eqs. (10) and (11) are conjunctively solved first to determine the hillslope-scale averaged flow depths and fluxes over interrill areas and in rills. The results are taken as inputs of Eqs. (18) and (19), which in turn are conjunctively solved for each time step. Eq. (18) is solved to calculate the hillslope-scale averaged sediment loading into rills and Eq. (19) is then solved to calculate the hillslope-scale averaged sediment loading from the rills into the stream located at the bottom end of the hillslope. In order to determine the total sediment loading from a hillslope into the stream, the probability of rill occurrence λ over the hillslope needs to be estimated using a high-resolution (e.g., 10-m) DEM (Govindaraju et al. 1992; Govindaraju and Kavvas 1992; and Kavvas and Govindaraju 1992). The hillslope-scale averaged rill sediment loading into the stream is multiplied by λ, and the hillslope-scale averaged interrill-area sediment loading into the stream is multiplied by (1-λ). These products are then summed up to get the total sediment loading from the hillslope into the adjunct stream.

The geometrics of rills and interrill areas can be determined using the DEM. The solution of Eqs. (18) and (19) requires that the width of a hillslope be greater than the ergodic length scale (i.e., 6 to 8 m) but be smaller than the terrain length scale. Also, the solution assumes that the geometry of a rill is fixed throughout a simulation period, but each rill's geometry may be different from the others'. This assumption is usually valid for rainstorms with a moderate or smaller intensity.

Table 3-3. Geometrics of the rills over the experimental hillslope in northern California[1]

Distance (m)	Expected Spatial Rill Density (%)	Mean Rill Depth (cm)	Mean Rill Width (cm)
9.0	0.1	7.0	10.0
10.5	0.2	8.0	14.0
12.0	0.2	9.0	16.0
13.5	25.0	9.5	19.0
15.0	28.0	11.0	22.0
16.5	0.3	11.5	22.5
18.0	33.0	12.0	23.5
19.5	0.4	12.0	24.0
21.0	38.0	13.0	25.0
22.5	0.4	13.0	24.5
24.0	38.0	13.0	25.5

[1] The data are from Govindaraju et al. (1992).

The solution was used to simulate the flows and sediment loadings from a cut bare hillslope located near Buckhorn Summit in Northern California in the United States of America (USA). The simulation results were compared with the corresponding observed values of the rainfall-runoff experiment conducted by Govindaraju et al. (1992) (Figs. 3-17 and 3-18). The hillslope has an average slope of about 67%. The lower portion of the hillslope, which is about 15 m long and 10 m wide, was subjected to intense rainfall of 152 mm h^{-1} for a duration period of 10 minutes. The sediment laden flow was collected at the downstream along the width of the slope after steady state had been achieved. The sediment loading was measured using a Parshall flume. The rill geometrics were surveyed using a tape measure and a ruler at 11 locations along the slope spaced at 1.5 m (Table 3-3). Govindaraju et al. (1992) determined that the hillslope has a Chezy roughness coefficient of $C_z = 16.6\ m^{0.5}$

s^{-1}. A laboratory test using 23 soil samples indicated that the hillslope has a saturated hydraulic conductivity of K_s = 37.8 mm h^{-1}. The infiltration rate was estimated using the Horton's formula with a rate constant of k = 0.0014 s^{-1} and an initial infiltration rate of f_o = 127 mm h^{-1}. The sediment density was measured to be 2662 kg m^{-3} (Govindaraju et al. 1992).

The solution successfully predicted both the flows (Figure 3-17) and the sediment loadings (Figure 3-18). The good predicted was further indicated by the low absolute errors of 11.07 L min^{-1} for flows and 0.382 kg s^{-1} for sediment loadings. Thus, the solution is judged to be capable in simulating the flow and sedimentation processes over rilled hillslopes.

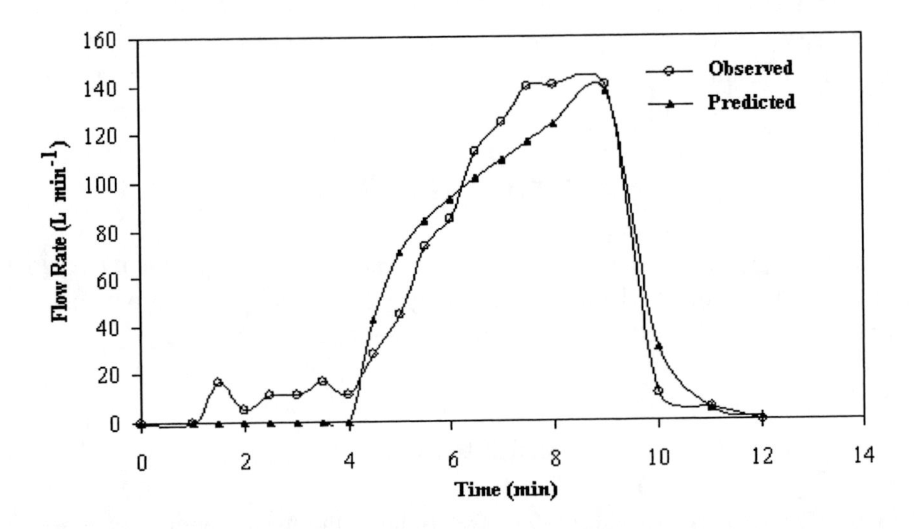

Figure 3-17. The observed versus predicted flows from the experimental hillslope near Buckhorn Summit in northern California. (After Tayfur 2007).

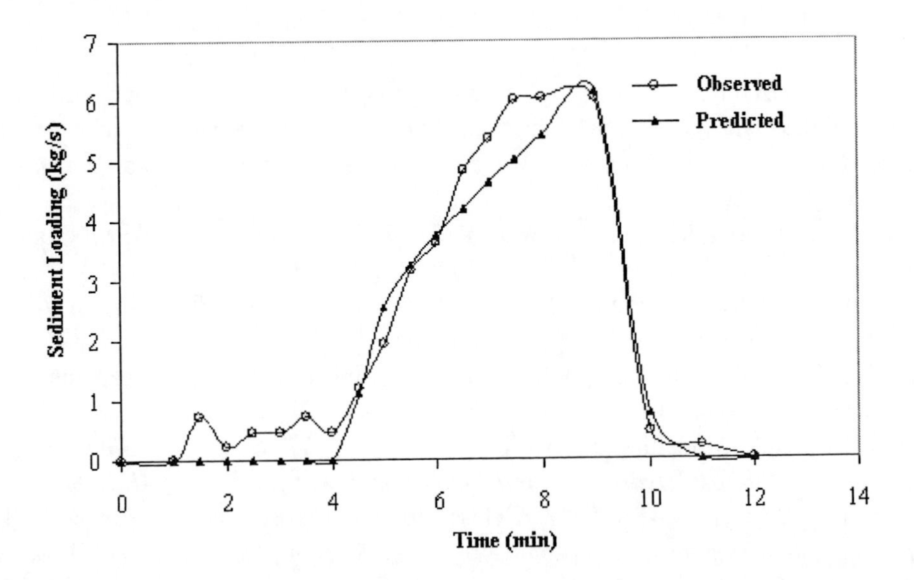

Figure 3-18. The observed versus predicted sediment loadings from the experimental hillslope near Buckhorn Summit in northern California. (After Tayfur 2007).

CONCLUSIONS

Understanding the physical processes of flow and sedimentation at hillsope scale is essential for watershed modeling. This chapter reviews the improved modeling approaches that enable considering the important effects of microtopographic features (e.g., local slope, rill, interrill, and variable roughness) on the processes. The results indicate that although the hillslope-scale areal averaged equations use easily-available data, they did very good jobs in predicting the flows and sediment loadings in the experiment plot S3R2A and the experiment hillslope in northern California in USA. The improved modeling approaches are expected to increase the analysis accuracy in comparison with the conventional methods that neglect microtopographic features.

ACKNOWLEDGMENTS

The author is grateful to B.J. Barfield and D.E. Storm of the Department of Agricultural Engineering, College of Agriculture, University of Kentucky, for providing observed hydrographs and three-dimensional picture of the experiment plot S3R2A.

REFERENCES

Abrahams, A.D. and Parsons, A.J. (1990). Determining the mean depth of overland flow in field studies of flow hydraulics. *Water Resour. Res.*, 26, 501-503.

Abrahams, A.D., Parsons, A.J. and Luk, S.-H. (1989). Distribution of depth of overland flow on desert hillslopes and its implication for modeling soil erosion. *J. Hydrology*, 106, 177-184.

Barfield, B.J., Barnhisel, R.I., Powell, J.L., Hirschi, M.C. and Moore, I.D. (1983). Erodibilities and eroded size distribution of Western Kentucky mine spoil and reconstructed topsoil. *Institute for Mining and Minerals Research Final Report,* Univ. of Kentucky, Lexigton, KY.

Bulygin, S.Y., Nearing, M.A. and Achasov, A.B. (2002). Parameters of interrill erodibility in the WEPP model. *Eurasian Soil Sci*, 35(11), 1237-1242.

Emmett, W.W. (1978). Overland Flow. *In: M.J. Kirkby (ed.) Hillslope Hydrology*, John Wiley and Sons, New York, N.Y., 145-176.

Engman, E.T. and Gurney, R.J. (1991). Remote sensing in hydrology, Chapman and Hall, London, UK.

Foster, G.R. (1982). Modelling the erosion process. *In : C.T. Haan, H.P. Johnson and D.L. Brakensiek (Editors), Hydrologic modelling of small watersheds'. ASAE*, 295-380.

Foster, G.R. and Meyer, L.D. (1972). A closed-form soil erosion equation for upland areas. *Sedimentation Symposium to Honor Prof. H.A. Einstein*, H.W. Shen ed., Fort Collins, Colorado, 12.1-12.9.

Gessler, J. (1965). The beginning of bedload movement of mixtures investigated as natural armoring in channels. E.A. Prych, translator, W.M. Keck Laboratory of Hydraulics and Water Research, CIT, Pasedana, California.

Govindaraju, R.S. and Kavvas, M.L. (1991). Modelling the erosion process over steep slopes: approximate analytical solutions. *J. of Hydrology*, 127, 279-305.

Govindaraju, R.S. and Kavvas, M.L. (1992). Characterization of the rill geometry over straight hillslopes through spatial scales. *J. Hydrology*, 130, 339-365.

Govindaraju, R.S., Jones, S.E. and Kavvas, M.L. (1988), On the Diffusion Wave Model for Overland Flow, 1, Solution for Steep Slopes. *Water Resources Research*, 24(5), 734-744.

Govindaraju, R.S., Kavvas, M.L., Tayfur, G. and Krone, R.B. (1992). Erosion control of decomposed granite at Buckhorn Summit. *Final Report*. California Department of Transportation.

Hairsane, P.B. and Rose, C.W. (1992a). Modelling water erosion due to overland flow using physical principles, 1. sheet flow. *Water Resour. Res.*, 28(1), 237-243.

Hairsane, P.B. and Rose, C.W. (1992b). Modelling water erosion due to overland flow using physical principles, 2. rill flow. *Water Resour. Res.*, 28(1), 244-250.

Hairsane, P.B., Sander, G.C., Rose, C.W., Parlange, J.-Y, Hogarth, W.L., Lisle, I. and Rouhipour, H. (1999). Unsteady soil erosion due to rainfall impact: a model of sediment sorting on the hillslope. *J. Hydrology,* 220, 115-128.

Kavvas, M.L. and Govindaraju, R.S. (1992). Hydrodynamic averaging of overland flow and soil erosion over rilled hillslopes. *Erosion, Debris Flows and Environment in Mountain Regions, Proceedings of the Chengdu Symposium,* IAHS Publ: 209.

Kilinc, M. and Richardson, E.V. (1973). Mechanics of soil erosion from overland flow generated by simulated rainfall. *Hydrology Papers, Colorado State University*, Fort Collins, Paper 63.

Li, R.M., Ponce, V.M. and Simons, D.B. (1980). Modeling rill density. *J. Irrig. and Drain. Div.*, ASCE, 106(1), 63-67.

Lisle, I.G., Rose, C.W., Hogarth, W.L., Hairsine, P.B., Sander, G.C. and Parlange, J.-Y. (1998). Stochastic sediment transport in soil erosion. *J. Hydrology*, 204(1-4), 217-230.

Loch, R.J. (1984). Field rainfall simulator studies on two clay soils of the Darling Downs, Queensland, III, An evaluation of current methods of deriving soil erodibilities (K factors). *Aust. J. Soil Res.*, 22, 401-412.

Loch, R.J. and Donnollan, T.E. (1983a). Field rainfall simulator studies on two clay soils of the Darling Downs, Queensland, I, The effects of plot length and tillage orientation on erosion processes and runoff erosion rates. *Aust. J. Soil Res.*, 21, 33-46.

Loch, R.J. and Donnollan, T.E. (1983b). Field rainfall simulator studies on two clay soils of the Darling Downs, Queensland, II, Aggregate breakdown, sediment properties and soil erodibility. *Aust. J. Soil Res.*, 21, 47-58.

Mankin, K.R., Koelliker, J.K. and Kalita, P.K. (1999). Watershed and lake water quality assessment: An integrated modeling approach. *J. Am. Water Resour. Assoc.*, 35(5), 1069-1088.

Meyer, L.D., Foster, G.R. and Romkens, M.J.M. (1975). Source of soil eroded from upland slopes. *Proc. 1972 Sediment Yield Workshop, U.S. Dept. Agric. Sediment Lab.,* Oxford, Mississippi, ARS-S-40, USDA, 177-189.

Moore, I.D. and Foster, G.R. (1989). Hydraulics and Overland Flow, *Process studies in hillslope hydrology,* John Wiley&Sons, Sussex, England, UK, 1-34.

Mosley, M.P. (1974). Experimental study of rill erosion. *Trans. Am. Soc. Agric. Eng.*, 17, 909-913.

Moss, A.J. (1979). Thin flow transportation of solids in arid and non-arid areas: A comparison of processes. IAHS-AISH Publ., 128, 435-445.

Moss, A.J., and Walker, P.H. (1978). Particle transport by continental water flows in relation to erosion, deposition, soil and human activities. *Sediment,* Geology, 20(2), 81-139.

Moss, A.J., Green, P. and Hutka, J. (1982). Small channels: Their experimental formation, nature and significance. *Earth Surf. Process. Landforms*, 7, 401-415.

Moss, A.J., Walker, P.H., and Hutka, J. (1980). Movement of loose, sandy detritus by shaloow water flows: An experimental study. *Sediment. Geol.,* 25, 43-66.

Negev, N. (1967). A sediment model on a digital computer. Tech. Rep. No. 76, Stanford University, California, 109 pp.

Parlange, J.-Y., Hogarth, W.L., Rose, C.W., Sander, G.C., Hairsine, P. and Lisle, I. (1999). Addendum to unsteady soil erosion model. *J. Hydrology*, 217(1-2), 149-156.

Rudra, R.P., Dickinson, W.T., Abedini, N.J. and Wall, G.J. (1999). A multi-tier approach for agricultural watershed management. *J. Am. Water Resour. Assoc*., 35(5), 1059-1070.

Sander, G.C., Hairsine, P.B., Rose, C.W., Cassidy, D., Parlange, J.-Y., Hogarth, W.L. and Lisle, I.G. (1996). Unsteady soil erosion model, analytical solutions and comparison with experimental results. *J. Hydrology*, 178(1-4), 351-367.

Sharma, P.P., Gupta, S.C. and Foster, G.R. (1993). Predicting soil detachment by raindrops. *Soil Sci. Soc. Am. Journal*, 57, 674-680.

Singh, V.P. and Woolhiser, D.D. (2002). Mathematical modeling of watershed hydrology. *J. Hydrologic Engrg*. ASCE, 7(4), 270-292.

Tayfur, G. (2001). Modelling two dimensional erosion process over infiltrating surfaces. *J. Hydrologic Engrg,* ASCE, 6(3), 259-262.

Tayfur, G. (2002). Applicability of sediment transport capacity models for non-steady state erosion from steep slopes." *J. Hydrologic Engrg,* ASCE, 7(3), 252-259.

Tayfur, G. (2007). Modeling sediment transport from bare rilled hillslopes by areally averaged transport equations. *Catena*, 70, 25-38.

Tayfur, G. and Kavvas, M.L. (1998). "Areal averaged overland flow equations at hillslope scale." *Hydrological Sciences J.,* IAHS, 43(3):361-378.

Tayfur, G. and Singh, V.P. (2004). Numerical model for sediment transport over non-planar, non-homogeneous surfaces. *J. Hydrologic Engrg*. ASCE , 9(1), 35-41.

Tayfur, G., and Kavvas, M.L. (1994). Spatially averaged conservation equations for interacting rill-interrill area overland flows . *J. Hydraulic Engrg,* ASCE, 120(12), 1426-1448.

Tayfur, G., Kavvas, M.L., Govindaraju, R.S., and Storm, D.E. (1993). Applicability of St.Venant equations for two-dimensional overland flows over rough infiltrating surfaces. *J. Hydraulic Engrg*, ASCE, 119(1), 51-63.

Wurbs, R.A. (1998). Dissemination of generalized water resources models in the United States. *Water Int.,* 23, 190-198.

Zhang, W. and Cundy, T.W. (1989). Modeling of two dimensional overland flow. *Water Resources Research,* 25(9), 2019-2035.

In: Modeling Hydrologic Effects…
Editor: Xixi Wang, pp. 59-82

ISBN 978-1-61668-628-4

Chapter 4

SIMULATING EFFECTS OF MICROTOPOGRAPHY ON WETLAND SPECIFIC YIELD AND HYDROPERIOD

David M. Sumner*

U.S. Geological Survey (USGS), Florida Integrated Science Center,
12703 Research Parkway, Orlando, Florida 32826, USA

ABSTRACT

Specific yield and hydroperiod have proven to be useful parameters in hydrologic analysis of wetlands. Specific yield is a critical parameter to quantitatively relate hydrologic fluxes (e.g., rainfall, evapotranspiration, and runoff) and water level changes. Hydroperiod measures the temporal variability and frequency of land-surface inundation. Conventionally, hydrologic analyses used these concepts without considering the effects of land surface microtopography and assumed a smoothly-varying land surface. However, these microtopographic effects could result in small-scale variations in land surface inundation and water depth above or below the land surface, which in turn affect ecologic and hydrologic processes of wetlands. The objective of this chapter is to develop a physically-based approach for estimating specific yield and hydroperiod that enables the consideration of microtopographic features of wetlands, and to illustrate the approach at sites in the Florida Everglades. The results indicate that the physically-based approach can better capture the variations of specific yield with water level, in particular when the water level falls between the minimum and maximum land surface elevations. The suggested approach for hydroperiod computation predicted that the wetlands might be completely dry or completely wet much less frequently than suggested by the conventional approach neglecting microtopography. One reasonable generalization may be that the hydroperiod approaches presented in this chapter can be a more accurate prediction tool for water resources management to meet the specific hydroperiod threshold as required by a species of plant or animal of interest.

Keywords: Specific yield, hydroperiod, wetlands, microtopography

* Hydrologist. E-mail: dmsumner@usgs.gov

INTRODUCTION

Specific yield and hydroperiod are measures of wetland hydrologic status and function. Specific yield is a measure of the relation between water storage and water level. Hydroperiod is a measure of temporal patterns of land-surface inundation. These hydrologic measures are widely used in hydrologic description, simulation, and interpretation. Microtopography can confound traditional forms of specific yield and hydroperiod that assume locally flat topography. This chapter describes enhanced, more physically-based, forms of specific yield and hydroperiod that honor the effects of land-surface microtopography on these hydrologic measures. A case study using data from wetlands of the Florida Everglades is used to contrast the proposed methods with traditional approaches.

Hydrologic models are useful in understanding and quantifying wetlands hydrology. Models provide quantitative estimates of wetland water levels and flows under proposed water management strategies or under conjectured climatic scenarios. Hydrologic models, linked to biogeochemical data or models, can quantify the transport of constituents such as nutrients and toxins or the viability of flora and fauna. Accurate hydrologic models are needed because small errors in simulated water levels can produce large errors in simulated hydroperiod of the low relief wetlands environment.

Hydrologic models for wetlands require a means to quantify the relation between imposed or calculated flows (including precipitation, evapotranspiration, surface- and ground-water flows) and the resulting change in water level. Specific yield is a parameter that has frequently been used to accomplish this goal (e.g., U. S. Geological Survey MODFLOW ground-water flow model, Harbaugh 2005). Additionally, specific yield is often used in conjunction with measured water-level changes to estimate ground-water recharge (Healy and Cook 2002, Crosbie et al. 2005, and Jaber et al. 2006) and evapotranspiration (Owen 1995, Rosenberry and Winter 1997, Lott and Hunt 2001, and Loheide et al. 2005) using the water table fluctuation method. Specific yield (S_y) is defined as the change in volume of water within a unit area associated with a unit change in water level. With this definition, a change in water level (Δh) is related to a depth-equivalent flow (D) or volume of flow per unit area as:

$$\Delta h = \frac{D}{S_y} \tag{1}$$

Many flow terms (evapotranspiration, precipitation, and deep percolation) are frequently reported in terms of a depth-equivalent flow (e.g., mm) over a given time period. Other flow terms (e.g., stream flow) are often reported in terms of volume (e.g., m^3) over a given time period and will require normalization by the area of interest to convert to a depth-equivalent flow.

The value of specific yield is frequently treated as a two-part function of water level composed of two constant values with a step transition between the two values (e.g., Evans 2000, Sun et al. 2000, Nair et al. 2001, and Wilsnack et al. 2001). In this traditional approach, a value equal to the depth-equivalent of drainable or fillable soil water per unit change in water level is used if the water level is below land surface and a value of one is used if the water level is above land surface. Johnson (1967) reports that average values for specific yield appropriate to subsurface water levels fall within the range of 0.21 to 0.27 for soils that are fine to coarse sands, and generally are less than 0.10 for soils that are silts or finer. For peatland soils, Johnson (1967) reports an average specific yield value of 0.44 and Letts et al. (2000) report values of 0.66, 0.26, and 0.13 for fibric, hemic, and sapric peats, respectively. A value of one for specific yield under inundated land surface conditions implies the absence of any standing vegetation. Vegetation can diminish the volume available for surface-water storage and, thereby, reduce open water specific yield below one.

Use of Eq. (1) to relate flows to changes in water levels is complicated by the influence of the capillary fringe on the magnitude of the drainable or fillable subsurface volume (Childs 1960, dos Santos and Youngs 1969, Duke 1972, and Gilham 1984). Figure 4-1 illustrates a generalized soil moisture profile at equilibrium above a water table. Capillary suction increases (and soil moisture decreases) with height above the water table under hydrostatic (equilibrium) conditions. Soil moisture is constant at saturation (θ_s) from the water table to a height corresponding to the air-entry or bubbling head (h_b), above which soil moisture decreases asymptotically with height above the water table to residual moisture content (θ_r). The capillary fringe is often defined as the height above the water table below which soil moisture exceeds residual moisture content. As described by Duke (1972), if the water table is sufficiently deep (e.g., land surface at height A in Figure 4-1) such that soil moisture at land surface approaches residual moisture content, the relatively large range in soil moisture content within the subsurface profile allows for a correspondingly large subsurface storage capacity. Under this condition, the specific yield is equal to the difference of saturation and residual moisture content. If the water table is shallower (e.g., land surface at height B in Figure 4-1), less range in soil moisture exists within the subsurface profile and specific yield is correspondingly lower than the previous case. If the water table is at a depth shallower than the air entry head (e.g., land surface at height C in Figure 4-1), no variation in soil moisture exists within the subsurface profile (subsurface is completely saturated) and, therefore, no subsurface storage capacity exists which indicates a specific yield equal to zero. As values of specific yield decrease with an increasingly shallow water table, the change in water level produced by a given addition or extraction of water increases as indicated by Eq. (1). In particular, when the water table is sufficiently shallow (within a distance less than the air entry suction of land surface), so that the specific yield is zero, large water-level changes can occur with relatively small additions or extractions of water (i.e., the reverse Wieringermeer effect, Healy and Cook 2002).

A capillary fringe causes specific yield to vary continuously with water-table depth in shallow water-table conditions. Duke (1972) developed a quantitative description (Eq. 2) of the variation in specific yield with water table depth under equilibrium (hydrostatic) conditions. Combining this description with the Brooks and Corey (1964) soil-water characteristic curve model given by Eq. (3), Duke (1972) showed that specific yield is given by Eq. (4) for the case of water levels below land surface.

$$S_y(H) = \theta_s - \theta(H) \tag{2}$$

$$\theta(H) = \begin{cases} \theta_r + (\theta_s - \theta_r)(h_b / H)^\lambda & (\text{for} H > h_b) \\ \theta_s & (\text{for} H \le h_b) \end{cases} \tag{3}$$

$$S_y(H) = \begin{cases} (\theta_s - \theta_r)(1 - (h_b / H)^{\lambda} & (\text{for} H > h_b) \\ 0 & (\text{for} H \leq h_b) \end{cases} \tag{4}$$

where θ is soil moisture content [L^3L^{-3}], θ_s is saturation moisture content [L^3L^{-3}], θ_r is residual moisture content [L^3L^{-3}], h_b is air-entry or bubbling suction [L], λ is the pore size distribution index [dimensionless], and H is water-table depth below land surface [L].

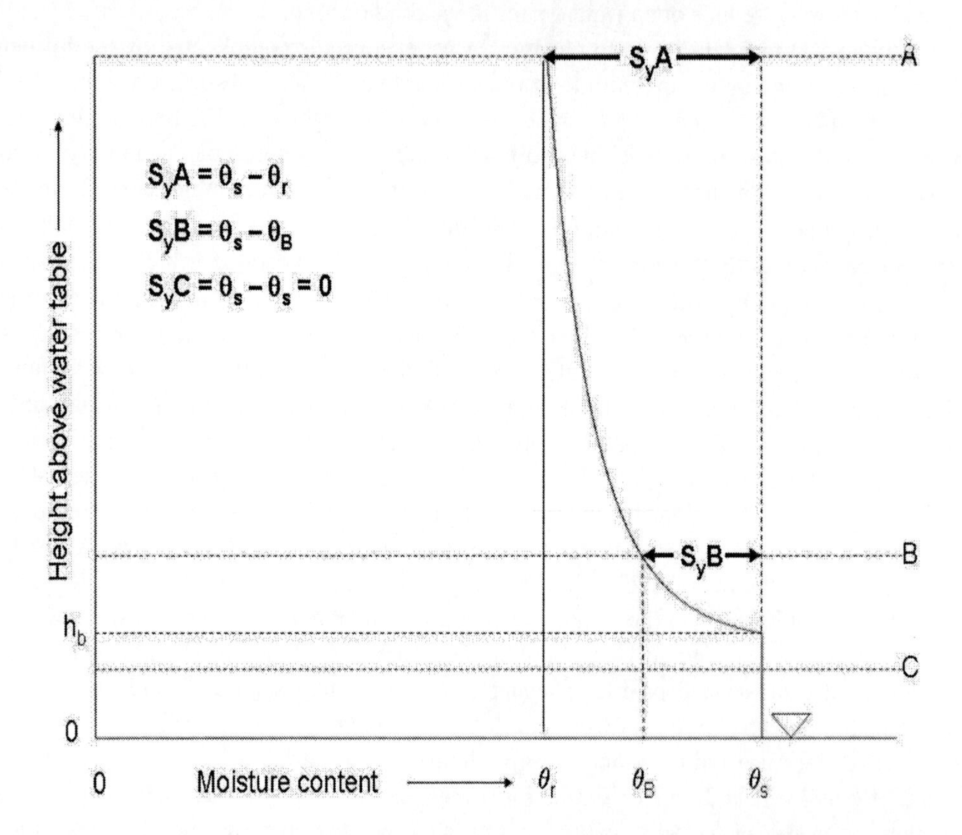

Figure 4-1. Schematic showing the equilibrium moisture contents above water table for land surface positions A, B, and C. h_b is the air-entry bubbling suction, θ_r is the residual moisture content, θ_s is the saturation moisture content, and S_y is the specific yield.

This formulation for specific yield is dependent on definition of the soil-moisture characteristic curve (Eq. (3)) relating capillary suction (or distance above water table if hydrostatic conditions prevail) to volumetric soil moisture content by either laboratory analysis of soil samples or by *in situ* methods (Hillel 1980). Hysteretic behavior often is observed in which the soil moisture content is higher for drainage than for wetting conditions for a given capillary suction, introducing non-uniqueness to the soil-moisture characteristic curve. Other factors that contribute to non-uniqueness in the soil-moisture characteristic curve are air entrapment during soil wetting (Peck 1969, Nachabe et al. 2004), microbial generation of gases below the water table, dissolution of gas bubbles, shrinking and swelling of the soil matrix (Kennedy and Price 2004), and soil water temperature variations (Hopmans and Dane 1986). Additionally, the formulation for specific yield given by Eq. (1) assumes that an equilibrium soil moisture profile exists above the water table at all times, an assumption that is increasingly violated as the depth to the water table increases and as the soil hydraulic conductivity decreases. In order to solve this problem, Nachabe (2002) developed a closed-form expression for specific yield to account for the case of transient soil drainage.

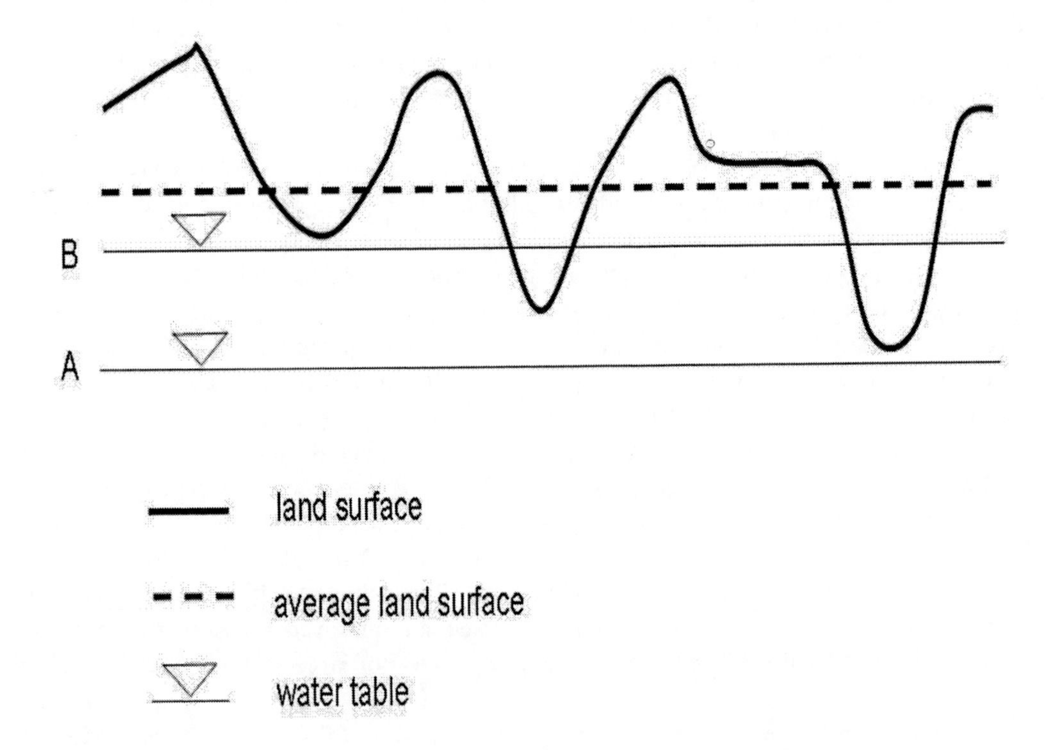

Figure 4-2. Schematic showing variations in land surface, water level, and surface inundation, of typical wetlands.

Another complicating factor in the use of Eq. (1) within hydrologic models is the effect of topographic variations in land surface as noted by Choi et al. (2003). Wetlands topographic variability is low relative to most other landscapes. However, the microtopography that does exist can be critical to determining the variability of water depths within wetlands. Microtopographic variations of only a few centimeters can greatly influence the extent of wetlands inundation. Hydrologic models often are discretized into finite sub-areas or grid cells. The variability in system properties within individual grid cells is composited into "effective" properties for each grid cell. A conceptualization of the topographic variation within a given grid cell (Figure 4-2) illustrates the resulting intra-cell variation in inundation and depth to water table. If the water table is at position A, the water table is below the average land surface elevation and the effective specific yield for this cell traditionally would be specified to a constant, global subsurface value.

This traditional approach neglects the effect of capillarity on specific yield, including the ramifications of intra-cell variations of water-table depth on the capillarity effect. If the water table is at position B in Figure 4-2, the water table is again below the average land surface elevation and this cell traditionally would again be specified by a constant, global subsurface value of specific yield. However, examination of intra-cell variability in microtopography reveals that the cell is partially inundated in this case. The effective specific yield for the cell is best estimated by compositing the specific yield for the inundated areas of the cell (a value near one) with appropriate values for subsurface specific yield in the areas of the cell with water levels below local topography. Microtopographic variations within a grid cell introduce a dependence of specific yield on water level through the variation in extent of inundation with a changing water level. Additionally, microtopographic variations modify the impact of the capillarity effect on specific yield by introducing a distribution, rather than a single value, of water-table depths within a cell.

Previous studies (e.g., Kushlan, 1990; Casanova and Brock, 2000) revealed that the particular biotic community of a wetland is highly influenced by hydroperiod. Hydroperiod is often categorized based on the frequency and duration of wetland flooding. For example, the St. Johns River Water Management District in Florida (SJRWMD, 2007) uses a range of hydroperiod categories, including "intermittently exposed", "temporarily flooded", "seasonally flooded", "semi-permanently flooded", and "typically saturated," while the University of New Hampshire Cooperative Extension (Tarr and Babbitt, 2009) categorizes hydroperiod as either "short", "intermediate", or "long" . The utility of hydroperiod categorization is handicapped by the coarseness of this method and the great variety of sometimes vaguely-defined categories used by different research and management agencies. Alternatively, hydroperiod is sometimes presented more quantitatively as a time series of inundation states (flooded or dry) at a particular location (Tarr and Babbitt, 2009) or as a single value or a range reflecting the typical number of days that a location is inundated over a year (USGS, 1999). Lee et al. (2009) presented an improved measure of fractional hydroperiod for isolated wetlands, incorporating the effects of large-scale topography within the wetland.

Microtopography complicates the available measures of hydroperiod by producing fractional inundation to a local area – low points may be inundated while slightly higher nearby points remain above the water. Wetland hydrologic processes and biotic diversity can be strongly impacted by microtopographic variations (Werner and Zedler, 2002; Bruland and Richardson, 2005; Tsuyuzuki, 2006; and Moser et al., 2007). Traditional measures of hydroperiod ignore fractional inundation and assume a binary state of either wet or dry at a location. However, the presence of standing water in small-scale topographic depressions in an otherwise dry wetland or the presence of dry areas in small-scale topographic highs in an otherwise wet wetland can serve as desirable habitats important to the viability of biota. Additionally, water depth varies locally in inundated areas because of microtopographic variations and water depth can be an important control on ecosystem diversity and productivity (Kushlan, 1990; Coops et al., 1996; and Miller and Zedler, 2003). Traditional hydrologic simulations often ignore this micro-variation in water depth, assuming that land surface, and therefore water depth, within an individual hydrologic model cell is single-valued. However, a microtopographically-generated distribution of water depths within wetlands provides more possibilities for biotic viability for those species that are sensitive to water depth and can impact overland flow of water (Tweedy et al., 2001; Choi et al., 2003). Alternative measures of hydroperiod incorporating the impact of microtopography on fractional inundation of wetlands are presented in this chapter.

MATERIALS AND STUDY METHODS

The Three Wetland Sites in the Florida Everglades

The three sites that were evaluated are located in areas of peatland soils in the Florida Everglades (Figure 4-3). The peat soils in these areas generally are greater than 1 m thick and are underlain by limestone. The microtopography of these sites was defined by previous researchers. Choi et al. (2003) measured variations in land-surface elevation within areas of approximately 4 hectares at sites F1 and U3 in a ridge and slough environment. Brandt et al. (2006) measured variations in land-surface elevation at site TI2 within a pop-up tree island of about 0.1 hectare. The cumulative frequency distributions of land-surface elevations are shown in Figure 4-4.

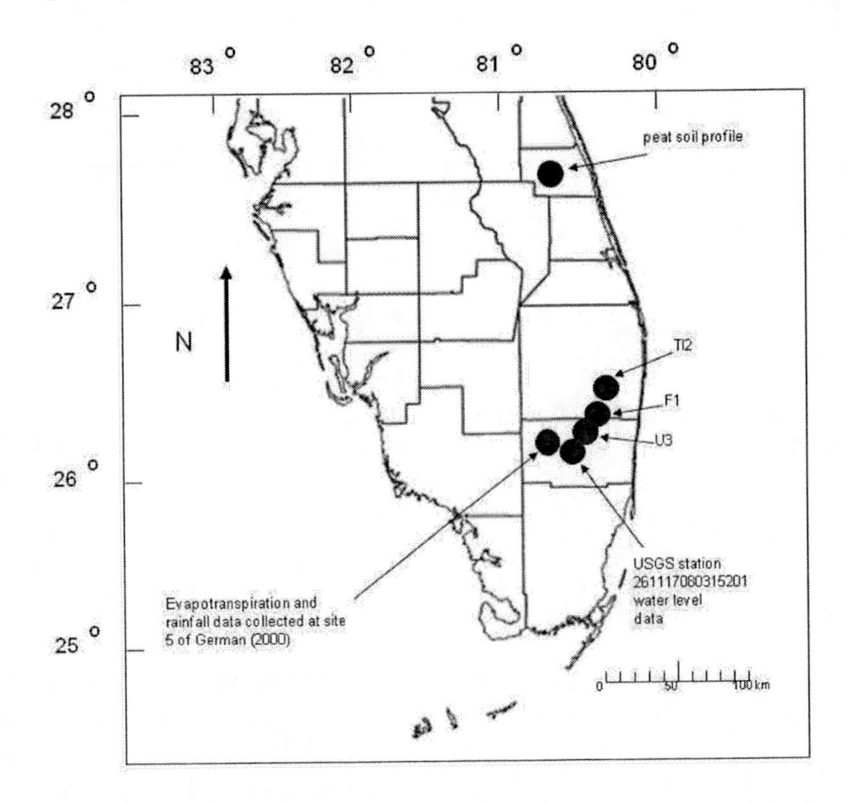

Figure 4-3. Map showing the locations of the three wetland sites (U3, F1, and TI2) in the Florida Everglades, the peat soil sampling site, the evapotranspiraiton/rainfall station, and the water level station.

Soil-moisture characteristic curve data were not available for the site locations. These data were estimated based on an analysis (Gator muck soil profile number 12 in Indian River County, University of Florida Institute of Food and Agricultural Sciences 2006) of a comparable saprist peat soil located about 170 km north of the study sites (Figure 4-3). The discrete values of soil moisture determined previously in the suction range of 3.5 to 200 cm of water were least squares fit to the Brooks and Corey (1964) model of the soil-moisture characteristic curve (Eq. 3) to define the model parameters. Eq. (4) was used to define capillarity-dependent specific yield curves. This approach is based on an assumption that a homogeneous soil profile exists within the range of water-table fluctuations.

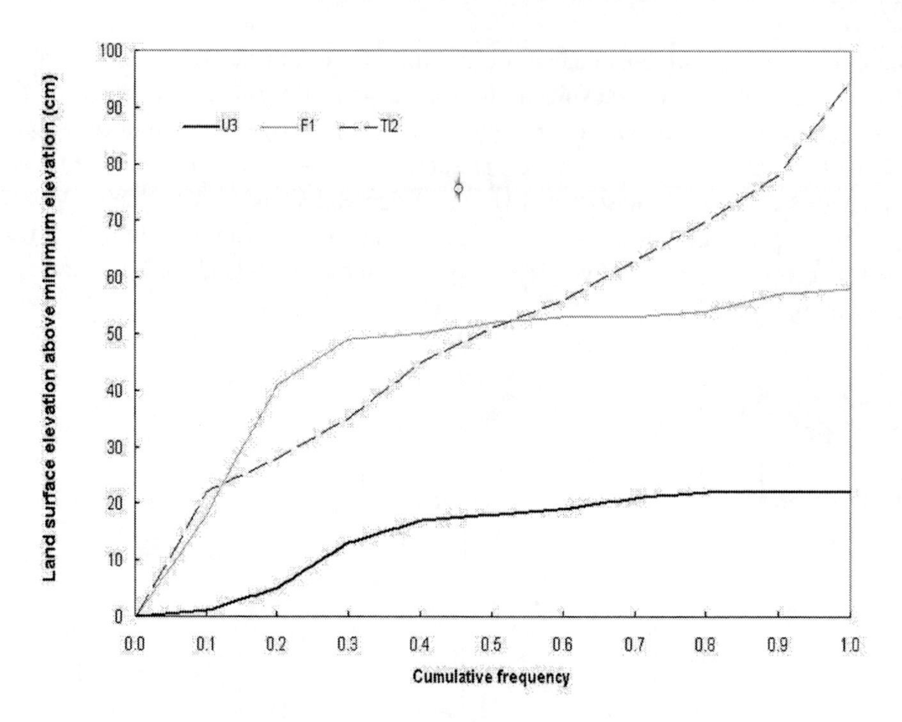

Figure 4-4. The cumulative frequency distribution (CFD) curves of the land surface elevations of the three wetland sites (U3, F1, and TI2) in the Florida Everglades. (After Choi et al. 2003 and Brandt et al. 2006).

Computation of Specific Yield

The relation of effective specific yield to water level was estimated for the three wetland sites under four conceptualizations: 1) the traditional approach of ignoring microtopography and capillarity, 2) an approach incorporating only the effects of capillarity, 3) an approach incorporating only the effects of microtopography, and 4) the most physically-based approach incorporating effects of both microtopography and capillarity.

A traditional measure of specific yield, assuming a flat land surface and neglecting capillarity, was computed as a two-part function of water level. Land surface elevation was assumed equal to the median elevation at each site. If the water level was above land surface, specific yield was set equal to a value near unity (indicative of surface-water storage). In this study, a value of 0.98 was used for specific yield when water level was above land surface based on field measurements by Choi et al. (2003) of vegetation biomass and density and a computed range of surface-water specific yield of 0.97 to 0.99. If the water level was below the flat land surface, specific yield was set equal to a value substantially less than unity (indicative of subsurface storage).

Selection of a traditional, constant value of specific yield appropriate to conditions of subsurface water levels is subjective. Sometimes, this value is estimated as the ultimate specific yield (i.e., the difference of saturation and residual moisture content) indicative of deep water levels. However, because of capillarity, use of the ultimate specific yield will over-estimate the optimal constant specific yield if water levels generally range close to the surface. In this study, the constant value of subsurface specific yield was selected based on a commonly used value within hydrologic models for the study area and a rule-of-thumb procedure based on the soil-moisture characteristic curve (Letts et al. 2000). Specific yield, incorporating capillarity but neglecting microtopography, was computed based on Eq. (4) for water levels below the assumed flat land surface. For water levels above land surface, specific yield was set equal to 0.98.

Microtopographic variations can produce spatial variability in specific yield as a result of spatial variability in land surface inundation and depth to water table (Figure 4-2). An effective specific yield for a site or model grid cell requires compositing areally variable values of specific yield. Assuming that changes in water level with time are spatially uniform within an area *A*, a simple mass balance approach leads to Eq. (5) which can be rearranged into Eq. (6). The comparison of Eqs. (1) and (6) indicates that area-weighted average of specific yield is the appropriate measure of effective specific yield (S_{ye}) for an area (Eq. 7):

$$D \cdot A = \iint_A \Delta h S_y(x,y)\,dx\,dy \tag{5}$$

$$\Delta h = \frac{D}{\dfrac{\iint_A S_y(x,y)\,dx\,dy}{A}} \tag{6}$$

$$S_{ye} = \frac{\iint_A S_y(x,y)\,dx\,dy}{A} \tag{7}$$

where *D* is the depth-equivalent input or output of precipitation, evapotranspiration, surface water, or ground water for area *A*; Δh is a spatially uniform change in water level within area *A*; and *x* and *y* are areal spatial coordinates.

Given that the only assumed areal variation within the site area is topographic, the areal average of Eq. (7) can be expressed as shown in Eq. (8) for a given water level *h*:

$$S_{ye} = \frac{\iint_A S_y(z(x,y),h)\,dx\,dy}{A} \tag{8}$$

where *h* is water level elevation and *z* is land surface elevation. Given the availability of the cumulative probability distributions of z shown in Figure 4-4, Eq. (8) was approximated as:

$$S_{ye}(h) \approx \frac{\sum_{i=1}^{n} S_y(z_i, h)}{n} \tag{9}$$

where n (specified to be 100) is the number of surface elevations sampled at equal probability intervals from the cumulative probability distributions (Figure 4-4).

Specific yield computations incorporating microtopography were performed using Eq. (9). In this equation, $S_y(z, h)$ was set to be 0.98 if the water level was above the local land surface. For the case of specific yield computed without consideration of capillarity, $S_y(z, h)$ was set equal to the traditional, constant estimate of subsurface specific yield if the water level was below the local land surface. For the case of specific yield computed with consideration of capillarity, $S_y(z, h)$ was determined using the depth of an assumed flat water level below the local topography and Eq. (4). Limited by the available data, this study did not consider effects of hysteresis, air entrapment, microbial generation of gases, dissolution of gases, shrinking/swelling of soil matrix, and a non-equilibrium soil moisture profile, on specific yield.

The effect of variations in specific yield conceptualization on simulated water levels was evaluated using a simple water balance for conceptualized one-dimensional systems at the most (site TI2) and least (site U3) topographically variable sites. These simulations were not intended to reproduce measured water levels at the site, but rather to exemplify the sensitivity of simulated water levels to the complexity of specific yield conceptualization. Measured hourly rainfall and evapotranspiration data (Figure 4-5) collected at a nearby marsh site (Figure 4-3) throughout 1997 were used as prescribed fluxes (D) in Eq. (1) to simulate water level changes over a one-year period. Surface- and ground-water inputs were neglected because this wetland system is mainly controlled by the relatively large atmospheric fluxes of precipitation and evapotranspiration. Based on an assumed initial water level (h^0), water levels (h^i) were computed for each time step i using Eq. (10). The computation was implemented for a variety of initial water levels.

$$h^{i+1} = h^i + \Delta h^i = h^i + \frac{D^i}{S_{ye}(h^i)} \tag{10}$$

Site water levels were computed using a discrete time step, initially set equal to the hourly resolution of the atmospheric flux data. As shown in Eq. (10), specific yield for a given time step was computed based on the water level at the beginning of the time step. As noted by Nachabe (2002), because of the dependence of specific yield on water level, this approach can introduce linearization error to computed water levels if water level changes are large over a given time step. Uniform subdivision of the hourly fluxes into smaller time steps was used to evaluate and minimize this linearization error.

Under the specific yield conceptualization that incorporates capillarity, but ignores microtopography, specific yield was set to zero for water levels between land surface and a depth below land surface equal to the bubbling pressure because all pore spaces are filled (Figure 4-1). For these situations in which specific yield (the denominator of Eq. 1) is zero, the computed water level was adjusted up to land surface when precipitation for the time step exceeded evapotranspiration, but the level was adjusted down to a subsurface depth 1 cm below the bubbling pressure when evapotranspiration exceeded precipitation.

Computation of Hydroperiod

The key to obtaining enhanced measures of hydroperiod, incorporating microtopography and the resulting fractional inundation, lies with quantification of microtopographic variations. Next, this microtopographic information can be coupled with known or simulated water level fluctuations, producing multiple measures of hydroperiod described below.

To chronologically present hydroperiod data, a time series of fractional inundation is a desirable alternative to traditional hydroperiods composed of a binary (wet or dry) time series. The cumulative frequency distribution of microtopography serves as a means for translating a given water level to a fractional inundation; as an operational example, in Figure 4-4, simply replace the x-axis annotation "cumulative frequency" with "fraction inundated" and the y-axis annotation "land surface elevation above minimum land surface" with "water level above minimum land surface." This operation serves to produce a one-to-one relation between water level and fraction of wetland inundated, allowing generation of a time series of fractional inundation. As an example, the data of Figure 4-4 are re-presented in Figure 4-6 as a water level to fractional inundation relation; the curve has been transposed to indicate that water level is the independent (x-axis) variable and fractional inundation is the dependent (y-axis) variable.

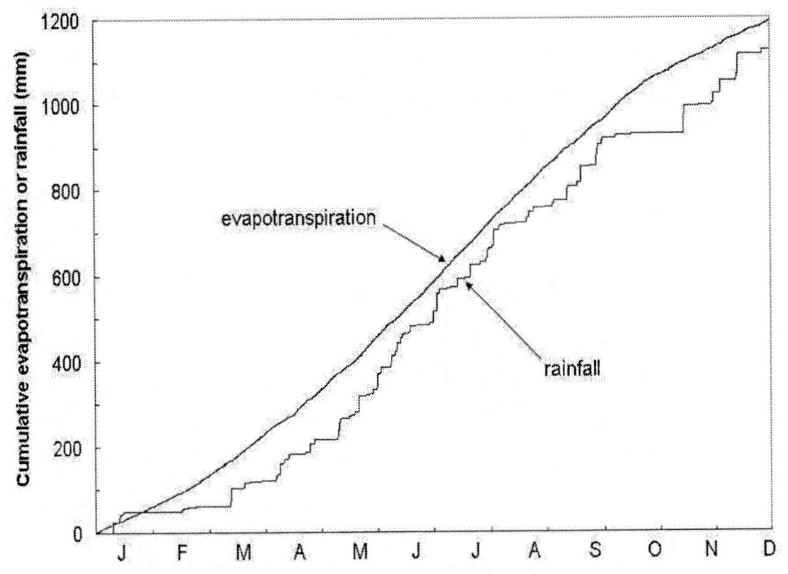

Figure 4-5. The evapotranspiration and rainfall data collected at site 5 shown in Figure 4-3 in 1997. (After German 2000).

As a statistical composite of hydroperiod over a period of time, exceedance curves provide a much greater wealth of information than the traditional measure of mean number of days inundated. Exceedance curves show the statistical likelihood, based on the selected data record (e.g., a particular season or the entire period of record), that any given fractional inundation is equaled or exceeded. An exceedance curve is computed using the available record of water level data (ideally collected at equal time intervals) and the one-to-one relation between water level and fractional inundation derived based on the microtopographic cumulative frequency distribution. Specifically, the full range of computed fractional inundation (zero to one) is divided into discrete intervals and the time series values within each interval are counted and normalized by the total number of values in the full record to estimate the probability that fractional inundation falls within any particular interval. Finally, the interval probabilities are accumulated to compute the probability that any particular fractional inundation interval is equaled or exceeded and fractional inundation is graphed versus the probability that fractional inundation is equaled or exceeded (for example, the "variable topography" curve in Figure 4-7).

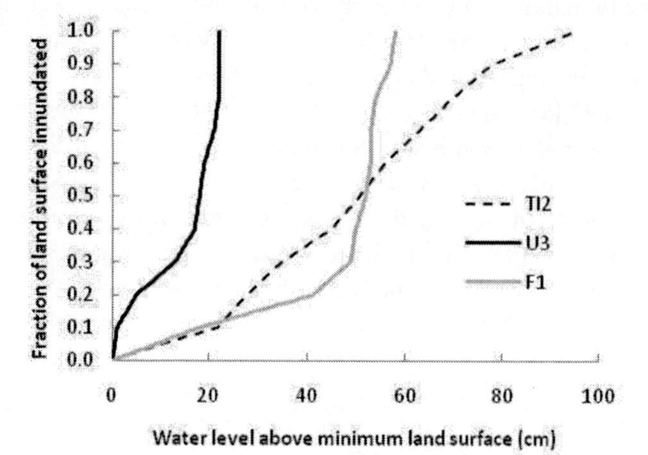

Figure 4-6. Plots showing fractional inundation versus water level at the three wetland sites (U3, F1, and TI2) in the Florida Everglades.

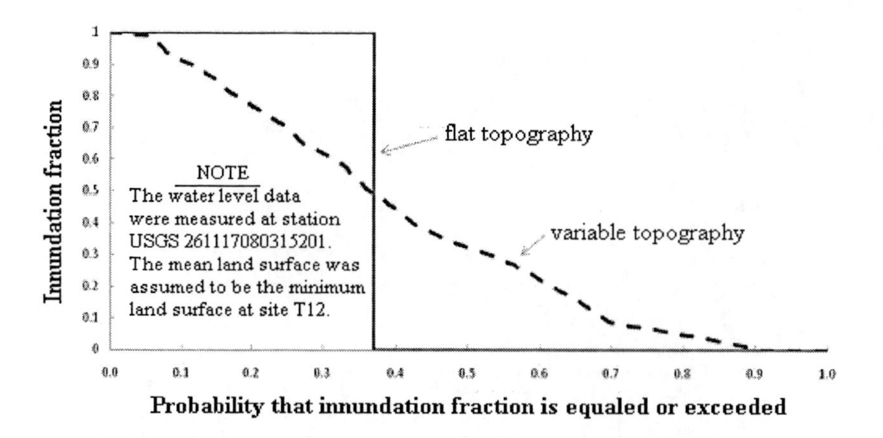

Figure 4-7. Exceedence curve of the inundation fraction for site TI2 in the Florida Everglades.

Knowing the distribution of water depths (above and below land surface), in addition to inundation status, can also be beneficial to ecological and hydrologic analyses. The cumulative frequency distribution of water depths for any given water level elevation is easily obtainable by subtraction of the land surface elevation of the microtopographic cumulative frequency distribution from the given water level. For example, the cumulative frequency distribution of water depths for a water level of 15 cm at site U3 of Figure 4-4 is shown in Figure 4-8. The frequency of a particular water level depth range can then be computed based on differencing the cumulative frequency over that water level range.

The alternative hydroperiod measures described here are examined in a case study using data from the wetlands of the Florida Everglades (Figure 4-3). The data include the microtopographic elevation distributions shown in Figure 4-4 and a time series of water level fluctuations shown from a long-term (over 17 years) monitoring station (USGS station 261117080315201 – Site 63 in Water Conservation Area 3A). These water level data are presented in Figure 4-9 relative to the mean elevation of the surrounding ridge and slough environment at the station. To relate the elevations of the measured water level to that of the topographic distributions, it was assumed that the mean land surface at the water level station was equal to the median land surface at the ridge-and-slough sites (U3 and F1). However, given that the tree islands are higher elevation features within the ridge-and-slough matrix of the Florida Everglades, the mean land surface of the ridge-and-slough water level station was assumed to be the minimum land surface at the tree island site (TI2).

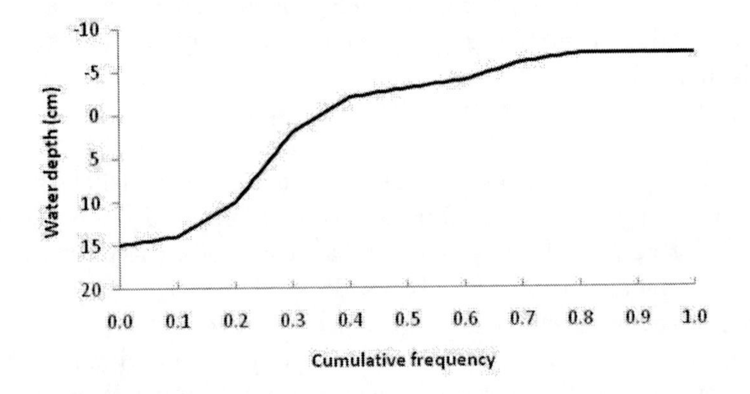

Figure 4-8. The cumulative frequency distribution (CFD) curve of the water depth at site U3. A positive depth signifies inundation, whereas, a negative depth signifies that the water level is below the local land surface.

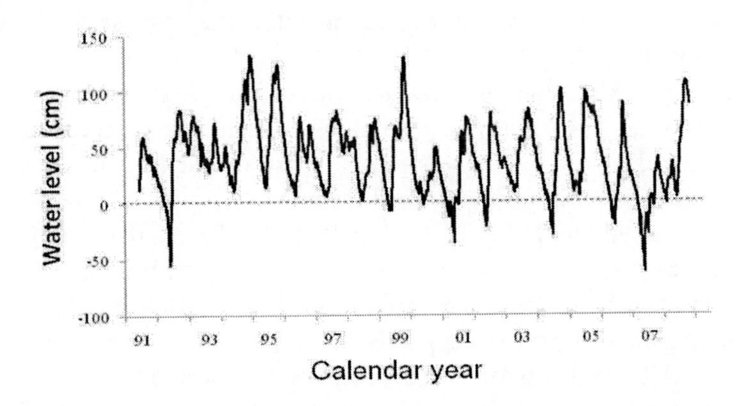

Figure 4-9. The water levels observed at station USGS 261117080315201. The datum is the average of the land surface elevation in the ridge-and-slough environment of the station location in the Florida Everglades.

RESULTS

Computed Specific Yield

The least squares fit of the discrete soil-moisture characteristic curve data for the peat soil (Figure 4-3) to the Brooks and Corey model was effective (Figure 4-10). Estimated Brooks and Corey parameters were: $h_b = 11$ cm; $\lambda = 0.20$; $\theta_r = 0.20$ cm^3 cm^{-3}; and $\theta_s = 0.81$ cm^3 cm^{-3}. The difference of volumetric moisture content at 0 and 0.1 bars (0.81 cm^3 cm^{-3} to 0.59 cm^3 cm^{-3}) provided a traditional estimate (Letts et al., 2000) of subsurface specific yield of 0.22, which is similar to the value of 0.2 used by others (e.g., Evans 2000, Nair et al. 2001, and Wilsnack et al. 2001) in simulating wetlands hydrology in south Florida. The method of Nachabe (2002) indicated that the soil moisture profile would reach a new equilibrium following an instantaneous 10 cm change in water table depth within 80 minutes for water table depths less than 100 cm and within 7 hours for water table depths less than 200 cm. These calculations were based on the Brooks and Corey parameters θ_r, θ_s, λ, and h_b determined for the peat soil and a measured value of saturated hydraulic conductivity for this soil of 71 cm/hr (University of Florida Institute of Food and Agricultural Sciences 2006). This estimated rapid re-equilibrium supports the assumption of an equilibrium soil moisture profile in this study.

The computed relations of specific yield with water level at the three study sites are shown in Figure 4-11. Specific yield relations computed incorporating both microtopography and capillarity can be considered most realistic. Specific yield relations based on simpler conceptualizations vary substantially from the more physically-based relations, particularly when water level is between the minimum and maximum land-surface elevations at each site. Additionally, the traditional estimate of specific yield was not consistent with the most physically-based approach throughout the range of water levels considered. Although microtopographic variations may strongly affect specific yield during partial inundation of land surface, topographically-induced variations in depth to water show little effect on the impact of capillarity on specific yield. A notable difference between specific yield conceptualizations incorporating microtopography and other conceptualizations that were considered is that the former conceptualizations result in values of specific yield that vary continuously with water level, without the step discontinuities of those conceptualizations not considering microtopography.

The impact of varying conceptualizations of specific yield on water levels simulated using the simple water approach along with prescribed fluxes of evapotranspiration and rainfall is exemplified in Figure 4-12. In these examples, none of the simpler conceptualizations of specific yield produced hydrographs at either site that were consistent with the most physically-based approach (Table 4-1). The relative utility of the simpler specific yield conceptualizations varied with the assumed initial water level at each site. For example, although the specific yield conceptualization that considered capillarity but ignored microtopography generally performed poorly at site TI2 for the relatively wet conditions shown in Figure 4-12, this conceptualization generally showed the greatest utility of the simpler approaches when the assumed initial water level was deeper (Table 4-1). The results shown in Figure 4-12 and Table 4-1 should not be taken as a generalization of the relative utility of simpler conceptualizations of specific yield, and are meant only as particular examples. The results of this comparison would be expected to vary with other scenarios of precipitation and evapotranspiration, with alternative initial water levels, and with incorporation of surface- and ground-water flows.

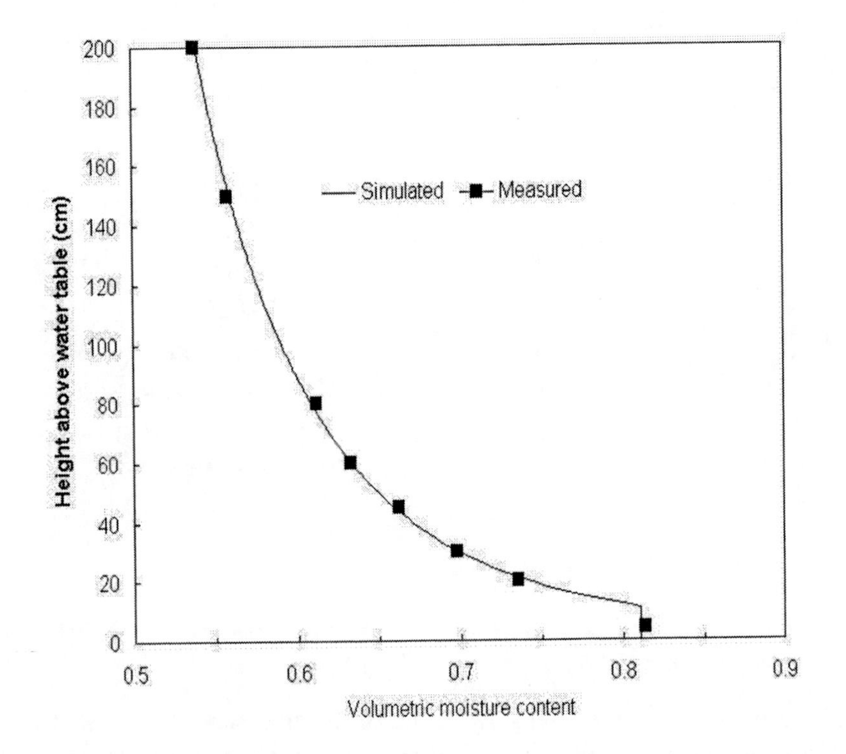

Figure 4-10. Plot showing equilibrium moisture content versus height above water level at the peat soil sampling point shown in Figure 4-3.

Table 4-1. Relative errors of the simulated hourly water levels for the conceptualizations as compared with those for the conceptualization that takes microtopography and capillarity into account[1]

Conceptualization	Site TI2				Site U3			
	h_0 = 55 cm		h_0 = 20 cm		h_0 = 20 cm		h_0 = 10 cm	
	e_a (cm)	e_{max} (cm)	e_a (cm)	e_{max} (cm)	e_a (cm)	e_{max} (cm)	e_a (cm)	e_{max} (cm)
Topography	3.3	16.8	12.6	16.3	17.1	35.7	23.5	30.7
Capillarity	36.2	55.1	2.0	23.7	25.5	45.9	9.3	22.3
Traditional	5.2	11.7	9.4	12.7	11.6	27.1	18.2	24.7

[1] h_0 is the initial water level; e_a is the average absolute relative error; and e_{max} is the maximum absolute relative error.

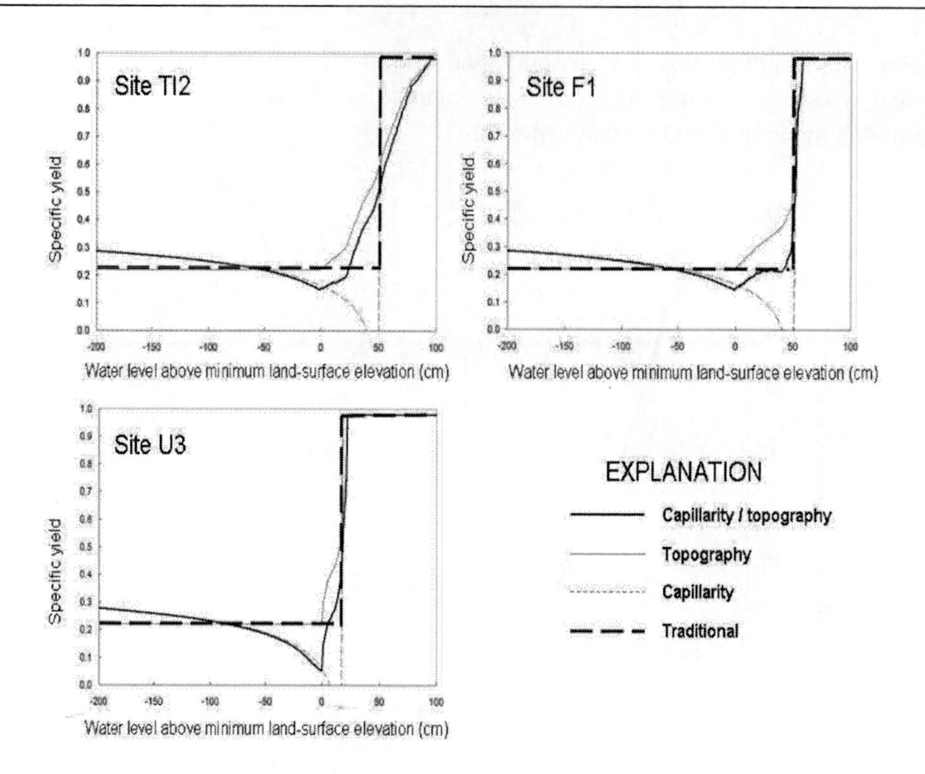

Figure 4-11. The predicted specific yields for the four conceptualizations at the three wetland sites (U3, F1, and TI2) in the Florida Everglades.

Simulated water levels were sensitive to length of time step. A time step of one hour generally was insufficient to avoid linearization errors in simulated water levels and mass balances associated with using the specific yield appropriate to the beginning of a given time step throughout the time step. Linearization error was particularly evident in simulations in which the computed water level quickly transited (generally coincident with a large rainfall event) a strong nonlinearity in the specific yield-to-water level relation. The strongest non-linearities in the specific yield-to-water level relations considered in this study occur in the two conceptualizations (traditional and capillarity alone) that ignore microtopography. These two conceptualizations exhibit step transitions in specific yield for water levels at land surface and simulations incorporating these conceptualizations often exhibited substantial linearization error (even at time steps of 20 minutes as shown for site TI2 in Figure 4-13). Simulations based on the less non-linear, topographically-variable conceptualizations of specific yield generally exhibited little linearization error for time steps of 20 minutes as exemplified in Figure 4-13. Linearization error is minimized in hydrologic simulation by either using a uniform and sufficiently fine temporal resolution, with a dynamic time step that is refined based on the non-linearity of the specific yield at the ambient water level, or by applying a mass-balance correction after each time step (Ross et al. 2004).

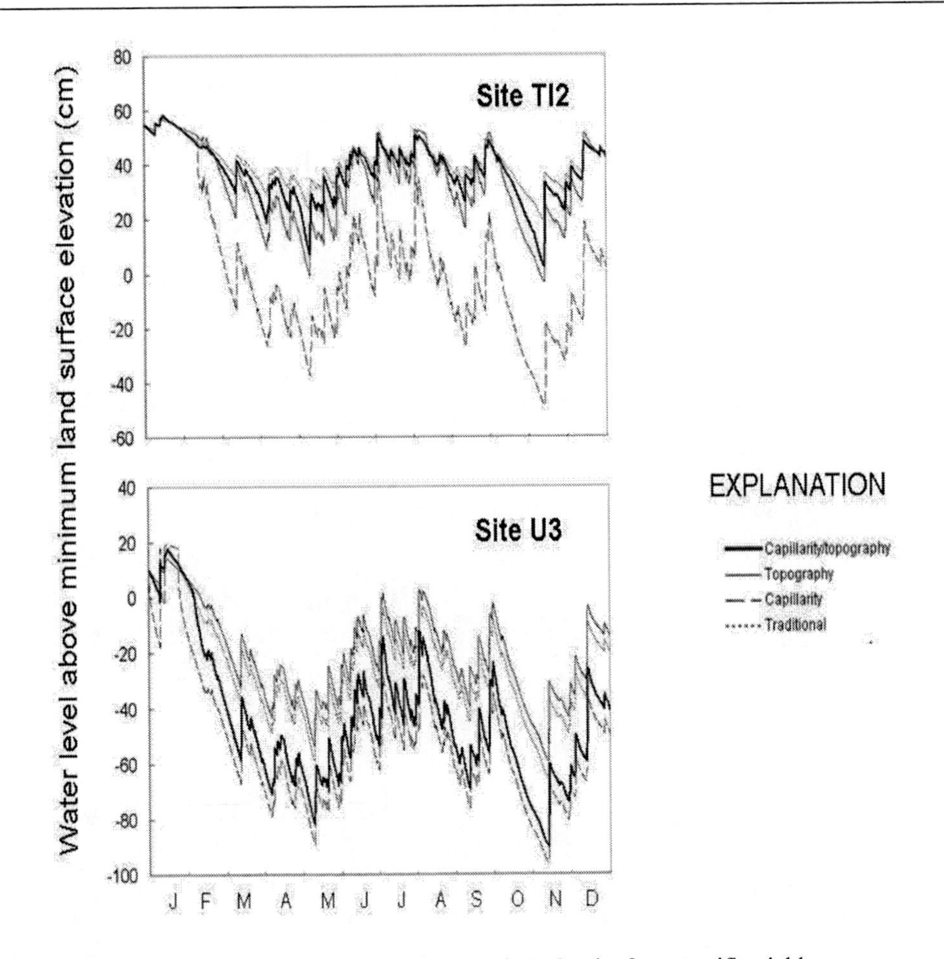

Figure 4-12. The predicted water levels at sites TI2 and U3 for the four specific yield conceptualizations.

Computed Hydroperiod

The time series of fractional inundation presented in Figure 4-14 are the result of combining the water level time series of Figure 4-9 with the water level-to-fractional inundation relations of Figure 4-6. The traditional binary (wet or dry) hydroperiod time series based on a flat, mean land surface is also presented in Figure 4-14. The fractional inundation time series conveys much greater information on the continuum of inundation status than does the traditional binary approach. For example, the traditional approach indicates completely dry conditions during several low water level periods during which the computed fractional inundation indicates residual water in microtopographic depressions. Likewise, during high water level periods (particularly at the tree island site), the traditional approach often indicates completely wet conditions when computed fractional inundation indicates that dry areas remain.

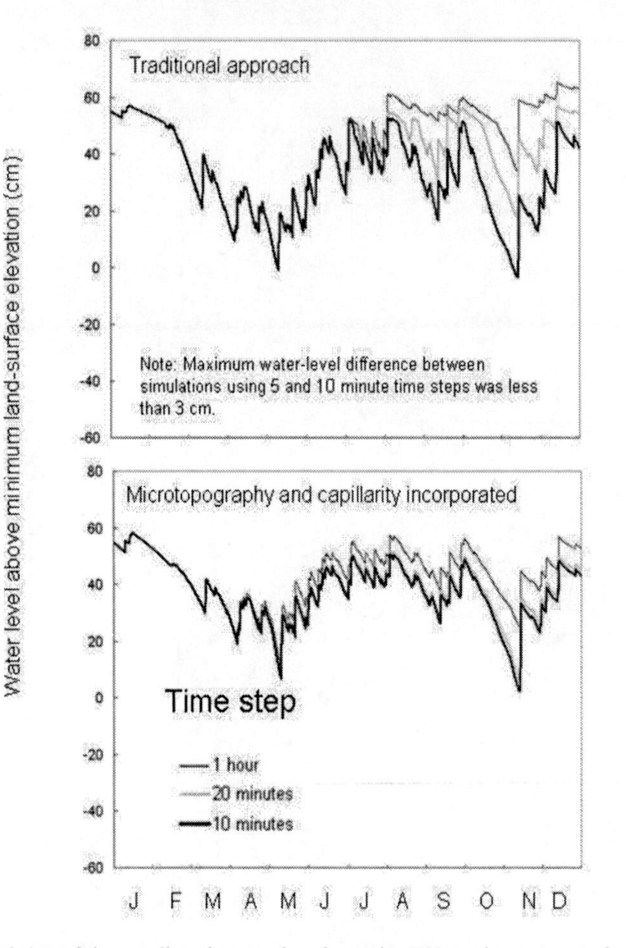

Figure 4-13. The sensitivity of the predicted water levels at site TI2 to the computation time step, for the conceptualizations of "traditional approach" and "microtography and capillary incorporated."

The exceedance curves presented in Figure 4-7 represent statistical composites of the time series for the tree island (TI2) site (Figure 4-3). The exceedance curve for the traditional binary inundation approach simply indicates that there is a 37 percent probability (135 days per year) that land surface is inundated. However, the exceedance curve for fractional inundation is much more descriptive of the statistical nature of inundation status. For example, the fractional inundation exceedance curve indicates that a fractional inundation of zero is exceeded 91 percent of the time (implying that the wetland is completely dry only 9 percent of the time), that fractional inundations of 0.25, 0.50, and 0.75 would be exceeded 58, 36 and 22 percent of the time, respectively, and that a fractional inundation of one is reached 6 percent of the time, implying that the wetland has at least some dry areas 94 percent of the time.

The water depth frequency distributions presented in Figure 4-15 summarize the variability of water depths at sites TI2 and U3 for given water levels. These graphs illustrate the wealth of water depth information obtained by consideration of microtopography compared to that of a single mean water depth provided by traditional approaches. Water depth frequency distributions can describe the variability of water depths (in inundated areas) and water-table depth below local land surface (in dry areas), both of which can be important to ecologic and hydrologic processes.

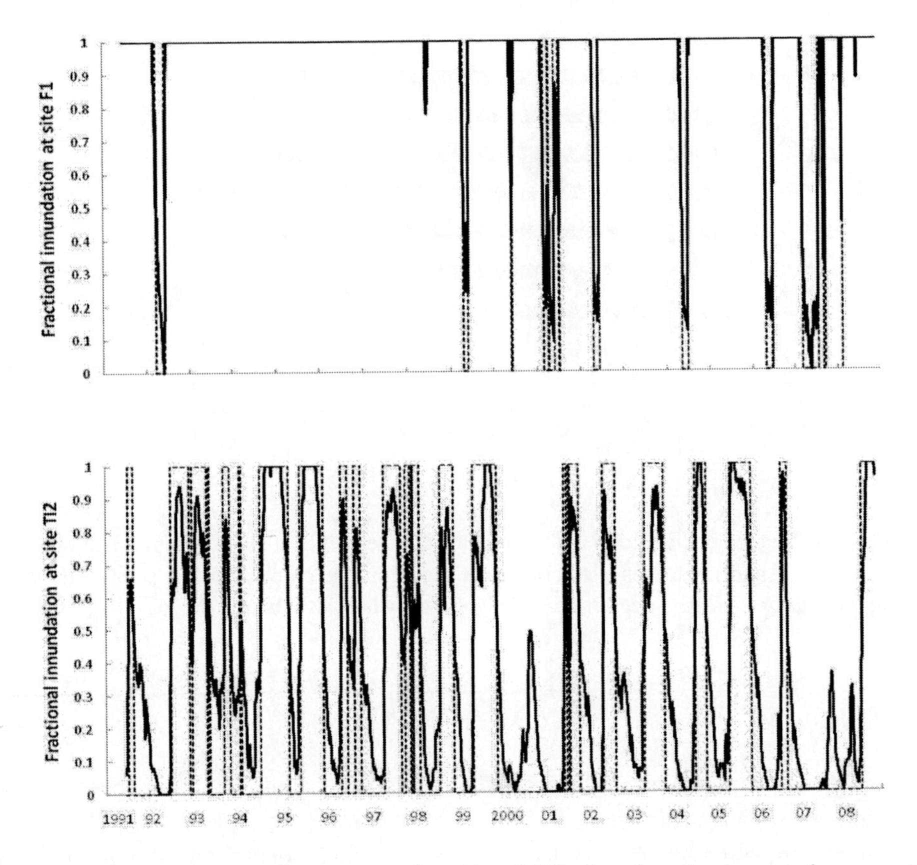

Figure 4-14. Predicted inundation fractions at sites F1 and TI2 when the microtopography was not considered (dashed lines) and when the microtopography was considered (solid lines).

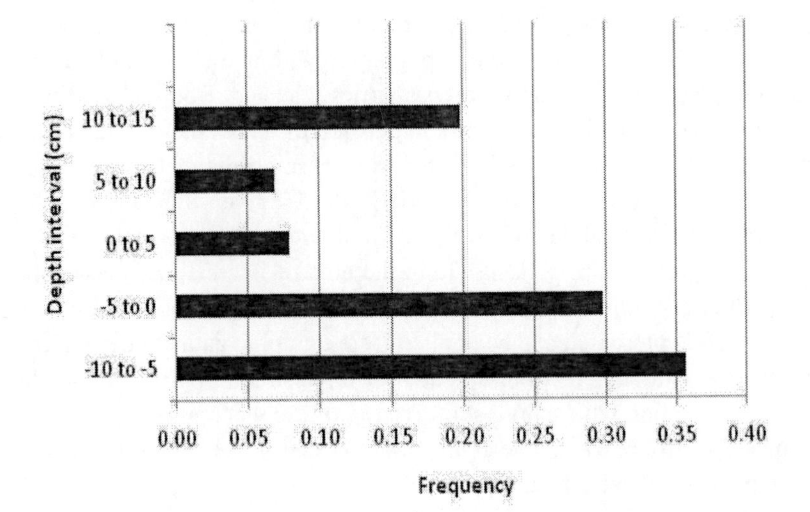

Figure 4-15. (Continued)

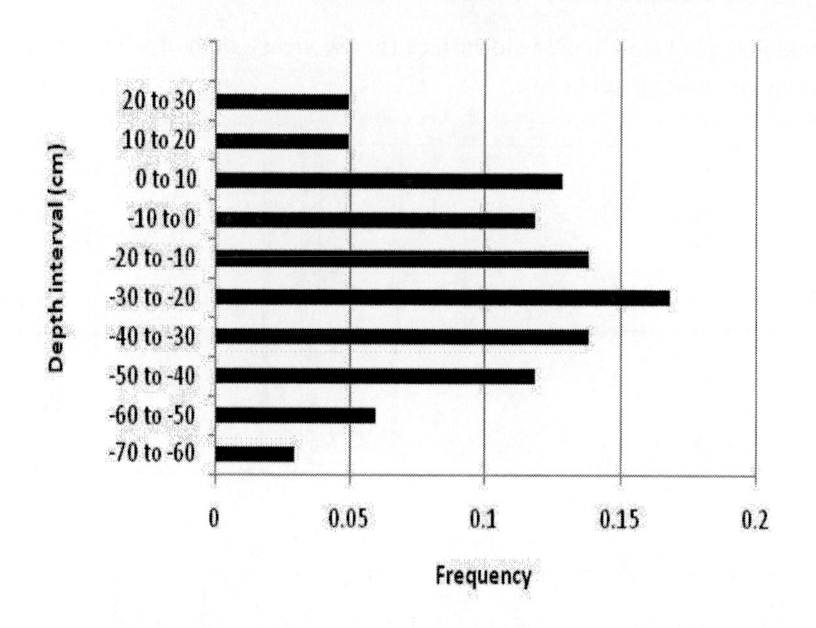

Figure 4-15. The frequency distribution of the predicted water depth at site TI2 (upper) and U3 (lower), at water levels of 30 and 15 cm, respectively, below the minimum land surface. A positive depth signifies inundation, whereas, a negative depth signifies a water level below the local land surface.

DISCUSSION

Critical to implementation of the proposed methods for computation of specific yield and hydroperiod is an adequate characterization of the microtopographic variations in land surface. Microtopography is most easily quantified probabilistically because of the enormous data burden of a deterministic description. An adequate characterization of the statistical distribution of microtopography can be obtained through traditional surveying techniques or a remote sensing approach such as airborne Light Detection and Ranging (LIDAR) (Rosso et al. 2006) performed during dry periods, when maximum exposure of land surface is available. With both techniques, a subset of the topography within the wetland of interest would typically be sampled and the statistical properties gleaned from the subset would then necessarily be assumed to be indicative of a larger part of the wetland. The proposed methods for computation of specific yield and hydroperiod assume that the water level is uniform within the area of interest (for example, a model cell) at a given time and that land surface is statistically stationary throughout the wetland of interest. If large-scale, non-flat trends in either water or land surface are present within the area of interest, these assumptions are not met. However, it is common that large-scale trends in water and land surface are parallel; in this case, the methods presented can be applied if topographic data are detrended and a mean, uniform water level is assumed within the area of interest. Non-parallel trends in water level and land-surface elevation will require sub-division of the wetland into smaller zones, within each of which the difference between the mean water level and the mean land surface elevation is approximately constant, if possible.

Hydrologic models may require modification to implement the continuously-varying specific yield versus water level relations (similar to Figure 4-11) that are a result of a more physically-based specific yield conceptualization. The Regional Simulation Model (South Florida Water Management District 2005a and 2005b) is an example of a hydrologic model that incorporates this capability. Computation of hydroperiod measures suggested in this chapter can be performed through post-processing of cell-by-cell simulated water level data along with cell-by-cell microtopographic data.

Implementation of the more physically-based conceptualization of specific yield requires soil-moisture characteristic curves data over the wetland of interest. Soil-moisture characteristic curves can be developed using laboratory analysis of soil cores (Hillel, 1980), preferably using multiple cores to examine spatial variability in soil properties. Alternatively, *in situ* paired measurements of moisture content and soil-water tension (or height above the water table, if an equilibrium soil-moisture profile is present) can define discrete points on the soil-moisture characteristic curve.

CONCLUSIONS

Three commonly-used specific yield conceptualizations for wetlands, which incorporate microtopography only, soil capillarity only, and neither microtopography nor capillarity, respectively, were compared with a more physically-based conceptualization scheme which incorporates both microtopography and capillarity for three peatland sites in the Florida Everglades. The comparisons were made in terms of the specific yield versus water level relations determined using the computational approaches developed in this study, and the water levels simulated by a simple water balance model in which the conceptualizations were embedded. The results indicate that it is important to consider microtopography and capillarity to adequately define the specific yields of wetlands. Linearization errors in hydrologic models associated with the abrupt step change in the traditional implementation of specific yield can be reduced by taking microtopography into account. In addition, the results indicate that the physically-based approach can better capture the variations of specific yield with water level, in particular when the water level falls between the minimum and maximum land surface elevations.

A variety of measures of hydroperiod were introduced and shown to convey an improved description of the partial areal inundation and variable depth distribution induced by microtopography. A time series of fractional inundation was proposed as more descriptive of inundation history than the traditional binary (wet or dry) time series. Statistical exceedance curves were proposed as a method to describe the probability of any given level of fractional inundation, providing a wealth of information on inundation beyond that of a single probability of inundation based on flat topography. Computation of the distribution of water depths in a microtopographic environment was suggested as a preferred measure of water depth over the single-valued depth associated with an assumed flat topography. Comparison of these enhanced hydroperiod measures with parallel measurements of ecologic form and function could provide improved understanding and predictive capability of the response of an ecosystem to hydrologic changes.

LITERATURE CITED

Brandt, L. A., Martin, G. L., & Mazzoti, F. J. (2006). Topography of pop-up bayhead tree islands in Arthur R. Marshall Loxahatchee National Wildlife Refuge. *Florida Scientist, 69*(1), 19-35.

Brooks, R. H., & Corey, A. T. (1964). Hydraulic properties of porous media. Colorado State University Hydrology Paper No. 3, Fort Collins, Colorado, USA.

Bruland, G.L., & Richardson. C.J. (2005). Hydrologic, edaphic, and vegetative responses to microtopographic reestablishment in a restored wetland. *Restoration Ecology, 13* (3), 515-523.

Casanova, M. T., & Brock, M. A. (2000). How do depth, duration and frequency of flooding influence the establishment of wetland plant communities?. *Plant Ecology 147*(2), 237-250.

Childs, E. C. (1960). The nonsteady state of the water table in drained land. *Journal of Geophysical Research 65*, 780-782.

Choi, J., Harvey, J., & Newlin, J. T. (2003). Significance of microtopography as a control on surface-water flow in wetlands. Greater Everglades Ecosystem Restoration Conference, April 13-18, 2003, Palm Harbor, Florida.http://sflwww. er.usgs.gov/ geer/2003/ posters/ microtopo_cntrl /microcntrl.pdf

Coops, H., van den Brink, F. W. B., & van den Velde, G. (1996). Growth and morphological responses of four helophyte species in an experimental water-depth gradient. *Aquatic Botany, 54*, 11-24.

Crosbie, R. S., Binning, P. & Kalma, J. D. (2005). A time series approach to inferring groundwater recharge using the water table fluctuation method. *Water Resources Research, 41*, W01008, doi: 10.1029/2004WR003077.

dos Santos, A. G., & Youngs, E. G. (1969). A study of the specific yield in land-drainage situations. *Journal of Hydrology 8*, 59-81.

Duke, H. R. (1972). Capillary properties of soils—Influence upon specific yield. *Transactions of the American Society of Agricultural Engineers, 15*(4), 688-691.

Evans, R. A. (2000). Calibration and verification of the MODBRANCH numerical model of south Dade County. Florida, U. S. Army Corps of Engineers, Jacksonville District, Jacksonville, Florida, USA.

German, E. R. (2000). Regional evaluation of evapotranspiration in the Everglades. U. S. Geological Survey Water-Resources Investigations Report 00-4217, Tallahassee, Florida, USA.

Gilham, R. W. (1984). The capillary fringe and its effect on water-table response. *Journal of Hydrology, 67*, 307-324.

Harbaugh, A. W. (2005). MODFLOW-2005. The U. S. Geological Survey modular ground-water model—the Ground-Water Flow Process. U. S. Geological Survey Techniques and Methods 6-A16, Washington D. C.

Healy, R. W., & Cook, P. G. (2002). Using groundwater levels to estimate recharge. *Hydrogeology Journal 10*, 91-109.

Hillel, D., 1980. *Fundamentals of soil physics*. Orlando, Florida: Academic Press, Inc.

Hopmans, W. J., & Dane, J. H. (1986). Temperature dependence of soil water retention curves. *Soil Science Society of American Journal, 50*, 562-567.

Jaber, F. H., Shukla, S., & Srivastava, S. (2006). Recharge, upflux and water table response for shallow water table conditions in southwest Florida. *Hydrological Processes, 20*, 1895-1907.

Johnson, A. I. (1967). Specific yield—Compilation of specific yields for various materials. U. S. Geological Survey Water-Supply Paper 1662-D, Washington D. C.

Kennedy, G. W., & Price, J. S. (2004). Simulating soil water dynamics in a cutover bog. *Water Resources Research, 40*, W12410, doi: 10.1029/2004WR003099.

Kushlan, J. A. (1990). Freshwater marshes. In R. L. Myers and J. J. Ewel (Eds.), *Ecosystems of Florida* (324-363), Orlando, Florida: University of Central Florida Press.

Lee, T.M., Haag, K.H., Metz, P.A., & Sacks, L.A. (2009). Comparative Hydrology, Water Quality, and Ecology of Selected Natural and Augmented Freshwater Wetlands in West-Central Florida: U.S. Geological Survey Professional Paper 1758, 152 p.

Letts, M. G., Roulet, N. T., Comer, N. T., Skarupa, M. R., & Verseghy, D. L. (2000). Parameterization of peatland hydraulic properties for the Canadian land surface scheme. *Atmosphere-Ocean 38*(1), 141-160.

Loheide, S. P., Butler, J. K., & Gorelick, S. M. (2005). Estimation of groundwater consumption by phreatophytes using diurnal water table fluctuations: A saturated-unsaturated flow assessment. *Water Resources Research, 41*, W07030, doi: 10.1029/2005WR003942.

Lott, R. B., & Hunt, R. J. (2001). Estimating evapotranspiration in natural and constructed wetlands. *Wetlands, 21*(4), 614-628.

Miller, R. C. & Zedler, J. B. (2003). Responses of native and invasive wetland plants to hydroperiod and water depth. *Plant Ecology, 167*, 57-69.

Moser, K., Ahn, C., and Noe, G. (2007). Characterization of microtopography and its influence on vegetation patterns in created wetlands. *Wetlands, 27*(4), 1081-1097.

Nachabe, M. H. (2002). Analytical expressions for transient specific yield and shallow water table drainage. *Water Resources Research, 38*(10), 1193, doi: 10.1029/2001WR001071.

Nachabe, M., Masek, C., & Obeysekera, J. (2004). Observations and modeling of profile soil water storage above a shallow water table. *Soil Science Society of American Journal, 68*, 719-724.

Nair, S. K., Montoya, A. M., Wilsnack, M. M., Zamorano, L. M., Obeysekera, J., Switanek, M., Herr, J. & Restrepo, J. I. (2001). South Palm Beach County ground water flow model. South Florida Water Management District, Hydrologic Systems Modeling Division, West Palm Beach, Florida.

Owen, C. R. (1995). Water budget and flow patterns in an urban wetland. *Journal of Hydrology, 169*, 171-187.

Peck, A. J. (1969). Entrapment, stability, and persistence of air bubbles in soil water. *Australian Journal of Soil Research, 7*, 79-90.

Rosenberry, D. O., & Winter, T. C. (1997). Dynamics of water-table fluctuations in an upland between two prairie-pothole wetlands in North Dakota. *Journal of Hydrology, 191*, 266-289.

Ross, M, Geurink, J., Aly, A., Tara, P., Trout, K., & Jobes, T. (2004). Integrated Hydrologic Model (IHM) volume I: Theory manual. Tampa Bay Water and Southwest Florida Water Management District, March 2004.

Rosso, P. H., Ustin, S. L., & Hastings, A. (2006). Use of LIDAR to study changes associated with *Spartina* invasion in San Francisco Bay marshes. *Remote Sensing of Environment, 100*(3), 295-306.

St. Johns River Water Management District (2007). St. Johns River Water Management District, Chapter 40C-8, *Florida Administrative Code*, Minimum Flows and Levels. Revised May 24, 2004.

South Florida Water Management District (2005a). Regional Simulation Model (RSM) – Theory manual. South Florida Water Management District, West Palm Beach, Florida.

South Florida Water Management District (2005b). Regional Simulation Model (RSM) – Hydrologic Simulation Engine (HSE) user's manual. South Florida Water Management District, West Palm Beach, Florida.

Sun, G., Riekerk, H., & Kornhak, L. V. (2000). Ground-water table rise after forest harvesting on cypress-pine flatwoods in Florida. *Wetlands, 20*(1), 101-112.

Tarr, M., & Babbitt, K. J. (2008). The importance of hydroperiod in wetlands assessment, University of New Hampshire Cooperative Extension. http://extension. unh.edu/ resources/representation/Resource000812_Rep847.pdf.

Tsuyuzuki, S. (2006). Plant establishment patterns in relation to microtopography on grassy marshland in Ruoergai, central China. *Lyonia, 11*(2), 35-41.

Tweedy, K. L., Scherrer, E., Evans, R. O., & Shear, T. H. (2001). Influence of micro topography on restored hydrology and other wetland functions, 2001 American Society of Agricultural Engineers Annual International Meeting, July 30-August 1, 2001, Sacramento, California.

USGS (United States Geological Survey). (1999). South Florida Restoration Science Forum, South Florida Information Access (SOFIA). http://sofia.usgs.gov/ sfrsf/rooms/ hydrology/ water/wheretoday.html.

University of Florida Institute of Food and Agricultural Sciences (2007). Florida Soil Characterization Data Retrieval System. http://flsoils.ifas.ufl.edu/index.asp.

Werner, K. J., & Zedler, J. B. (2002). How sedge meadow soils, microtopography, and vegetation respond to sedimentation. *Wetlands, 22*(3), 451-466.

Wilsnack, M. M., Welter, D. E., Montoya, A. M., Restrepo, J. I., & Obeysekera, J. (2001). Simulating flow in regional wetlands with the MODFLOW wetlands package. *Journal of the American Water Resources Association, 37*(3), 655-674.

In: Modeling Hydrologic Effects...
Editor: Xixi Wang, pp. 83-102
ISBN 978-1-61668-628-4

Chapter 5

EFFECTS OF LAND MANAGEMENT AND TOPOGRAPHY ON LAND SURFACE ENERGY FLUXES OF ECOSYSTEMS AT DIFFERENT LATITUDES

Assefa M. Melesse[1,*], Xixi Wang[2,†], Michael McClain[1,3,‡]
[1]Department of Earth and Environment, Florida International University, Miami, Florida 33199, USA
[2]Department of Engineering and Physics, Tarleton State University, BOX T-0390, Stephenville, Texas 76401, USA
[3]Department of Water Engineering, UNESCO-IHE Institute for Water Education, Delft, The Netherlands

ABSTRACT

Energy (e.g., latent and sensible heats) fluxes are important components of the land-atmosphere processes governing the hydrologic cycle. Understanding these energy fluxes as a function of topography and land management for ecosystems at different latitudes is essential to estimating water-energy exchanges between land surface and atmosphere. This chapter presents the dependence of energy fluxes on physical, geographical, and temporal factors. In addition, this chapter also discusses the estimation and evaluation of latent heat, sensible heat, evaporative energy flux, and non-evaporative energy flux for three selected sites at different latitudes. Microtopography can regulate soil moisture content and its spatial distribution, as indicated by a good correlation between topographic index and latent heat flux. The results indicate that vegetation controls the partition of energy fluxes in the three sites considered in this analysis located at different

[*] E-mail: melessea@fiu.edu
[†] E-mail: xxqqwang@gmail.com
[‡] E-mail: m.mcclain@unesco-ihe.org

latitudes. Remote sensing can be very useful for evaluating the moisture availability, vegetative cover in wetland and grassland ecosystems, and microtopographic effects.

Keywords: energy flux, evaporative flux, Glacial Ridge, Kissimmee River, Landsat, Mara River, MODIS, topographic index, vegetation cover, wetland

INTRODUCTION

Energy (i.e., latent heat, sensible heat, and soil heat) fluxes and land surface environment variables (e.g., management practices and topography) determine land surface characteristics and availability of moisture, and thus are commonly used to understand land surface-atmosphere interactions. The partition of net radiation into latent, sensible and soil heat fluxes is mainly controlled by vegetation cover and temperature gradient, vapor pressure difference, and wind speed. These energy fluxes have been used for estimating irrigation water demand (e.g., Kustas 1990; Bastiaanssen 2000; Kustas et al. 2004; Melesse and Nangia 2005), evaluating wetland restoration (e.g., Loiselle et al. 2001; Mohamed et al. 2004; Melesse et al. 2006; Oberg and Melesse 2006; Melesse et al., 2007), and understanding effects of land management on vegetation cover (e.g., Kustas et al. 1994; Kustas and Norman 1999; French et al. 2000; Hemakumara et al. 2003; Kustas et al. 2004). This chapter examined effects of topography, vegetation cover, and land management on the energy fluxes of net radiation, latent heat, and sensible heat for three selected sites: the Mara River basin (1°23′38"S, 35°11′24"E) in Kenya/Tanzania; the Kissimmee River basin (27°44'42"N, 81°14'43"W) in Florida of the United States (USA); and the Glacial Ridge wetland (47°41'25"N, 96°16'53"W) in northwestern Minnesota of USA.

In this chapter, the different factors that govern available solar radiation at the land surface are discussed, followed by the descriptions and analysis of energy fluxes at the three study sites. In the last section, four examples, which use surface energy fluxes determined from remotely-sensed data to evaluate wetland restoration and quantify moisture or water deficit, are presented. Also, the effects of micro-topography on latent heat for crop fields are shown.

SOLAR RADIATION SPECTRUM

Understanding the surface energy budget requires the consideration of the solar radiation spectrum and its interactions with the atmosphere and surface of the earth. Solar radiation reaches the earth's surface through atmospheric processes of transmission, scattering, diffusion, and reflection. About 50% of incoming shortwave solar radiation is reflected back into space, mainly by clouds and snow covered surfaces, and the remaining half is absorbed at the earth's surface and re-radiated as thermal infrared (i.e., longwave) radiation (NSIDC, 2009; Acra et al., 1989). Depending on albedo, emissivity, and surface and air temperatures, a certain percentage of the incoming solar energy can be reflected back as outgoing radiation. The remaining percentage, defined as net radiation, can be further partitioned into fluxes of latent heat, sensible heat, and soil heat.

The intensity of solar radiation at the earth's surface is dependent on factors of latitude, geographic location, season, cloud cover, air pollution, altitude, and inclination angle. The earth's axis is tilted at an angle of 23.5° affecting the angle of incidence of solar radiation on the earth's surface and in turn causing diurnal, seasonal, and latitudinal variations of radiation. These variations result in differences in levels of solar insolation, at sea level in the visible and in the infrared spectrums (Acra et al., 1989).

Latitude is an important factor in determining the amount of insolation that a geographic location can receive. The global solar intensity is almost constant throughout a year irrespective of slight seasonal variations, with an annual average sunshine duration of about 2500 hours (NSIDC, 2009). Regions at 15 to 35°N, which encompass large parts of northern Africa and southern parts of Asia, have an annual average sunshine duration of over 3000 hours and low cloud coverage. In these regions, more than 90% of the incident solar radiation comes as direct radiation (NSIDC, 2009). In contrast, the equatorial belt at 0 to15°N has a higher humidity and cloudiness, leading to increased proportion of the scattered radiation. However, while the presence of water droplets in the atmosphere causes the dispersion (i.e., diffusion) of the incoming solar radiation, the solar radiation intensity in these latitudes is very high because the places have higher solar altitudes. Further, on landscapes at 35 to 45°N, the scattering of the solar radiation is noticeably increased because the landscapes have higher latitudes but a lower solar altitude. The solar radiation intensity on these landscapes tends to be sharply reduced as a result of elevated cloudiness and air pollution.

The solar radiation at higher latitudes spreads over a greater area and thus is less intense per unit surface area than that at lower latitudes (e.g., around the equator). In contrast to incoming solar radiation, the amount of infrared radiation leaving the earth-atmosphere system is independent of latitude. The places at 30°N or higher cool throughout a year because the outgoing longwave radiation exceeds the incoming shortwave radiation (NSIDC, 2009).

With a higher albedo, clouds can reflect more incoming radiation back to space, reducing the amount of radiation that reaches the earth's surface. On the other hand, clouds absorb and can re-radiate infrared (i.e., longwave) energy radiating from the earth's surface, moderating the temperature of the ambient atmosphere. Globally, clouds have a cooling effect on the earth-atmosphere system, in Polar Regions, however, clouds seem to have a net warming effect as the reduction in solar radiation is outweighed by the increase in longwave radiation (NSIDC, 2009).

Albedo, the ratio of the reflected to the incident solar radiation, is dependent on land surface properties and solar incidence angle. Its physical meaning is surface reflectivity. The incoming solar radiation that reaches the earth's surface is reflected in proportion to albedo. Surfaces with darker color have a lower albedo than those with lighter color. Also, the albedo of a surface is dependent on the incidence angle of solar radiation. That is, the amount of solar energy that a surface absorbs depends on the solar altitude.

SURFACE ENERGY BUDGET

The estimation of surface energy fluxes requires parameters on energy, soil moisture, vegetation, and surface microclimate (Norman et al, 1995, and French et al. 2000). These parameters can be determined using remote sensors at reasonable spatial and temporal scales.

In the absence of horizontally advective energy, the surface energy budget of land surface satisfies the law of conservation of energy and can be expressed as:

$$R_n - LE - H - G = 0 \tag{1}$$

where R_n is the net radiation at the surface, LE is the latent heat or moisture flux (the energy used to evaporate water), H is the sensible heat flux to the air, and G is the soil heat flux.

Eq. (1) is solved by using energy flux models to estimate the different terms separately. Remote sensing based models, such as the commonly used Surface Energy Balance Algorithms for Land (SEBAL; Bastiaanssen et al. 1998a, b) can take into account the spatial variability by computing values at the pixel level and the temporal variability by processing a series of instantaneously-acquired images. SEBAL uses thermal and visible spectrum data, as well as NIR data of sensors (e.g., Landsat, MODIS and ASTER) to estimate the terms in Eq. (1). R_n is determined by estimating albedo, emissivity, surface temperature, and NDVI, while H is determined using surface temperature and emissivity. G is computed as a function of R_n, NDVI, albedo, and surface temperature. LE is determined as R_n minus H and G. The details of using SEBAL to estimate surface energy budget can be found in Bastiaanssen et al. (1998a, b) and to assess agricultural and wetland ecosystems can be found in Melesse and Nangia (2005), Oberg and Melesse (2006), and Melesse et al. (2007). Because its direct measurement is difficult, time consuming, and costly, soil heat flux (G) is usually estimated using empirical equations.

Campbell and Norman (1998) defined *G* as the conduction rate of heat from its storage to ground surface, which is expressed as:

$$G = \lambda_s \frac{dT}{dz} \ (\text{W/m}^2) \tag{2}$$

where λ_s is the thermal conductivity of the soil, *dT* is the difference in temperature over depth *dz*.

λ_s is a function of the conductivities and volumetric fractions of soil constituents, including minerals, water, and air (Campbell and Norman 1998). G is normally equal to 5 to 20% of R_n during daylight hours, but, G cannot be measured remotely.

Bastiaasnssen et al. (1998b) used field measurements to correlate G with Index Normalized Difference Vegetation (NDVI) and R_n for areas of predominately vegetative surfaces. The correlation is expressed as:

$$G/R_n = 0.30\left(1 - 0.98\,NDVI^4\right) \tag{3}$$

Using data from irrigated agricultural regions in Turkey, Bastiaanssen (2000) developed another empirical relationship for *G* that is expressed as:

$$G/R_n = \frac{T_s}{\alpha}(0.0038\alpha + 0.0074\alpha^2)(1 - 0.98\,NDVI^4) \tag{4}$$

where T_s is the surface temperature (°C), α is the surface albedo, and *NDVI* is the normalized difference vegetation index.

Equations (3) and (4) are not applicable to water surfaces because the penetration of solar radiation into water is a function of water transparency, which is not accounted for by these two equations. Bastiaanssen et al. (2000) pointed out that the information on how to account for water transparency is lacking in literature.

Table 5-1. Description and location of the study sites

Site	Latitude	Longitude	Topography	Land Use	Climate
Mara River basin, Kenya/ Tanzania	1°23'38"S	35°11'24"E	Flat to ragged	Mixed: ag. & rangeland	Tropical
Kissimmee River basin, Florida, USA	27°44'42"N	81°14'43"W	Flat	Wetland, ag.	Sub-tropical
Glacial Ridge Wetland, Minnesota, USA	47°41'25"N	96°16'53"W	Flat	Wetland, ag.	Temperate

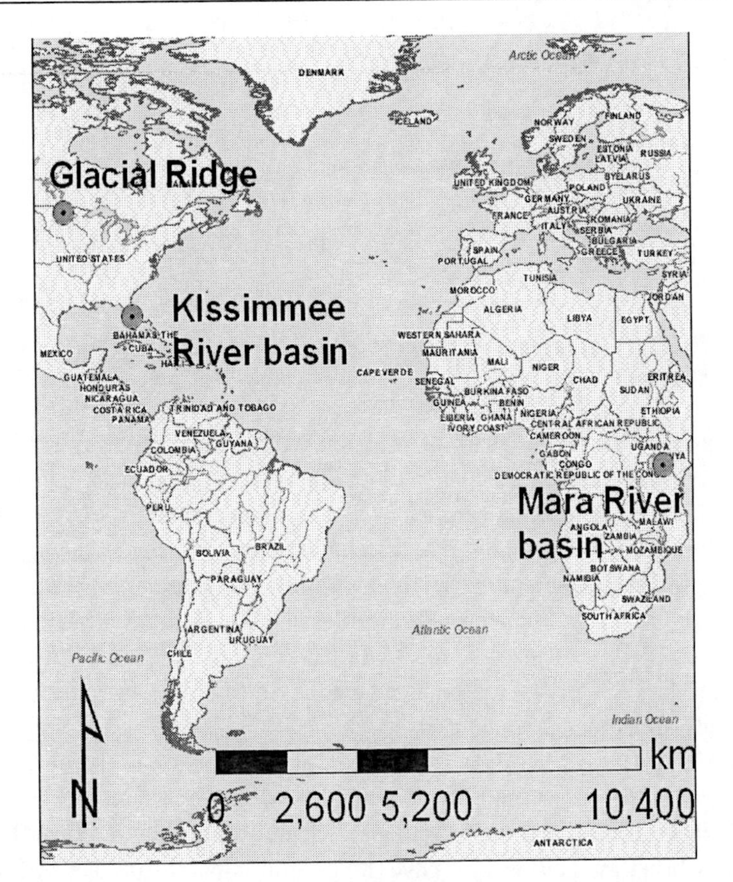

Figure 5-1. Map showing locations of the three study sites.

STUDY SITES

The effects of land cover (e.g., wetland and grassland) and topography on the terms of surface energy budget in Eq. (1) are examined for three sites located at different latitudes (Figure 5-1; Table 5-1). The sites are the Mara River basin (Figure 5-2), the Kissimmee River basin (Figure 5-6), and the Glacial Ridge Wetland (Figure 5-9).

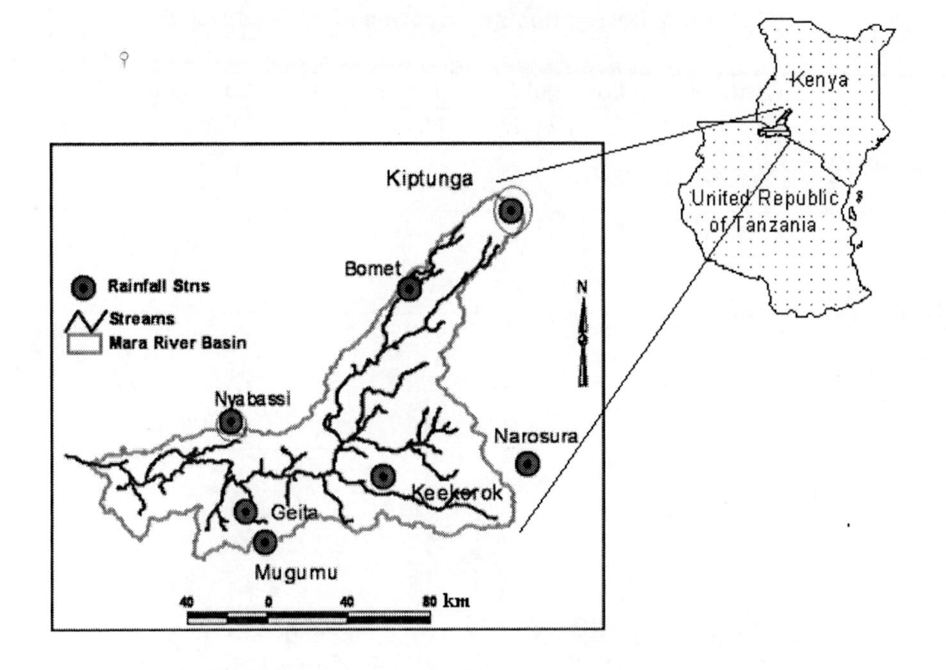

Figure 5-2. Map showing the location and boundary of the Mara River basin, superimposed by the drainage network and weather stations where data were used in this study.

The Mara River Basin

Head watered on the Mau Escarpment in Kenya, the Mara River menders 395 km with the elevation drop from 2920 m above sea level (a.s.l) to 1134 m a.s.l and flows into Mara Bay of Lake Victoria in Tanzania (Figure 5-2). The river drains 13,750 km^2, of which 8941 km^2 (65 %) is located in Kenya and the remaining drainage area in Tanzania. The Masai Mara Game Reserve of Kenya accounts for about 17% of the drainage area in Kenya (i.e., 11% of the total drainage area). Flowing through the Masai Mara Game Reserve in Kenya and the Serengeti National Park in Tanzania, the Mara River is the life line for the Mara-Serengeti ecosystem. The Mara River basin is bounded by the Soit Ololo Escarpment on the west and by the Loita and Sannia plains on the east.

The major land uses of the Mara River basin include forests in the upstream watershed, mixed agriculture in the upper central part of the basin, and grassland and protected areas for the Serengeti National Park and the Masai-Mara Game Reserve in the lower basin. Land use changes in the basin, which would have long-term impacts on the hydrology and the sustainability of other resources (e.g., fuel), are documented by Ottichilo et al (2001), Gereta et al (2002), Serneels et al (2001), and Onjala (2004).

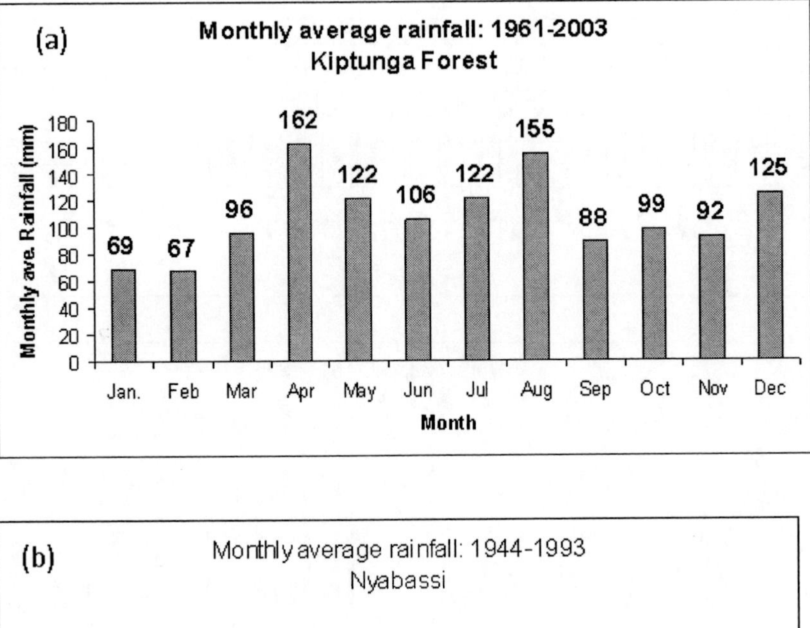

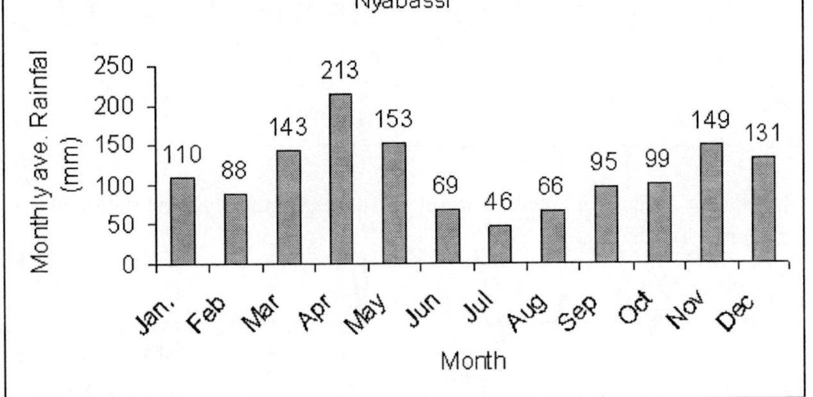

Figure 5-3. The long-term average precipitations at (a) the Kiptunga Forest station in Kenya and (b) the Nyabassi station in Tanzania.

Meteorological data from selected monitoring stations (Figure 5-2) indicate that precipitation varies greatly across the basin as well as from season to season. For example, most of the annual precipitation occurs in April and May, followed by dry months and the months from October to December with small rainfall events (Figures 3 and 4). The long-term average precipitation records from the Kiptunga Forest station in Kenya and the Nyabassi station in Tanzania indicate greater rainfalls occurred in the upstream part of the basin than in the downstream. Figure 5-5 shows the historical daily potential evapotranspiration at the Narok station adjacent to the basin.

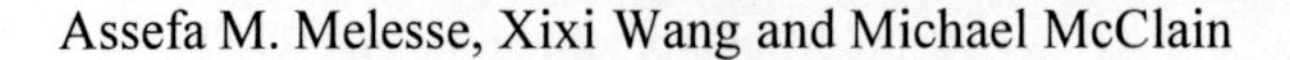

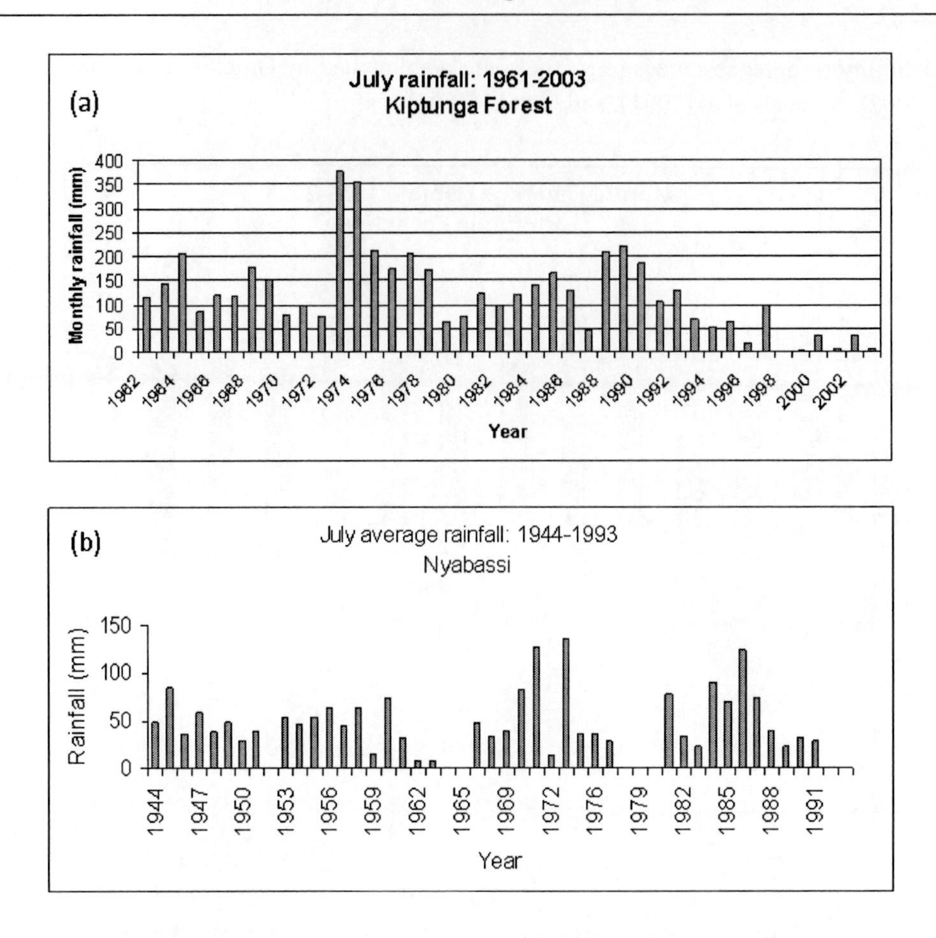

Figure 5-4. The monthly mean precipitations in July at (a) the Kiptunga Forest station in Kenya and (b) the Nyabassi station in Tanzania.

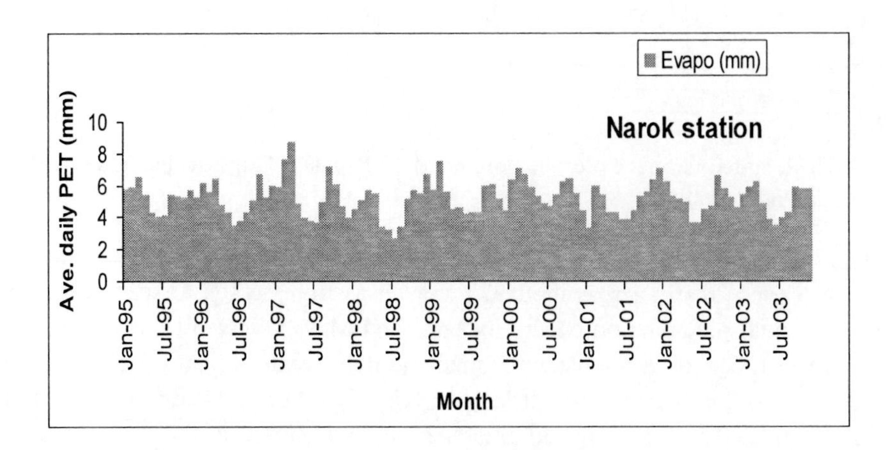

Figure 5-5. The average daily potential evapotranspiration (PET) at the Narok station adjacent to the Mara River basin in Kenya.

The Kissimmee River Basin

Since European colonization began, hundreds of thousands of kilometers of river corridors have been encroached and millions of hectares of wetlands have been drained and/or dredged throughout USA (Melesse et al. 2007). The ecosystems in southern Florida, in particular the Everglades and the Kissimmee River basin, have been incurring continuous ecological alterations since early 1900s. The noticeable alternations resulted from the conversion of wetlands into agricultural and/or urbanized areas as well as the channelization of the Kissimmee River between 1962 and 1970 for flood control purposes. Currently, restoration activities are underway to reverse the alterations.

The Kissimmee River basin is located north of Lake Okeechobee in southern Florida (Figure 5-6). The basin has a total drainage area of 7680 km^2, stretching from southern Orlando southward to Lake Okeechobee in central Florida. The average annual precipitation is 1273 mm, of which 833 mm occurs in the wet season (June through October), 434 mm occurs in the dry season (November through February), and 6 mm occurs in the three months of March through May. The minimum temperature occurs in January and has an annual average value of 8.9°C, while the maximum temperature occurs in July and has an annual average value of 33.3°C. Meteorological data (Figures 7 and 8) at the Ft. Lauderdale station adjacent to the basin verify these seasonal patterns of air temperature, evapotranspiration, and solar radiation.

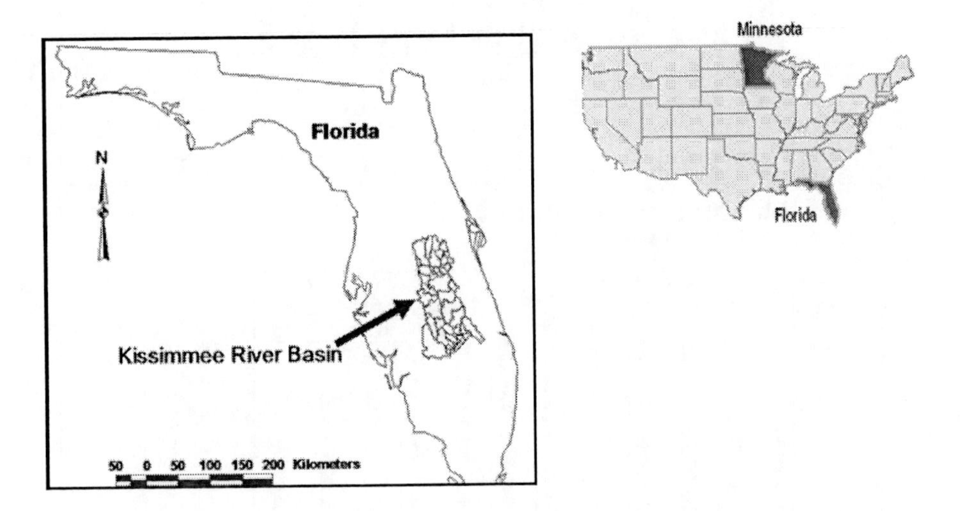

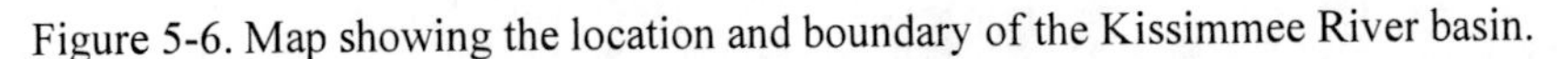

Figure 5-6. Map showing the location and boundary of the Kissimmee River basin.

The major land uses of the basin are wetlands, cropland, rangeland, and forests. The areal extents of these land uses have been greatly changed from the pristine conditions as a result of the development and wetland drainage. The dominant wetland types are broadleaf marsh, wet prairie, and wetland shrub. In wetlands where water is available throughout a year, the evaporative fluxes are controlled by the available evaporative energy (i.e., solar radiation). Figure 5-8 shows the peak solar radiations on 1 April and 1 August 2008 for comparison purposes.

Prior to the channelization, the Kissimmee River had a much larger sinuosity index and an over 180 km^2 floodplain. The channelization changed this river into a canal of 90 km long, 64 to105 m wide and 9 m deep. In addition, the channelization drained 120 to 140 km^2 of wetlands, which disturbed the ecological balance between the flora, fauna, and hydrology of this riverine ecosystem (Toth, 1996). The Kissimmee River restoration project began in 1997 and is working to reestablish the natural conditions of hydrology, vegetation, and biodiversity, and to recreate historical floodplains through the removal of flood control canal, water control structures, and levees (SFWMD, 2009). The specific restoration activities that have been implemented since 1999 include rechannelization, revegetation, and land acquisition.

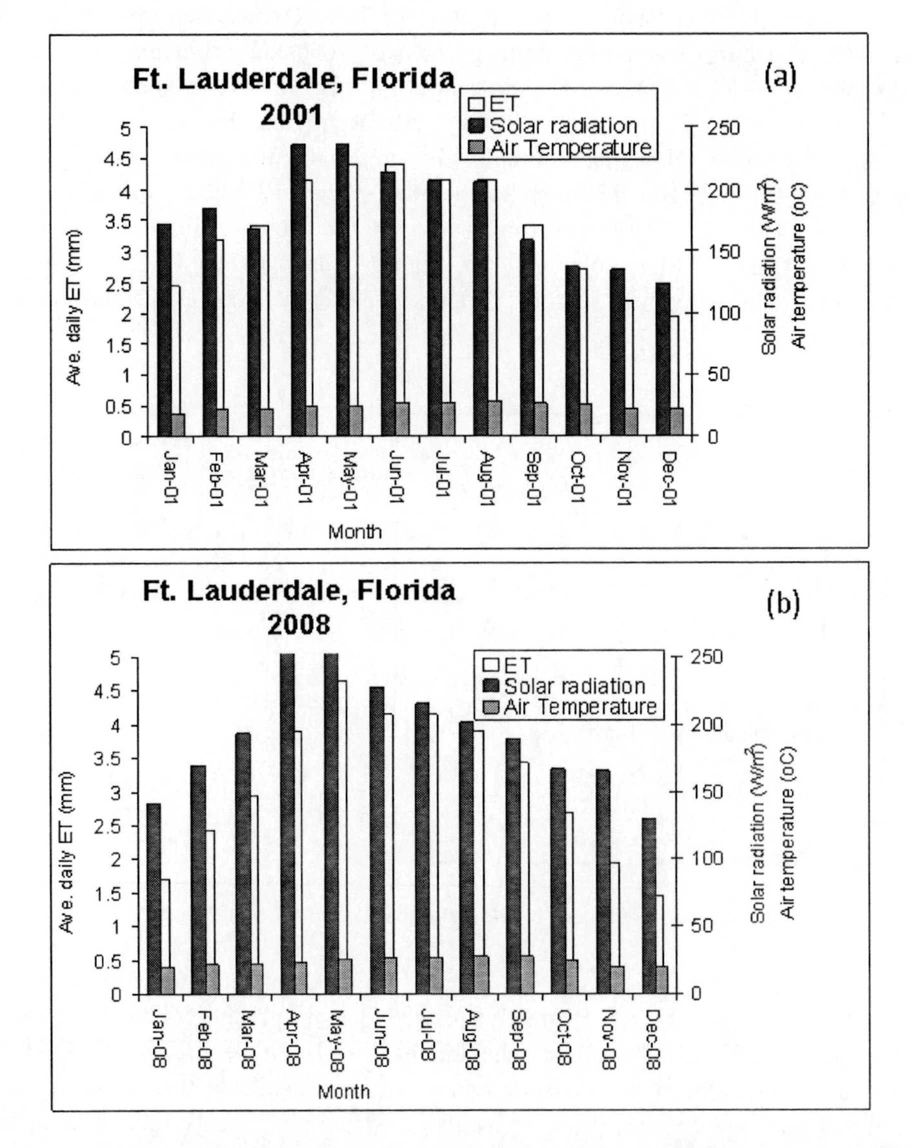

Figure 5-7. The average daily evapotranspiration (ET), solar radiation, and air temperature at the Ft. Lauderdale station in the year of (a) 2001 and (b) 2008.

The Glacial Ridge

Implemented by The Nature Conservancy (TNC, 2009) since late 2001, the Glacial Ridge prairie restoration project is located in northwest of Minnesota (Figure 5-9). The project covers 99.7 km^2 of prairie lands, of which about 20.2 km^2 are native prairie lands but the remaining are disturbed and have been used for gravel extraction, crop production, and cattle and sheep grazing. The prairie wetland complex supports a great diversity of plant species, including the threatened western prairie fringed orchid (TNC, 2009). Other vegetation types found at this site include wet and mesic tallgrass prairie and gravel prairie, willow thickets, mixed prairie, sedge meadow, aspen woodlands, and emergent marsh. The topography of the Glacial Ridge site consists of level to gently rolling tallgrass prairie and agricultural fields that are interspersed with wetlands. The 30-year average mean annual precipitation is 590 mm, of which 90 mm is in the form of snowfall. The mean maximum and minimum monthly temperatures are 10 and -2.3 °C, respectively. The maximum temperature occurs in July or August, while the minimum temperature occurs in January or February. Figure 5-10 shows the average monthly solar radiation, evaporation, and temperature of the Glacial Ridge site.

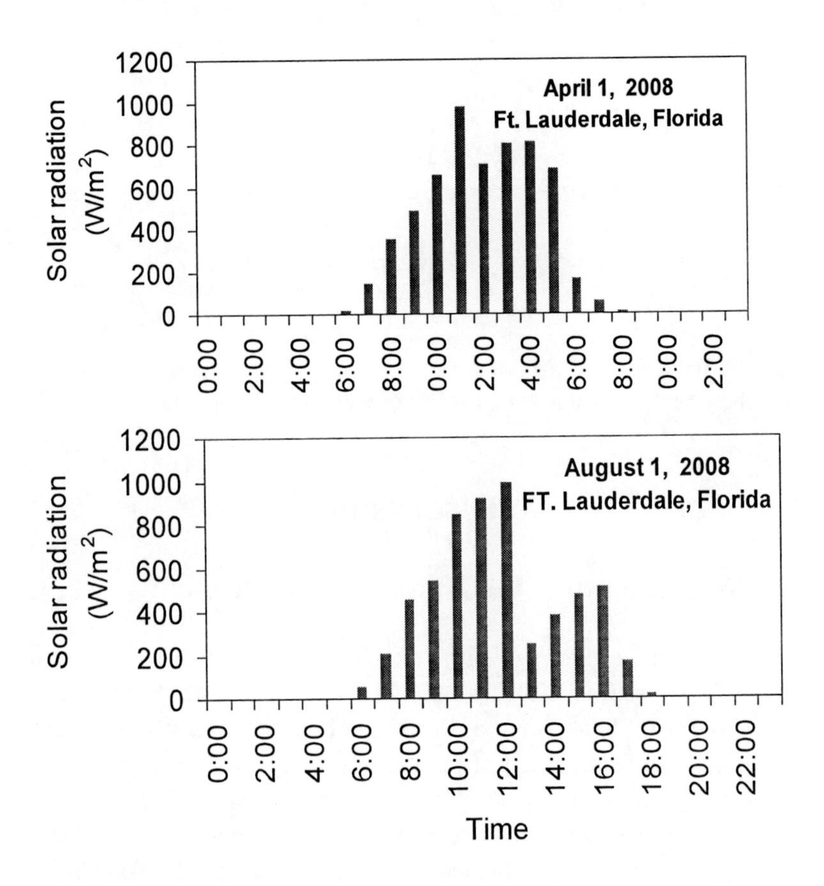

Figure 5-8. The hourly solar radiation at the Ft. Lauderdale station on (a) 1 April 2008 and (b) 1 August 2008.

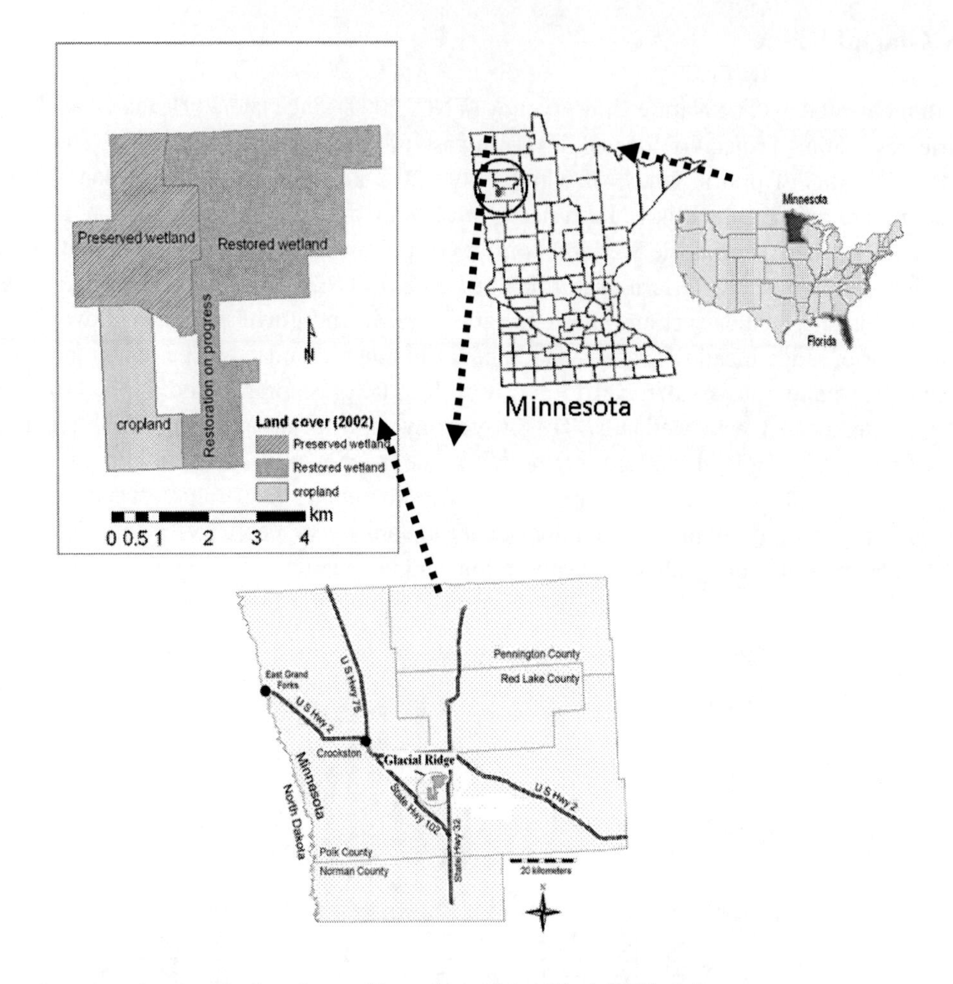

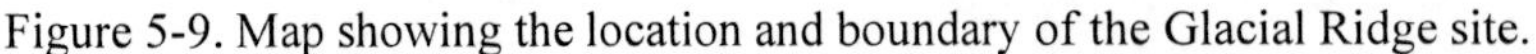

Figure 5-9. Map showing the location and boundary of the Glacial Ridge site.

DISCUSSION

Drought and Water Deficit in the Mara River Basin

To understand the onset of droughts in the basin, the spatially distributed non-evaporative energy flux (NEF) was estimated using remotely sensed data acquired in February, May, August, and October of 2004 from a Moderate Resolution Imaging Spectroradiometer (MODIS) sensor aboard Terra. In addition, data on daily surface temperature, NDVI, and albedo were obtained from the Land Processes Distributed Active Archive Center (LP DAAC, 2005) and used in the surface energy balance computation. Further, data on air temperature and wind speed were collected from the Kenyan and Tanzanian meteorological offices.

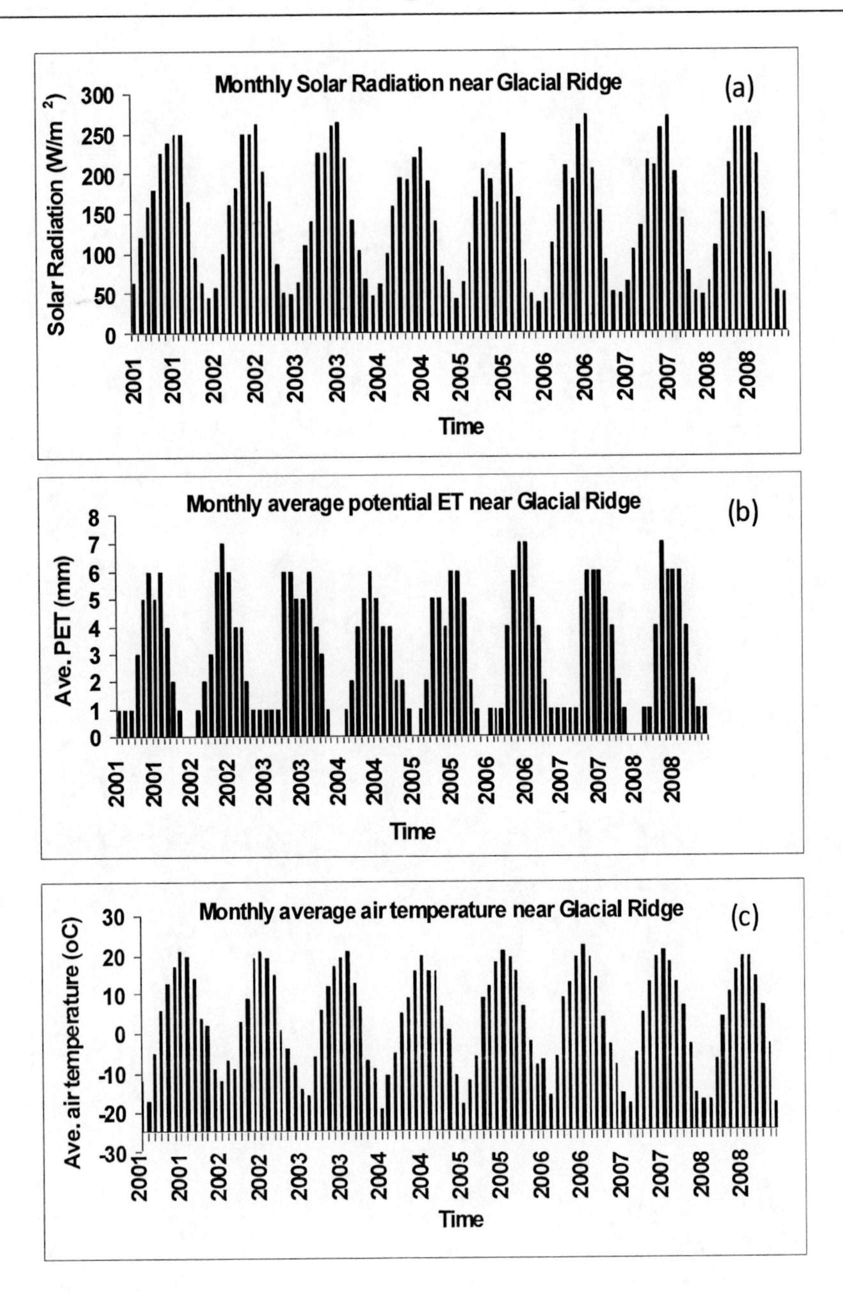

Figure 5-10. The Glacial Ridge monthly average (a) solar radiation, (b) potential evapotranspiration (PET), and (c) temperature.

Evaporative energy flux (EF) is complementary with NEF and is computed as 1- NEF. EF (Eq. 5) and NEF (Eq. 6) are functions of latent heat (LE), sensible heat (H), net radiation (R_n), and soil heat flux (G). EF indicates the availability of evaporable water: a high EF value indicates the presence of wet surface and green vegetative cover. In contrast, a high NEF value indicates water deficit and drought.

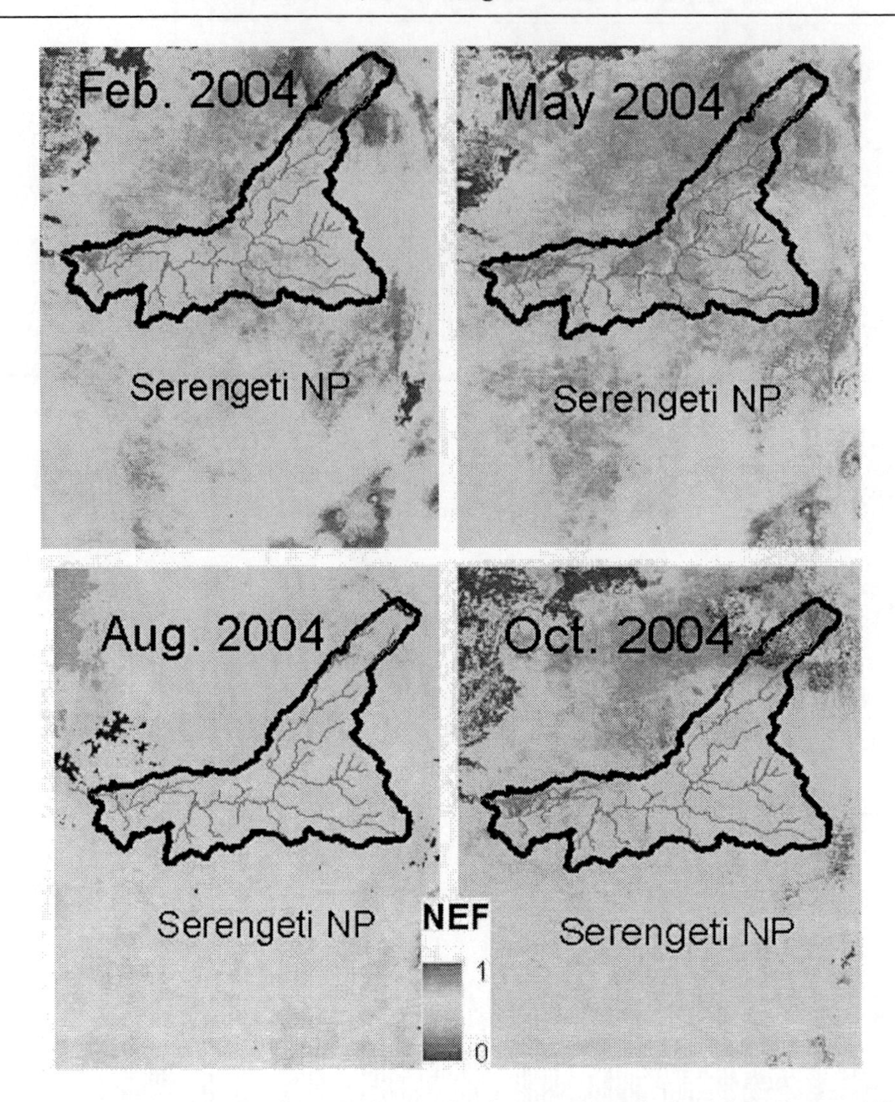

Figure 5-11. The non-evaporative flux (NEF) maps for the Mara River basin and the Serengeti National Park (SNP).

$$EF=\frac{LE}{LE+H}=\frac{LE}{R_n-G} \tag{5}$$

$$NEF=1-EF=\frac{H}{LE+H}=\frac{H}{R_n-G} \tag{6}$$

In terms of the management of the Masai Mara National Reserve (MMNR) and Serengeti National Park (SNP) in the Mara River basin, the critical factors are: 1) droughts; 2) dry season flows; and 3) grazing land. The MODIS-based NEF maps (Figure 5-11) were generated and used to examine the extent of moisture stressed areas for different seasons. The maps show higher NEF values for the months of February and August than for the months of May and October. In the SNP, the NEF values were higher for the months of February, May and August than for the other months indicating low vegetation cover and moisture in this period. In the months of May to October, the wildlife population is in the northern part of the SNP in search of grazing land and water along the Mara River. Since the MODIS data are available on daily basis, this information can be useful in generating fluxes at a daily and weekly basis to inform park managers and others on the status of vegetation and grassland cover.

Kissimmee River Basin Restoration and Evaporative Flux Dynamics

To understand the effects of restoration on evapotranspiration, the MODIS remote sensing data were used to compute the EF (Eq. 5) values in April for years 2002 to 2004, and the results are shown in Figure 5-12. The results indicate that the EF values vary spatiotemporally and are higher in the lower part of the basin along the restored areas in 2003 and 2004. The removal of flood control structures and rechannalization of the river to its natural course will increase the floodplain area and in turn higher latent heat flux and EF. It is shown that higher latent heat flux along the river can be attributed to the increased flood plain areas and vegetation cover.

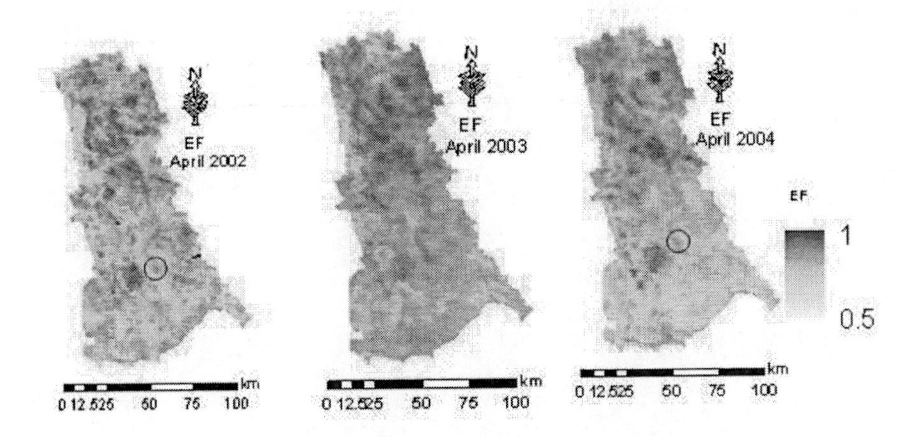

Figure 5-12. The evaporative flux (EF) maps for the Kissimmee River basin.

Evapotranspiration Dynamics in the Glacial Ridge Wetland Restoration Site

The effects of the prairie restoration on available surface water and hence evapotranspiration (ET) were evaluated using the Landsat-based ET values in July for the years 2001 to 2003 (Figure 5-13). The restoration began in late 2001 and the comparison provides a very good base for evaluating the effects of the restoration activities (e.g., planting of native species, drainage canal closure, and water diversion) on ET. The results show that the restoration areas had greater ET values in 2003 than 2001 and that the ET for the restored part was compatible with the pristine section of the Glacial Ridge (Figures 9 and 13).

Comparison of the monthly ET from the pristine and restored part of the Glacial Ridge is shown in Table 5-2. The analysis showed comparable ET values between the restored and preserved values for the months of July and August for the years 2001 and 2003. Since the restoration began in late October, the comparison between the 2001 and 2003 depicts the effect of restoration on spatial ET. In 2001, both sections of the wetland have similar ET but the spatial variability is indicated to be higher by high values of SD for the impacted part of the wetland (Table 5-1). The ET values between 2001 and 2003 increased by 11.5% in July and 86% in August.

Table 5-2. Comparison of monthly ET (mm) from the preserved and restored section of the wetland

Layer		**Preserved**		**Impacted (2001) Restored (2003)***		**Average**	
		Mean	STD	Mean	STD	Mean	STD
2001	**July**	117.8	17.8	119.0	28.0	84.9	22.9
	Aug.	130.1	11.3	127.8	20.9	89.7	16.1
2003	**July**	136.6	11.7	132.7	15.7	93.7	13.7
	Aug.	246.1	16.2	237.7	23.6	166.7	19.9
Ave		**157.7**	**14.3**	**154.3**	**22.0**	**112.5**	**19.3**

* In 2001 summer, this area was still impacted.

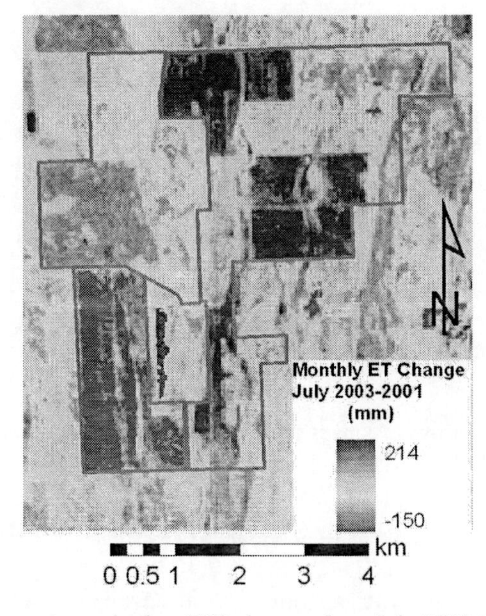

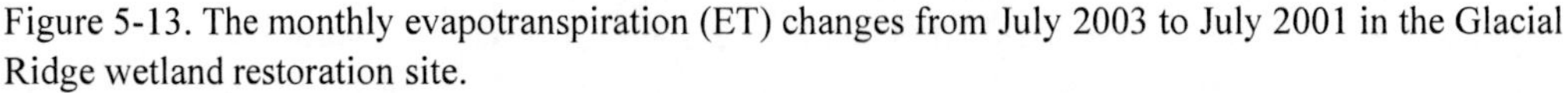
Figure 5-13. The monthly evapotranspiration (ET) changes from July 2003 to July 2001 in the Glacial Ridge wetland restoration site.

Effects of Microtopography on Latent Heat

Among other environment parameters, topography is a determinant for magnitudes and spatial distributions of water and energy fluxes over natural landscapes. The topographic configuration of a landscape is a control boundary condition for the hydrologic processes of surface runoff, evaporation, and infiltration, which take place at the ground-atmosphere interface. For example, wetness index (WI) provides a description of the spatial distribution of soil moisture in terms of topographic information. WI is computed as.

$$WI = Ln\left(\frac{A}{S}\right) \tag{7}$$

where A and S are the specific drainage (i.e., flow accumulation) area and slope, respectively.

As A increases and/or S decreases, WI becomes larger, indicating that soil moisture content will increase. Because WI takes into account local slope variations, it has proven to be a reasonable indicator for soil wetness, flow accumulation, saturation dynamic, water table fluctuation, evapotranspiration, soil horizon thickness, organic matter content, pH, silt and sand content, and plant cover density (Kulagina et al. 1995, Florinsky 2000).

The microtopography and latent heat flux are found to be well correlated for agricultural fields of wheat and soybean (Figure 5-14). This may be because the microtopography controls soil moisture content as well as its spatial distribution. Grids with higher WI values are identified as the areas receiving more overland flows (i.e., with greater flow accumulations) and having a smaller gradient. These areas have higher soil moisture but a higher evaporation rate than the areas with lower WI values. The correlation between WI and soil moisture is further verified by the observation that when water is a limiting factor of an agricultural field, the crop in the areas with higher WI values tends grow better than the crop in the areas with lower WI values. This can be attributed to more water availability for transpiration (i.e., latent heat demand) in areas with higher WI values. Figure 5-14 shows WI versus latent heat for wheat and soybean for selected seasonal dates in North Dakota. For all dates and both crops, the latent heat increases with the WI, indicating a positive correlation between evapotranspiration and soil moisture. The latent heat seems to increase at a greater rate when the WI is lower, which is partially due to the proportional relationship between available water for evapotranspiration and WI.

SUMMARY

The effects of land management and topography on surface energy fluxes were examined. Using remotely sensed data, the surface energy fluxes, the evaporative flux, and the non-evaporative flux were estimated and evaluated. The results indicate that these fluxes can be good spatial indicators of moisture deficits in the Mara River basin and the SNP, and of evapotranspiration (ET) changes as a result of ecological restorations in the Kissimmee and Glacial Ridge sites. For the latter two sites, ET (i.e., latent heat flux) tends to increase after restoration responding to the restorative activities. In addition, the microtopography can regulate the soil moisture content and its spatial distribution. For an agricultural field of wheat or soybean, the crop in the areas with higher WI values tends to have a greater latent heat than the crop in the areas with lower WI values.

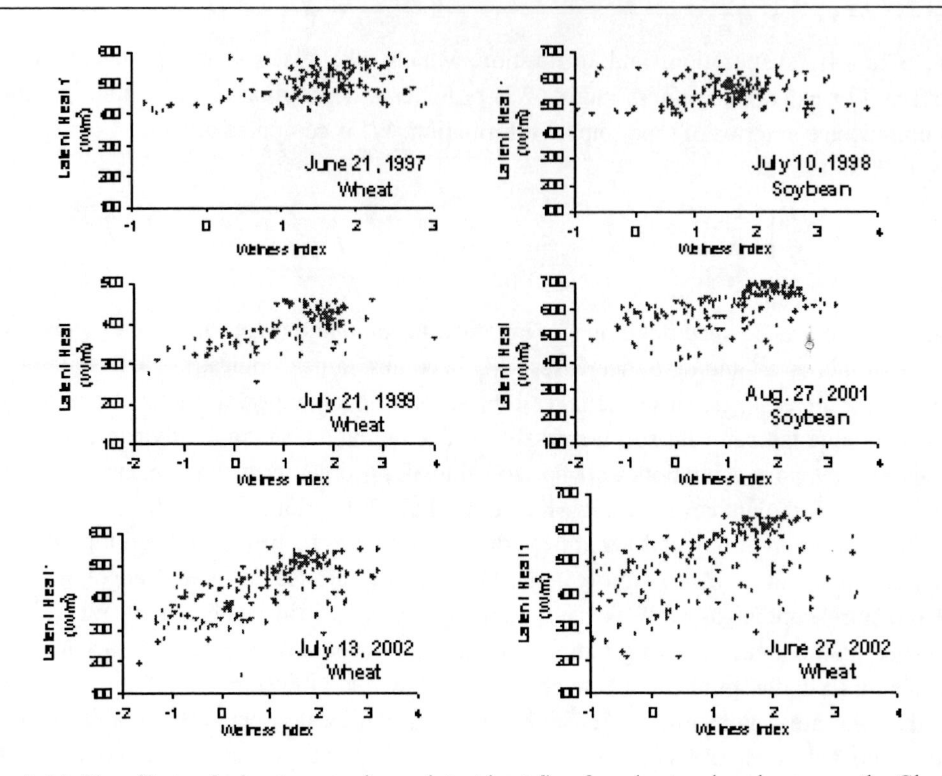

Figure 5-14. The effects of microtopography on latent heat flux for wheat and soybean near the Glacial Ridge site.

REFERENCES

Acra, A.; Jurdi, M.; Mu'allem, H.; Karahagopian, Y.; Raffoul, Z. (1989). *Water disinfection by solar radiation: assessment and application.* IDRC. 86 pp.

Bastiaanssen, W.G.M., Menenti, M., Feddes, R.A., and Holtslag, A.M. (1998a). The Surface energy balance algorithm for land (SEBAL): Part 1 formulation. *Journal of Hydrology*, 212-213, 198-212.

Bastiaanssen, W.G.M., Pelgrum, H., Wang, J., Ma, Y., Moreno, J., Roerink, G.J., and van der Wal, T. (1998b). The surface energy balance algorithm for land (SEBAL): Part 2 validation. *Journal of Hydrology,* 212-213, 213-229.

Bastiaanssen, W.G.M.(2000). SEBAL-based sensible and latent heat fluxes in the irrigated ediz Basin, Turkey. *Journal of Hydrology,* 229, 87-100.

Campbell, G. S., and Norman, J.M. (1998). *An Introduction to Environmental Biophysics (2nd ed).* Springer-Verlag: New York. pp 4-5.

Florinsky, I.V. (2000). Relationships between topographically expressed zones of flow accumulation and sites of fault intersection: Analysis by means of digital terrain modeling. *Environmental Modeling and Software,* 15, 87-100.

French, A.N., Schmugge, T.J., and Kustas, W.P. (2000). Estimating surface fluxes over the SGP site with remotely sensed data. *Physics and Chemistry of the Earth*, 25(2), 167-172.

Kulagina, T.B., Meshalkina, J.L., and Florinsky, I.V. (1995). The effect of topography on the distribution of landscape radiation temperature. *Earth Observation and Remote Sensing* 12: 448–458.

Gereta, E., Wolanski, E., Markus, B., and Serneels, S. (2002). Use of an ecohydrology model to predict the impact on the Serengeti Ecosystem of deforestation, irrigation and the proposed Amala Weir water diversion project in Kenya: Transboundary Issue. *Ecohydrology & Hydrobiology,* 2 (1-4), 127-134.

Hemakumara, H. M., Chandrapala, L., and Moene, A. F. (2003). Evapotranspiration fluxes over mixed vegetation areas measured from large aperture scintillometer. *Agricultural Water Management,* 58(2),109-122.

Kustas, W.P. (1990). Estimates of evapotranspiration with a one-and two-layer model of eat transfer over partial canopy cover. *Journal of Applied Meteorology,* 29, 704-715.

Kustas, W.P., Perry, E.M., Doraiswamy, P.C., and Moran, M.S. (1994). Using satellite remote sensing to extrapolate evapotranspiration estimates in time and space over a semiarid Rangeland basin. *Remote Sensing of Environment,* 49(3), 275-286.

Kustas, W.P, and Norman, J. (1999). Evaluation of soil and vegetation heat flux predictions using simple two-source model with radiometric temperatures for partial canopy cover. *Agricultural and Forest Meteorology,* 94, 13-29.

Kustas, W.P., Li, F., Jackson, T.J., Prueger, J.H., MacPherson, J.I., and Wolde, M. (2004). Effects of remote sensing pixel resolution on modeled energy flux variability of croplands in Iowa. *Remote Sensing of Environment,* 92(4), 535-547.

Land Processes Distributed Active Archive Center (LP DAAC). *Available at* http://lpdaac.usgs.gov/modis/dataproducts.asp, *Accessed on* 24 January 2005.

Loiselle, S., Bracchini, L., Bonechi, C., and Rossi, C. (2001). Modeling energy fluxes in remote wetland ecosystems with the help of remote sensing. *Ecological Modeling* 45 (2), 243-261.

Melesse, A., and Nangia, V. (2005). Spatially distributed surface energy flux estimation using remotely-sensed data from agricultural fields. *Hydrological Processes,* 19 (14), 2653-2670.

Melesse, A., Oberg, J., Beeri, O., Nangia, V., and Baumgartner, D. (2006). Spatiotemporal dynamics of evapotranspiration and vegetation at the Glacial Ridge prairie restoration. *Hydrological Processes,* 20(7), 1451-1464.

Melesse, A., Nangia, V., Wang, X., and McClain, M. (2007). Wetland restoration response analysis using MODIS and groundwater data. Special Issue: Remote Sensing of Natural Resources and the Environment. *Sensors,* 7, 1916-1933

Mohamed, Y.A., Bastiaanssen, W.G.M., and Savenije, H.H.G. (2004). Spatial variability of evaporation and moisture storage in the swamps of the upper Nile studied by remote sensing techniques. *Journal of Hydrology,* 289, 145-164.

Norman, J., Divakarla, M., Goel, N.A. (1995). Algorithms for extracting information from remote thermal-IR observations of the earth's surface. *Remote Sensing of Environment,* 51, 157-168.

NSIDC, (2009) NSIDC Arctic Climatology and Meteorology Primer (http:// nsidc.org/ rcticmet/), *Accessed* on 2 March, 2009.

Oberg, J. and Melesse, A.M. (2005). Wetland evapotranspiration dynamics vs. ecohydrological restoration: An energy balance and remote sensing approach. *Journal of American Water Resources Association,* 42(3), 565-582.

Onjala, J. (2004). Gross Margin Analysis of Various Crops Cultivated in the Mara River (Incomplete Draft Repot). World Wide Fund for Nature–East African Regional Office, Nairobi, Kenya.

Ottichilo, W., de Leeuw, K.J., and Herbert, H. T. P. (2001). Population trends of resident wildebeest [*Connochaetes taurinus hecki* (Neumann)] and factors influencing them in the Masai Mara Ecosystem, Kenya. *Biological Conservation*, 97, 271-282.

Serneels, S., Said, M.Y., and Lambin, E.F. (2001). Land-cover changes around a major East-African wildlife reserve: the Mara Ecosystem, Kenya. *International Journal of Remote Sensing,* 22(17), 3397-3420.

SFWMD, 2009. *Available a*t http://www.sfwmd.gov/org/erd/closer/ clkissim.pdf, *Accessed* on 30 April, 2009).

TNC. (2009). *Available at* http://nature.org/wherewework/northamerica/states/ minnesota/ preserves/art6943.htm, *Accessed on* 21 August 2009.

Toth, L.A. (1996). Restoring the hydrogeomorphology of the channelized Kissimmee River. pp. 369-383 in: Brookes, A., and F.D. Shields, Jr., (eds.). *River Channel Restoration: Guiding Principles for Sustainable Projects*. John Wiley & Sons, Ltd., Chichester, England.

In: Modeling Hydrologic Effects…
Editor: Xixi Wang, pp. 103-120
ISBN 978-1-61668-628-4

Chapter 6

EFFECTS OF DETENTION PONDS IN CONTROLLING STORMWATER QUALITY AND PEAK DISCHARGE FROM SITE DEVELOPMENT

Xiaoqing Zeng *
Stetson Engineers Inc., 2171 E. Francisco Blvd., Suite K,
San Rafael, CA 94901, USA

ABSTRACT

A new site development is typically required neither to increase the peak discharge nor to degrade the water quality, of the stormwater leaving site, as compared with the corresponding values for the predevelopment conditions. In practice, stormwater detention ponds have been widely used for new site developments to satisfy these requirements. However, very limited evaluations have been conducted to document the cumulative effects of such detention ponds on hydrology and water quality at the watershed scale. The objectives of this chapter are to: 1) overview the engineering practices of stormwater detention ponds in the State of California in the United States; and 2) evaluate the cumulative effects of detention ponds within the Corte Madera Creek watershed located about 20-km north of San Francisco in California. The results from the study watershed indicate that the detention ponds designed for peak flow reduction combined with runoff volume reduction measures can be an effective means to minimize watershed environmental impacts resulting from site developments. This is consistent with the findings revealed by the practices that are overviewed in this chapter.

Keywords: Best management practices; California; Detention ponds; Stormwater management; Watershed

* Ph.D, P.E., Supervising Engineer (415)457-0701, Fax (415)457-1638, email: xiaoqingz@stetsonengineers.com.

INTRODUCTION

Stormwater runoff is a component of natural hydrologic processes. However, the stormwater runoff resulting from site developments within a watershed can alter the characteristics of natural processes as well as transport nonpoint source (NPS) pollutants (e.g., phosphorus) downstream into the receiving water bodies. The development that converts pervious land covers into impervious (e.g., concrete-paved) surfaces can modify the rainfall-runoff mechanisms and thus change the watershed hydrology. The modifications, in particular as a result of new developments, tend to increase runoff volume, peak discharge, and duration and frequency of bankfull flow, which is likely to cause secondary problems such as degraded habitat, increased stream bed and bank erosion, and disturbed ecosystem. The degree of impacts is dependent on the aforementioned modifications and the resulted increase of stream power.

Urban runoff has been cited as a major nonpoint pollution source because it can cause the shrinkage of fishery and restrictions for swimming, and limit human's options to enjoy many of the other benefits that water resources could provide. While the natural stormwater runoff contains various kinds of constituents, their levels in the stormwater runoff resulting from site developments could be elevated to a level above which the constituents would become water quality concerns. The typical pollutants associated with the modified stormwater are sediment, nutrients, bacteria and viruses, oil and grease, metals, organics, pesticides, and trash (floatables). For site developments, the sediment is typically originated from the construction of roads and parking lots, the disturbance of landscapes, and the removal of vegetation covers. The organic compounds are secondary products of automotive fluids, pesticides, and herbicides, whereas, nutrients (i.e., nitrogen and phosphorus) are mainly from organic litter, fertilizers, food waste, sewage, and sediment. Sources of trace metals include motor vehicles, roofing and construction materials, and chemicals. Pet waste and solid waste disposal areas contribute bacteria and viruses and motor vehicles are the dominant source of oil and grease compounds. The modified, nutrient-enriched stormwater runoff, once it accumulates and stands for more than 72 hours, can be an attractive medium for vector production. In general, site developments can impact water quality as results of: 1) the erosion, sedimentation, and contamination during the construction period; and/or 2) the increased stormwater runoff generated from the engineered landscapes after the construction.

The Clean Water Act (CWA), enacted in 1972, authorized the U.S. Environmental Protection Agency (USEPA) to regulate point source pollutants to be discharged into waters (e.g., streams and lakes) across the nation. The 1987 amendments to the CWA as section 402(p) established a framework for regulating NPS pollutants transported by stormwater discharges. This framework is well known as the National Pollutant Discharge Elimination System (NPDES), which is only comprised of Phase I to regulate stormwater discharges from major industrial facilities, large and medium-sized municipal storm sewer systems (those serving more than 100,000 persons), and construction sites that disturb five or more acres of land.

In 1999, USEPA expanded the NPDES by including Phase II, which requires "small" municipal storm sewer systems (i.e., those serving populations less than 100,000) and construction sites that disturb 1 to 5 acres of land implement programs/practices to control contaminated stormwater runoff. The control is realized by developing and executing a site-specific plan, designated as the Storm Water Pollution Prevention Plan (SWPPP). The main purpose of Phase II is to prevent water quality from being impaired by the small municipal storm sewer systems and construction projects that are not under the regulation of Phase I. The most noticeable outcome of the NPDES has been the applications of Best Management Practices (BMPs) to control stormwater runoff. BMPs are combinations of a variety of abatement procedures, structural measures, nonstructural measures, regulatory restrictions, and management practices. BMPs are usually designed to minimize the adverse impacts of watershed activities and/or site developments on for instance riparian morphology.

The conventional BMPs were usually designed to control onsite (i.e., localized) sources. Among those BMPs, stormwater detention ponds have been widely used for site developments to satisfy the requirements of neither to increase the peak discharge nor to degrade the water quality, of the stormwater leaving site, as compared with the corresponding values for the predevelopment conditions. However, Emerson et al. (2005) showed that the stormwater detention ponds designed based on this conventional onsite control philosophy may not be able to reduce the peak discharge at watershed scale, but many even increase the discharge. In contrast, when designed and constructed to satisfy watershed-scale stormwater runoff control goals, detention ponds may be more cost-effective. Previous studies (e.g., Harrell and Ranjithan, 1997, 2003; Yeh and Labadie, 1997) examined the placement and sizing of detention ponds in terms of watershed protection. For example, Yeh and Labadie (1997) used a multi-objective genetic algorithm to optimize locations and sizes of detention ponds for flood control. Nevertheless, practical applications of detention ponds to achieve watershed-scale benefits are still facing a number of challenges. The objectives of this chapter are to: 1) overview the engineering practices of stormwater detention ponds in the state of California in the United States; and 2) evaluate the cumulative effects of detention ponds within the Corte Madera Creek watershed (Figure 6-5) located about 20-km north of San Francisco in California. In order to be consistent with their sources, some numbers hereinafter are reported in U.S. Customary Units. The conversions are 1 ft = 0.3048 m, 1 inch = 1 in = 25.4 mm, and 1 cfs = 0.02832 m^3/s.

Engineering Practices of Stormwater Detention Ponds in California

In California, the State Water Resources Control Board (SWRCB) and the Regional Water Quality Control Boards (RWQCBs) are responsible for administering the NPDES program. Under the NPDES program, an applicant is required to submit a Notice of Intent (NOI) with the SWRCB Division of Water Quality. The NOI includes general information on the types of construction activities that will be conducted on the site. The applicant is also required to prepare and implement a SWPPP to address stormwater management issues after the project is completed. The SWPPP usually includes a description of appropriate BMPs to maximize stormwater retention onsite and to minimize discharge of pollutants from the site. It is the responsibility of the applicant to obtain the NPDES permit authorization prior to initiating any site construction activities.

Detention, One of the Basic Elements of Stormwater BMPs

Many common site features can achieve stormwater management goals by incorporating one or more basic elements of BMPs, either alone or in combination, depending on the onsite conditions. The basic elements include infiltration, detention, and biofilters. Infiltration is the process where water enters the ground and moves downward through the unsaturated soil zone. Infiltration is ideal for management and conservation of runoff because it filters pollutants through the soil and restores natural flows to groundwater and downstream water bodies. A stormwater infiltration system aims to infiltrate the majority of runoff into the soil by allowing it to flow slowly over permeable surfaces. The slow flow of runoff allows pollutants to settle into the soil where they are naturally mitigated. The reduced volume of runoff that remains takes a long time to reach the outfall, and when it empties into a natural water body or storm sewer, its pollutant load would be noticeably reduced. Infiltration basins can range from a single shallow depression in a lawn, to an integrated swale, pond, and underground storage basin network. The basic design goal of infiltration systems is to provide opportunities for rainwater to enter the soil. Infiltration systems are often designed to capture the "first flush" storm event. They effectively remove suspended solids, particulates, bacteria, organics and soluble metals and nutrients through the vehicle of filtration, absorption and microbial decomposition. Site soil conditions generally determine if infiltration is feasible. In Soil Groups A and B infiltration is usually acceptable, but it is severely limited in Soil Groups C and D.

Detention systems differ from infiltration systems in terms of their intended usages. While infiltration systems are intended to percolate water into the soil, detention systems are designed primarily to store runoff for later release. Detention systems store runoff for one to two days after a storm and are dry until the next storm. Properly designed detention systems release runoff slowly enough to reduce downstream peak flows to their pre-development levels, allow fine sediments to settle, and uptake dissolved nutrients in the runoff where wetland vegetation is included. Detention systems are most appropriate for areas where soils percolate poorly.

Biofilters, also known as vegetated swales, are vegetated slopes and channels designed and maintained to transport shallow depths of runoff slowly over vegetation. Biofilters are effective if flows are slow and depths are shallow. This is generally achieved by grading the site and sloping pavement in a way that promotes sheet flow of runoff. For biofilter systems, features that concentrate flow, such as curb and gutter, paved inverts, and long drainage pathways across pavement must be minimized. The slow movement of runoff through the vegetation provides an opportunity for sediments and particulates to be filtered and degraded through biological activity. In most soils, the biofilter also provides an opportunity for stormwater infiltration, which further removes pollutants and reduces runoff volumes.

Sizing Criteria for Stormwater Detention Ponds

As one of the basic elements of stormwater BMPs, stormwater detention ponds are widely-used in California, aiming at managing stormwater quality and reducing the post-development peak flow rate to pre-development level. For flood protection, detentions are designed with capacity for the expected peak runoff volumes and flow rates of a given design storm size. This is known as the "peak runoff volume." Peak runoff volumes and flow rates are calculated for various design storm sizes, depending on local conditions, codes, and the potential damage that can be caused by flooding. Large drainage systems flood very infrequently, but they are expensive to construct. Therefore, drainage systems are typically sized to balance flooding risk and cost. Street drainage systems are typically designed for a 10-year storm, meaning that there is a 10 percent chance in any given year that a storm will be large enough to overwhelm the drainage system and flood the street. Since the flooding of a street once every ten years, on average, is a minor inconvenience, designing streets for a ten year storm represents a generally accepted balance of protection and cost. Homes and buildings suffer more severe damage from flooding, and are typically designed to remain protected in the 100-year storm, meaning that the probability of flooding is one percent in any given year.

The same need to balance costs and benefits applies to detention pond design for stormwater quality protection. Many pollutants may be carried by small, frequent storms. Because of this phenomenon, the water quality protection component of a drainage system can be designed to manage a much smaller volume and flow rate of water than the flood protection component. Also, because most rainfall occurs in small, frequent storms, water quality systems with relatively small capacities can have a large impact in minimizing overall runoff and preserving base stream flows. This amount of water that can be managed to protect water quality is called the "water quality volume."

As with flood control, there are a variety of standards and approaches for quantifying how to manage stormwater for water quality protection. The California Storm Water Quality Task Force, in its *Storm Water Best Management Practice Handbooks* (1993), and the Water Environment Federation/American Society of Civil Engineers in their jointly published *Urban Runoff Quality Management* (1998), each adopted an 80% annual capture rate as a standard of practice for the water quality volume. In the San Francisco Bay Area, this translates into approximately the first 0.50-1.25 inches of rain, or a two-year recurrence interval storm. The reason that designing stormwater treatment control BMPs for water quality protection targeting small, frequent storm events, instead of the larger, infrequent storms targeted for design of flood control facilities, can be seen by examination of Figure 6-1 and Figure 6-2. Figure 6-1 shows the distribution of storm events at San Jose, California where most storms produce less than 0.50 inches of total rainfall. Figure 6-2 shows the distribution of rainfall intensities at San Jose, California, where most storms have intensities of less than 0.25 inch/hour. The patterns at San Jose, California are typical of other locations throughout the state. Figures 1 and 2 show that as storm sizes increase, the number of events decreases. Therefore, when BMPs are designed for increasingly larger storms (for example, storms up to 1.0 inch versus storms of up to 0.5 inch), the BMP size and cost increase dramatically, while the number of additional treated storm events are small. Table 6-1 shows that doubling the design storm depth from 0.50 inch to 1.00 inch only increases the number of events captured by 23%. Similarly, doubling the design rainfall intensity from 0.25 inch/hour to 0.50 inch/hour only increases the number of events captured by 7%.

Due to economies of scale, doubling the capture and treatment requirements for a detention pond are not likely to double the construction cost, but the incremental cost per event will increase, making increases beyond a certain point generally unattractive. Typically, design criteria for stormwater quality control are set to coincide with the "knee of the curve", that is, the point of inflection where the magnitude of the event increases more rapidly than number of events captured. Figure 6-3 shows that the "knee of the curve" or point of diminishing returns for San Jose, California is in the range of 0.75 to 1.00 inch of rainfall. In other words, targeting design storms larger than this will produce gains at considerable incremental cost. The "knee of the curve" typically corresponds to the 75th to 85th percentile capture range of storm events for many sites in California.

Table 6-1. Incremental design criteria versus storms treated at San Jose in California[1]

Proposed BMP Design Target	Number of Historical Events in Range	Incremental Increase in Design Criteria	Incremental Increase in Storms Treated
Storm Depth 0.00 to 0.50 inch	1,067	+100%	+23%
Storm Depth 0.51 to 1.00 inch	242		
Rainfall Intensity 0.10 to 0.25 inch/hour	2,963	+100%	+7%
Rainfall Intensity 0.26 to 0.50 inch/hour	207		

[1]Source: California Stormwater BMP Handbook, 1993.

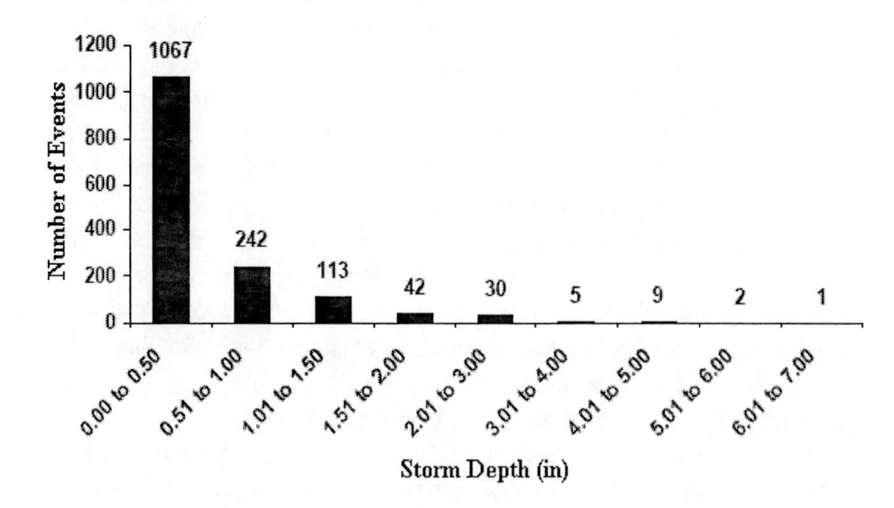

Figure 6-1. The historic (1948 to 2000) rain storms at San Jose in California. The unit of inch (in) is used to be consistent with the source (i.e., California Stormwater BMP Handbook, 1993). 1 in = 25.4 mm.

It is important to note that arbitrarily targeting large, infrequent storm events can actually reduce the pollutant removal capabilities of stormwater detention ponds. This occurs when outlet structures, detention times, and drain down times are designed to accommodate unusually large volumes and high flows. When the detention ponds are over-designed, the more frequent, small storms that produce the most annual runoff pass quickly through the over-sized detention ponds and therefore receive inadequate treatment. For example, a detention pond might normally be designed to capture 0.5 inch of runoff and to release that runoff over 48 hours, providing a high level of sediment removal. If the detention pond were to be oversized to capture 1.0 inch of runoff and to release that runoff over 48 hours, a more common 0.5 inch runoff event entering the detention pond would drain in approximately 24 hours, meaning the smaller, more frequent storm that is responsible for more total runoff would receive less treatment than if the detention were designed for the smaller event. Therefore, efficient and economical BMP sizing criteria are usually based on design criteria that correspond to the "knee of the curve" or point of diminishing returns.

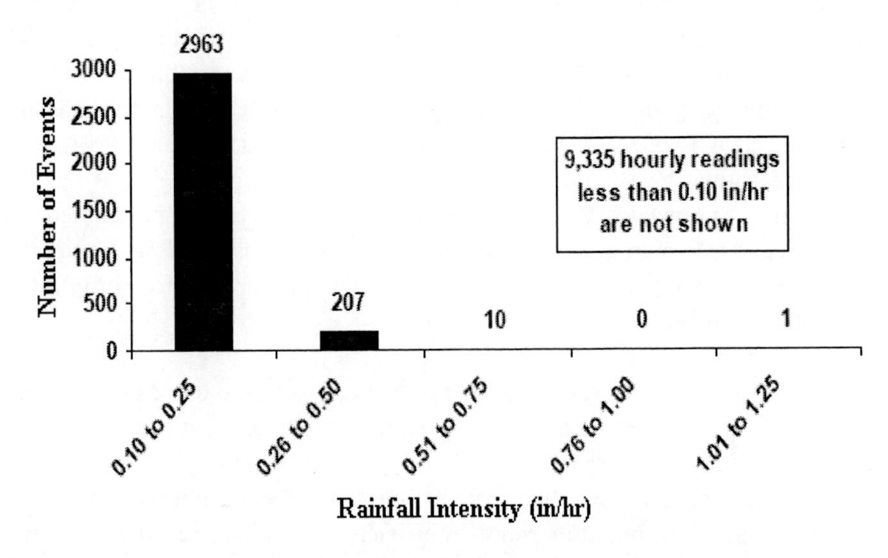

Figure 6-2. The historic (1948 to 2000) rainfall Intensity at San Jose in California. The unit of inch per hour (in/hr) is used to be consistent with the source (i.e., California Stormwater BMP Handbook, 1993). 1 in/hr = 25.4 mm/hr.

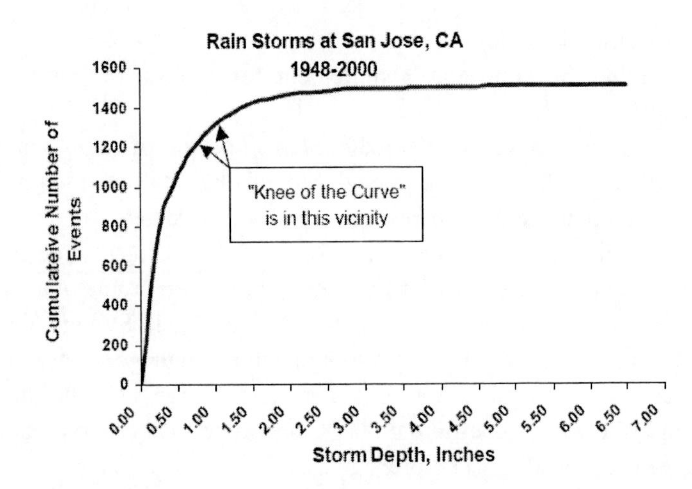

Figure 6-3. The historic (1948 to 2000) storm events at San Jose in California. The unit of inch (in) is used to be consistent with the source (i.e., California Stormwater BMP Handbook, 1993). 1 in = 25.4 mm.

California Stormwater BMP Handbook Approach for Determining Water Quality Volume of a Stormwater Detention Pond

The California Stormwater BMP Handbook approach for determining the water quality volume of a stormwater detention pond is based on results of a continuous simulation model, the Storage, Treatment, Overflow, Runoff Model (STORM), developed by the Hydrologic Engineering Center of the U.S. Army Corps of Engineers. STORM was applied to long-term hourly rainfall data at numerous sites throughout California, with sites selected throughout the state representing a wide range of municipal stormwater permit areas, climatic areas, geography, and topography. STORM translates rainfall into runoff, then routes the runoff through detention storage. The volume-based BMP sizing curves resulting from the STORM model provide a range of options for choosing a BMP sizing curve appropriate to sites in most areas of the state. Figure 6-4 shows an example of the volume-based BMP sizing curve at San Jose in California.

The California Stormwater BMP Handbook approach is easy to be applied, and uses commonly available information about a project. The following steps describe the use of the BMP sizing curves contained in the Handbook to determine the water quality volume of a stormwater detention pond.

Step 1. Identify the "BMP Drainage Area" that drains to the proposed detention pond. This includes all areas that will contribute runoff to the proposed detention pond, including pervious areas, impervious areas, and off-site areas, whether or not they are directly or indirectly connected to the detention pond.

Step 2. Calculate the composite runoff coefficient "C" for the area identified in Step 1.

Step 3. Select a capture curve representative of the site and the desired drain down time using. Curves are presented for 24 hour and 48 hour draw down times in the Handbook. The 48 hour curve should be used in most areas of California. Use of the 24 hour curve should be limited to drainage areas with coarse soils that readily settle and to watersheds where warming may be detrimental to downstream fisheries.

Step 4. Determine the applicable requirement for capture of runoff (Capture, % of Runoff).

Step 5. Enter the capture curve selected in Step 3 on the vertical axis at the "Capture, % Runoff" value identified in Step 4. Move horizontally to the right across capture curve until the curve corresponding to the drainage area's composite runoff coefficient "C" determined in Step 2 is intercepted. Interpolation between curves may be necessary. Move vertically down from for this point until the horizontal axis is intercepted. Read the "Unit Basin Storage Volume" along the horizontal axis. If a local requirement for capture of runoff is not specified, enter the vertical axis at the "knee of the curve" for the curve representing composite runoff coefficient "C." The "knee of the curve" is typically in the range of 75% to 85% capture.

Step 6. Calculate the required capture volume of the BMP by multiplying the "BMP Drainage Area" from Step 1 by the "Unit Basin Storage Volume" from Step 5 to give the water quantity volume of the proposed detention pond.

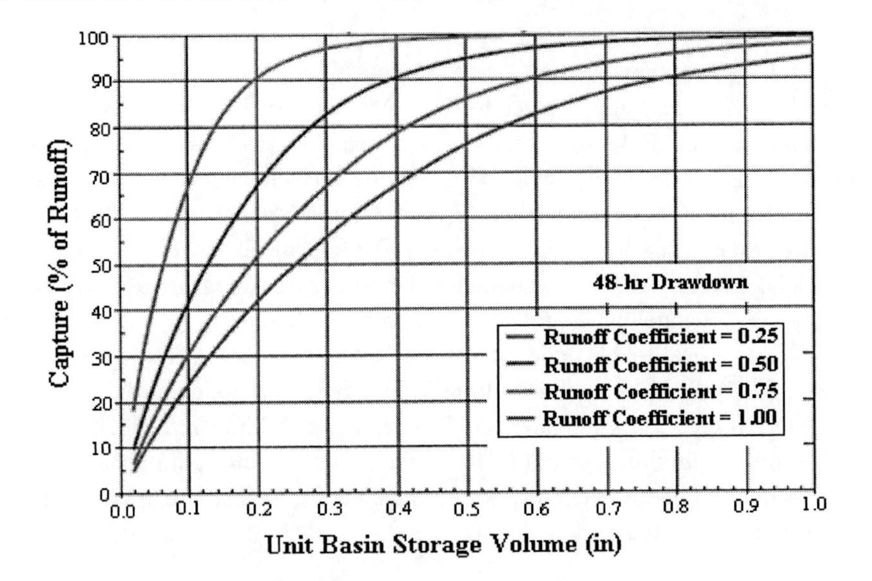

Figure 6-4. Capture/treatment analysis at San Jose in California. The unit of inch (in) is used to be consistent with the source (i.e., California Stormwater BMP Handbook, 1993). 1 in = 25.4 mm.

Maintenance of Stormwater Detention Ponds

Maintenance activities of a stormwater detention pond mainly consist of removal of sediment and trash and debris, vegetation management, and vector control. Typical activities and frequencies include:

- Schedule semiannual inspection for the beginning and end of the wet season for standing water, slope stability, sediment accumulation, trash and debris, and presence of burrows.
- Remove accumulated trash and debris in the detention pond and around the inlet of the discharge pipe during the semiannual inspections. The frequency of this activity may be altered to meet specific site conditions.
- Trim vegetation at the beginning and end of the wet season and inspect monthly to prevent establishment of woody vegetation and for aesthetic and vector reasons.
- Remove accumulated sediment and regrade the detention pond about every 10 years or when the accumulated sediment volume exceeds 10 percent of the pond volume. Inspect the pond each year for accumulated sediment volume.

Peak Reduction Effects of Detention Basins

The Corte Madera Creek Watershed

The 72-km^2 Corte Madera Creek watershed (Figure 6-5), also named the Ross Valley watershed, is located in the eastern part of central Marin County in California. The upper watershed has three major creeks: San Anselmo Creek, Fairfax Creek, and Sleep Hollow Creek. Fairfax Creek is connected to Corte Madera Creek by a manmade culvert. In the middle watershed, Phoenix Lake, drained to Corte Madera Creek by Ross Creek, functions as a de facto detention basin. From upstream to downstream, Corte Madera Creek has reaches of natural channel, concrete lined channel, and manmade earthen channel. The Ross streamflow gaging station (Ross gage) monitors streamflows in Corte Madera Creek at the Town of Ross. This station was established by the United States Geological Survey (USGS) in 1951 and is the only official station in the watershed. The 100-year flood in 31 December 2005 overtopped the main channel banks, forming "secondary" channels.

During the appraisal-level flood study for Corte Madera Creek in 2007 (Stetson Engineers Inc., 2007), five potential detention basin locations were identified. These locations are labeled as DB1 to DB5 in Figure 6-5. DB5 coincides with Phoenix Lake. These sites were selected based on: 1) proximity to the Corte Madera Creek mainstem or major tributary; 2) location upstream of a critical reach or flood prone area; 3) open site with few structures; and 4) large contributing drainage area.

Analysis Methods

The U.S. Army Corps of Engineers Hydrologic Engineering Center's Hydrologic Modeling System (HEC-HMS) was used to evaluate the cumulative effectiveness of the proposed detention basins in reducing peak flows. In flood control planning, HEC-HMS is often used to develop a streamflow hydrograph for the design storm event and, in particular, to estimate the magnitude of the peak discharge for the design event. The peak discharge serves as the design objective for channel improvements and other measures that aim at containing the peak flow within the channel to prevent breaching. With the design hydrograph known, HEC-HMS can be used further to evaluate the effectiveness of detention basins at lowering the peak flow to reduce or eliminate breaching.

HEC-HMS simulates the rainfall-runoff process using four sub-procedures of loss, transformation, baseflow, and routing. The loss sub-procedure computes excess rainfall (i.e., runoff volume), while the transformation sub-procedure transforms runoff volume to flow hydrograph at the sub-watershed outlet. The baseflow sub-procedure estimates baseflow. The routing sub-procedure routes the hydrograph along the drainage network from upstream to downstream. For each of the four sub-procedures, HEC-HMS provides several different methods (Table 6-2). Each method has its own applicability, assumptions, and limitations. A model development has to select an appropriate method for each sub-procedure and estimate the associated parameters.

In this study, the SCS CN method was selected for loss sub-procedure. The SCS CN loss method is a commonly used method for simulating excess rainfall. The method is well documented and widely accepted for use by professional communities in the U.S. as well as in the other countries. The method is simple, predictable, and stable. The method uses one single parameter of curve number (CN), which varies as a function of watershed characteristics of soil hydrologic group, land use, surface condition, and antecedent moisture condition.The Clark Unit Hydrograph transform method was used for transformation sub-procedure. This selection was predominately guided by model calibration and professional judgment. Baseflow is the sustained fair-weather runoff of prior precipitation that is stored temporarily in the watershed, plus the delayed subsurface runoff from the current storm. The Corte Madera Creek watershed flood modeling is an event modeling, the exponential recession baseflow method was found to be most appropriate. The Muskingum-Cunge method with 8-point cross section configuration was used for routing sub-procedure since this method allows considering overbank flows. As mentioned above, the 31 December 2005 flood caused significant overbank flows along Corte Madera Creek.

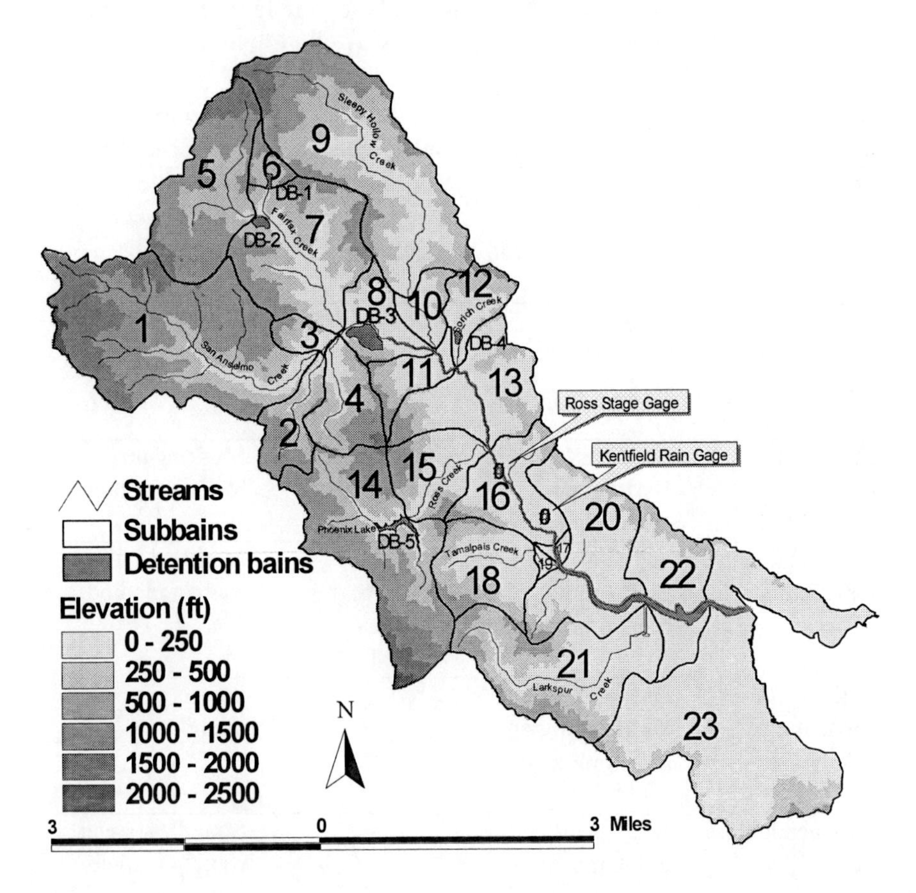

Figure 6-5. Map showing the locations of the potential detention basins within the Corte Madera Creek watershed. 1 ft = 0.3048 m, and 1 mile = 1.609 km.

The HEC-HMS hydrologic model for the Corte Madera Creek watershed employs watershed-related and physically-based characteristics as the model inputs. The model was first calibrated within reasonable parameter ranges for both the in-channel flows observed at the Ross gage and the peak stage at the spillway of Phoenix Lake for the 31 December 2005 flood event. The updated rating curve of Ross gage was used to convert the in-channel stages to stream discharges at Ross gage. The model calibration was based on the in-channel part of the rating curve only since the out-of-channel part of the rating curve has not been verified. It was assumed that the simulated out-of-channel hydrograph would approximate the actual out-of-channel hydrograph if the simulated in-channel hydrograph was well calibrated to the observed in-channel hydrograph in both flow magnitude and hydrograph shape. The estimated peak stage at the spillway of Phoenix Lake by high water mark survey was used to calibrate the subbasin upstream of the lake. The high water mark (debris line) along the upstream face of the dam was surveyed to estimate the maximum lake level above the spillway crest; 6.65 ft. The water flowing out of the spillway was calculated using the Phoenix Lake Spillway rating curve.

Table 6-2. The available HEC-HMS methods for the four sub-procedures (The methods used in this study are highlighted)

Loss Sub_procedure	Transform Sub-procedure	Baseflow Sub-procedure	Routing Sub-procedure
Initial and constant rate loss method	Clark Unit Hydrograph transform method	Constant monthly baseflow method	Kinematic wave routing method
Deficit and constant rate loss method	ModClark transform method	Exponential recession baseflow method	Muskingum-Cunge routing method
SCS Curve Number (CN) loss method	Snyder's Unit Hydrograph transform method	Linear reservoir baseflow method	Muskingum routing method
Green and Ampt loss method	SCS Unit Hydrograph transform method	Bounded recession baseflow method	Modified Puls routing method
Soil-moisture accounting (SMA) loss method	User specified Unit Hydrograph transform method		Lag routing method
Exponential loss method	User specified S-graph transform method		Straddle stagger routing method
	Kinematic wave transform method		

Figure 6-6 shows the calibrated in-channel hydrograph at Ross gage. The hydrograph calibration result gives an estimated peak flow of about 6,850 cfs at Ross gage during the 31 December 2005 storm event. Figure 6-7 shows the simulated stage hydrograph of Phoenix Lake and Figure 6-8 shows the simulated routing effect of Phoenix Lake during the 31 December 2005 storm event. Phoenix Lake has an ungated spillway with a crest elevation at 174 ft. The simulated peak elevation was 180.66 ft, which was 6.66 ft above the spillway crest. This is very close to the surveyed peak water of 6.65 ft above the spillway crest. Note that Phoenix Lake has an existing 30-inch low-level outlet. However, the low-level outlet has been kept closed since it was built. The lake was full prior to the storm event and this explains why the attenuation effect of the lake under existing conditions (Figure 6-8) was not significant.

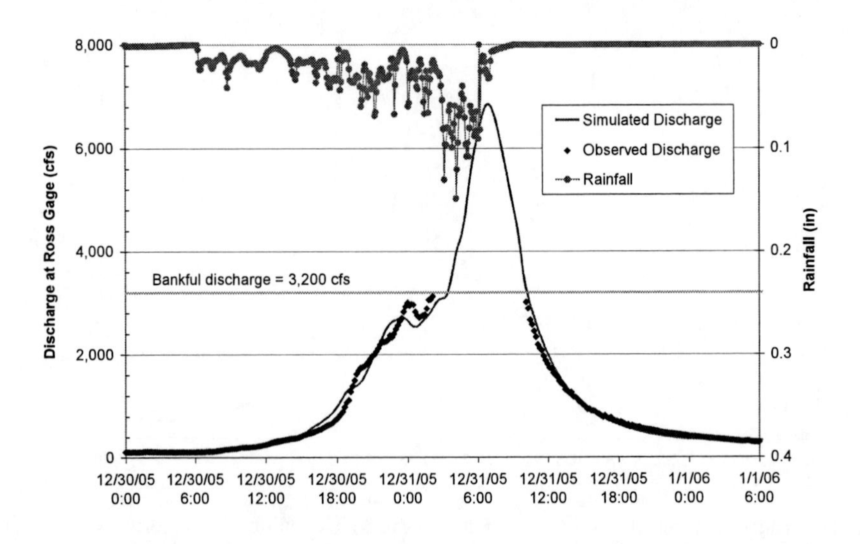

Figure 6-6. HEC-HMS model calibration results at Ross gage during the 31 December 2005 storm event. 1 cfs = 0.02832 m^3/s; 1 in = 25.4 mm.

The calibrated model was then applied to the 29 December 2003 storm event without any parameter change to verify the model's reliability. During this event, the peak flow was right at the bankfull level of the channel at Ross gage. The verification result is shown in Figure 6-9. It can be seen that the model worked satisfactorily. There are some mismatches at the rising limb and the peak flow. But the result is generally acceptable. The observed peak was 3,560 cfs, while the simulated peak was 3,300 cfs. This test verified that the calibrated model captured the dominant characteristics of the Corte Madera Creek watershed and thus can be used for reasonable scenario simulations without further changing the model parameters.

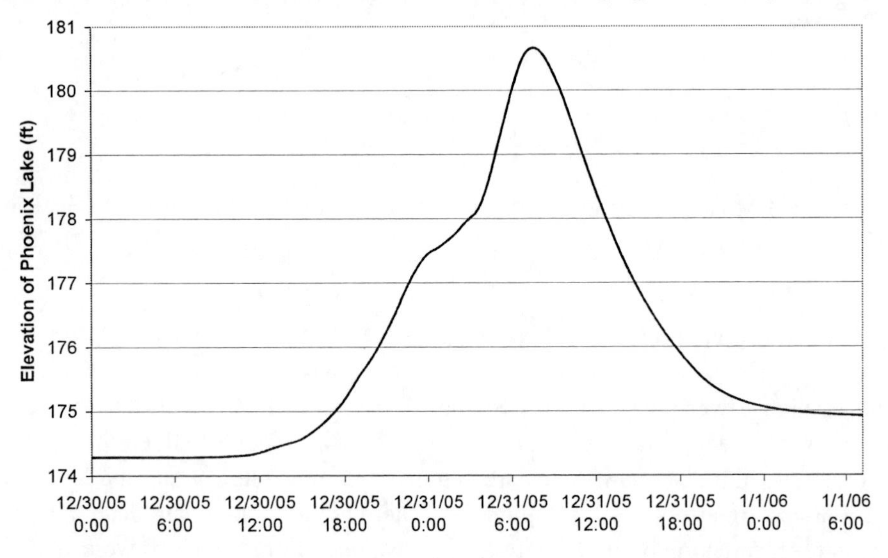

Figure 6-7. Simulated water level of Phoenix Lake during the 31 December 2005 storm event. 1 ft = 0.3048 m.

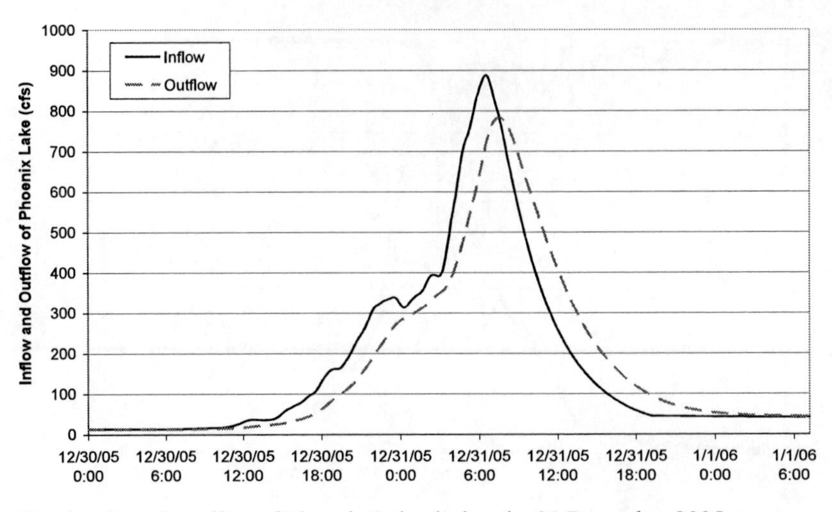

Figure 6-8. Simulated routing effect of Phoenix Lake during the 31 December 2005 storm event. 1 cfs = 0.02832 m^3/s.

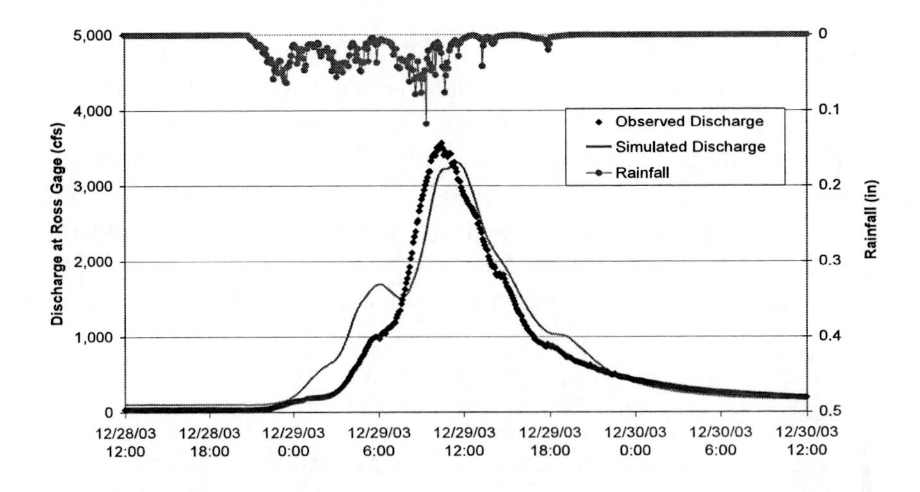

Figure 6-9. HEC-HMS model verification results at Ross gage for the 29 December 2003 storm event. 1 cfs = 0.02832 m^3/s; 1 in = 25.4 mm.

Operation Scenarios and Analysis Results

The calibrated/verified HEC-HMS model was used to evaluate the cumulative effectiveness of the proposed five detention basins in reducing flood peak flows. The characteristics of the detention basins are summarized in Table 6-3 and the locations of the detention basins are also shown in Figure 6-5. There are two on-stream detention basins and three off-stream detention basins. The following two operations scenarios were analyzed:

- Scenario 1: this scenario simulates the uncontrolled operations. Except for Phoenix Lake, all other detention basins' low-level outlets are kept open and no operations to the detention basins are needed. The low-level outlet of Phoenix Lake is kept closed as existing and all outflows from the lake will be through the ungated spillway.

- Scenario 2: this scenario simulates the controlled operations. The low-level outlets of the three off-stream detention basins are kept open but are closed during the flood. The low-level outlet of Phoenix Lake is kept closed as existing but is open to empty the lake prior to the storm, and then closed during the flood. This controlled operations scenario is intended to evaluate the benefit in peak flow reduction by retention of some runoff volume in the detention basins.

Table 6-3. Summary of the analyzed detention basins and their operation scenarios

Detention Basin	DB-1	DB-2	DB-3	DB-4	DB-5
Location	White Hill School - Upstream (Fairfax Creek)	White Hill School - Lefty Gomez Field (Fairfax Creek)	Marin Town & Country Club (San Anselmo Creek)	San Anselmo Memorial Park (Sorich Creek)	Phoenix Lake (Ross Creek)
Type	On-stream	Off-stream	Off-stream	Off-stream	On-stream
Drainage Area (mi^2)[1]	0.19	1.61	4.91	0.47	2.30
Features	Dry basin Dam Spillway Low level outlet	Dry basin Dam Spillway Low level outlet	Dry basin Dam Spillway Low level outlet	Dry basin Dam Spillway Low level outlet	Existing reservoir Raise existing dam and spillway by 15 ft 30-inch existing closed low level outlet
Storage Volume (ac-ft)[2]	54	71	395	26	680
Operations Scenario 1 - Uncontrolled Operations	Low level outlet kept open (Initial storage = 0)	Low level outlet kept open (Initial storage = 0)	Low level outlet kept open (Initial storage = 0)	Low level outlet kept open (Initial storage = 0)	Low level outlet kept closed as existing (Initial storage = 175 acre-ft)
Operations Scenario 2 – Controlled Operations	Low level outlet kept open (Initial storage = 0)	Low level outlet kept but closed during flood (Initial storage = 0)	Low level outlet kept open but closed during flood (Initial storage = 0)	Low level outlet kept open but closed during flood (Initial storage = 0)	Low level outlet kept closed but open to empty the lake prior to a storm, and then closed during the flood (Initial storage = 0)

[1] 1 mi^2 = 2.59 km^2

[2] 1 ac-ft = 1233.5 m^3

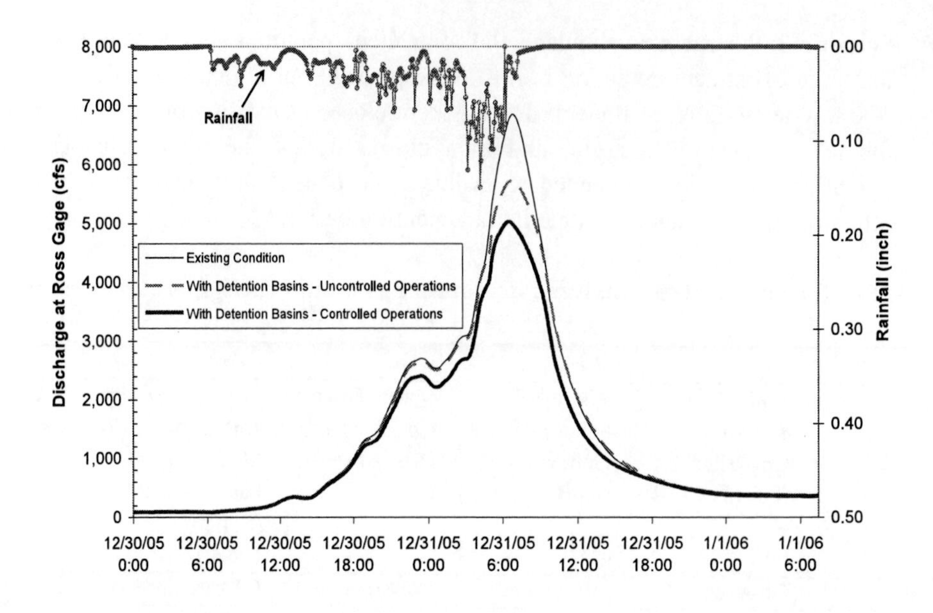

Figure 6-10. The simulated flow hydrographs for the existing condition and the two operation scenarios. 1 cfs = 0.02832 m^3/s; 1 in = 25.4 mm.

Figure 6-10 shows the simulated hydrographs for the 31 December 2005 storm event for the existing condition and the two operation scenarios of the proposed detention basins. The simulated peak discharges for the three conditions are approximately 6,850 cfs, 5,720 cfs, and 5,030 cfs, respectively. Under the uncontrolled operation scenarios, the peak flow reduction by the proposed detention basins is about 1,130 cfs (i.e., 16%). Under the controlled operation scenarios, the proposed detention basins could further reduce the peak flow by about 690 cfs (i.e., 10%). This indicates that the runoff volume control scenario by just modifying the detention basin operations is very effective in reducing the watershed peak discharges.

DISCUSSION

The analysis results for the existing condition and the two operations scenarios of the proposed detention basins indicate that runoff volume retention by modifying the detention basin operations is very effective in watershed-wide peak flow reduction. Although the results also show that appropriate design of detention basins could also be effective in watershed-wide peak flow reduction, however, as demonstrated in the study by Emerson et al. (2005), detention basins designed for attenuating individual on-site peak flow rates to not exceed prescribed standards may not be effective in collectively reducing watershed-wide peak flows. Detention basins are also ineffective at controlling peak water levels in downstream ponds and lakes that may drain slowly. During a storm, the level of these water bodies will not drop significantly during the time that detention facilities store water. Thus the increased volume of stormwater runoff associated with the introduction of impervious surfaces from site developments will lead to increased peak water levels downstream. The most extreme examples are ponds and lakes whose discharges are affected by high tides. However, even lakes that discharge to streams or rivers may drain slowly enough to negate the flood-reduction benefits of detention basins.

Both peak flow rate and runoff volume under post-development conditions are likely to increase compared to pre-development conditions. The detention concept for individual site development is solely focused on peak flow reduction for the site. It does not provide mechanisms for runoff volume attenuation. To effectively offset the impervious surface stormwater contributions in both peak flows and runoff volume, the BMPs that can promote retention of stormwater runoff volume would be more effective. Through these BMPs, such as bioretention facilities, stormwater can be filtered on-site, back into the ground, rather than contributing to downstream impacts. If properly designed, bioretention facilities can effectively mitigate urban and suburban stormwater impacts that detention basins cannot. They control the volume of storm runoff and enhance groundwater recharge. In addition, they enhance overall stormwater treatment. As a developing trend, stormwater regulators are beginning to require runoff volume control in addition to peak flow reduction and stormwater quality treatment. The application of low impact development (LID) techniques (e.g., bioretention) in project design may maximize the benefits of detention ponds/basins.

CONCLUSIONS

Impervious areas resulting from site developments within a watershed can change the land form, adversely affecting the natural runoff processes and water quality. In practice, it is important to minimize the impacts by preserving the pre-development flow hydrographs. However, when implementing their favorite site development plans, the upstream communities within the watershed may not acknowledge what would happen for the downstream communities. For this reason, stormwater regulators tend to use a variety of means, including law enforcements, political measures, and economic incentives, to promote applications of LID techniques, a new concept for practical engineers. The LID design concept requires peak discharge, runoff volume, and water quality treatment be taken into account. Most of the existing detention ponds/basins were not designed and constructed from watershed perspective. As a result, these detention ponds/basins are unlikely to play roles in sustaining the watershed-scale environmental integrity while they may have attenuation effects of localized storm peaks. The analysis results from the Corte Madera Creek watershed indicated that the detention ponds designed for peak flow as well as runoff volume reductions can be a more effective means to minimize watershed-scale environmental impacts resulting from site developments.

REFERENCES

CASQA (California Stormwater Quality Association), 2003. California *Stormwater Best Management Practices Handbook.*

Emerson, C. H., Welty, C., and Traver, R. G., 2005. Watershed-Scale Evaluation of a System of Storm Water Detention Basins. *Journal of Hydrologic Engineering,* 10(3), 237-242.

Harrell, L.J. and Ranjithan, S.R., 1997. Generating Efficient Watershed Management Strategies Using a Genetic Algorithm-Based Method. *Proceedings of the 24th Annual Water Resources Planning and Management Conference,* ASCE, Houston, TX, 1997.

Harrell, L. J. and Ranjithan, S. R., 2003. Detention Pond Design and Land Use Planning for Watershed Management. *Journal of Water Resources Planning and Management,* 129(2), 98-106.

Stetson Engineers Inc., 2007. *Appraisal-Level Flood Study for Corte Madera* Creek, CA.

Yeh, C. H. and Labadie, J. W, 1997. Multi-objective Watershed Level Planning on Storm Water Detention System. *Journal of Water Resources Planning and Management,* 123(6), 336-343.

In: Modeling Hydrologic Effects...
Editor: Xixi Wang, pp. 121-206

ISBN 978-1-61668-628-4

Chapter 7

EVALUATE EFFECTS OF THE WAFFLE® ON FLOOD REDUCTION IN THE RED RIVER OF THE NORTH BASIN USING COUPLED SWAT AND HEC–RAS MODELS

Xixi Wang[*]
Hydrology and Watershed Management Program, Department of Engineering and Physics, Tarleton State University, BOX T-0390, Stephenville, Texas 76402, USA

ABSTRACT

The Red River of the North borders North Dakota and Minnesota and flows north toward Lake Winnipeg in Manitoba, Canada. The river is susceptible to flooding because of the synchrony of its discharge with spring thaw and ice jams, its shallow and sinuous channel, its low gradient, and the decrease in its gradient downstream. As a result, the property adjacent to the river is subject to frequent, damaging inundation from minor and major flood events, with a truly devastating flood about every decade. To mitigate flooding, various structural and nonstructural measures have been employed. However, the extensive flooding in 1997 necessitated reexamination of these measures and exploration of innovative concepts to augment traditional approaches. Hydrologic and hydraulic models play a key role in evaluating and identifying economical and feasible measures for flood reduction in this complex river system. In terms of complexity and modeling objective, the models developed in the past two decades can be categorized as 1) explanatory analyses, 2) floodplain and floodway management analyses, 3) land planning and management analyses, 4) flood mitigation engineering design analyses, 5) flood forecasting, and 6) miscellaneous. While a few of these models were used for some

[*] Assistant Professor and Coordinator,Tel. (254) 968-9164, E-mail: xxqqwang@gmail.com

initial flood reduction analyses, they were developed mainly for other purposes. In addition, the models insufficiently address inflows from ungauged areas and overland flows, which significantly affect their calibration, verification, and application. This chapter discusses a conceptual model scheme applied to study the impacts of various storage scenarios on flood reduction in the Red River. Under this scheme, a coupled hydrologic–hydraulic model was developed by integrating two decades of modeling achievements with new algorithms specially designed for this study, employing updated modeling techniques and utilizing improved spatial and temporal data. Furthermore, this model was used to analyze storage scenarios necessary to mitigate 1997-type floods and the probable maximum flood (PMF) in the Red River.

Keywords: Hydro modeling, flood mitigation, nonstructural measures

INTRODUCTION

The Red River of the North Basin

The Red River of the North originates in Minnesota, forms the boundary between North Dakota and Minnesota, and enters Canada at Emerson, Manitoba, where it continues northward to Lake Winnipeg, Manitoba (Figure 7-0). It meanders approximately 548 mi (883 km) through the flat and fertile valley of former glacial Lake Agassiz, forming the 45,000 mi^2 (116,500 km^2) Red River Basin. Both the river channel and the basin are intersected by the international border between the United States and Canada, with approximately 75% of the Red River Basin located in the United States. The basin is remarkably flat, as reflected by the 0.5 ft per mile gradient of the Red River (International Joint Commission, 1997). Over 66% of the basin is conducive to agriculture and productive cropland because of the fertile, black, fine-grained soils (Stoner et al., 1993). In addition to farmland, several major population centers (cities) are located on the banks of the Red, including Wahpeton–Breckenridge with a population of 12,000, Fargo–Moorhead with 100,000, Grand Forks–East Grand Forks with 60,000, Selkirk with 9800, and Winnipeg with 670,000 (International Joint Commission, 2000).

The Red River Basin is subject to frequent damaging inundation from major flood events. Since official record keeping began in 1882, major floods affecting large areas of the basin have occurred once about every 4 to 6 years, with a truly devastating flood occurring, on average, about every decade (LeFever et al., 1999; International Joint Commission, 1997). Major historical floods occurred in 1826, 1897, 1950, 1966, 1969, 1975, 1978, 1979, and 1997. Based on historical accounts, the 1826 flood is the worst flood known. Unfortunately, there are no detailed data on this flood. While all of the other floods caused severe damage, the 1997 flood is the worst in the official record and forced the evacuation of many families in most of the cities mentioned above (International Joint Commission, 1997).

To mitigate flooding loss, various measures have been taken (International Joint Commission,1997, 2000; Red River Basin Board, 2000; Kingery et al., 1999; Dyhouse, 1995). LeFever et al. (1999) categorized the various flood protection measures proposed for the Red River Basin as structural and nonstructural. Construction of levees and reservoirs, enlargement and straightening of channels, and installation of channel bypasses are structural measures, whereas nonstructural measures consist of regulating development, improving operation of the structures, establishing policies for emergency and loss protection, and employing new concepts such as mitigating flooding at its source. Both types of measures are generally needed for most of the basin, as nonstructural measures augment the flood mitigation capacities of structural measures (International Joint Commission, 1997, 2000).

Figure 7-0. Map showing the Red River of the North Basin.

After the 1997 flood, the Energy & Environmental Research Center (EERC) recognized the need for alternative methods of flood protection to augment existing flood protection measures. This sentiment was mirrored by other major organizations and agencies in the Red River Basin, and it was determined that innovative concepts of nonstructural measures should be explored to augment the design capacities of structural measures planned to protect against future floods similar in scope to, or greater than, the 1997 flood (International Joint Commission, 2000). As a result, the EERC proposed to investigate the technical feasibility of utilizing the Red River Basin's existing gridlike infrastructure of fields bounded by raised roads as a means to temporarily hold water for both flood and drought mitigation. This project became known as "The Waffle®" because of the resemblance of farm fields bounded by raised roads to the squares on breakfast waffles and the ability of waffle squares and/or roads to retain fluid. Other potential storage areas include ditches, wetlands, or any type of natural depression. These storage areas, supplemented by roads and drainage structures, could act as a network of channels and control structures to hold and slowly release water into the Red River as the flood crest passes. Although the main goal of this concept is flood and drought mitigation, it is also recognized that this approach could have multiple environmental benefits, including improvements to water quality, decreased soil erosion, and increased groundwater recharge.

The Integrated Modeling Approach

In the Waffle® project, an integrated modeling approach was used to evaluate the effects of Waffle® on reducing the 1997-type flood in the U.S. portion of the Red River of the North Basin (RRB). Hereinafter, RRB referred to the U.S. portion of this basin. RRB comprises 28 USGS 8-digit hydrologic unit codes (HUCs), 27 of which are drained into the RR mainstem (Figure 7-1 and Table 7-1). For description purposes, each HUC was designated a watershed.

Figure 7-2 shows the general framework of this approach, which uses the SWAT (Soil and Water Assessment Tool) models to determine the flow reductions at the outlets of the watersheds, and a HEC-RAS (Hydrologic Engineering Center River Analysis System) model to simulate the corresponding flood crest reductions along the mainstem. For a watershed that is located either in Minnesota or North Dakota, one SWAT model was set up. In contrast, two SWAT models were set up for a watershed that encompasses a segment of the Red River mainstem, one model for the North Dakota portion and the other for the Minnesota portion.

Description of SWAT

SWAT has been widely used to predict impacts of land management practices on water, sediment, and agricultural chemical yields in large complex watersheds with varying soils, land use, and management conditions over long periods of time. It is composed of three major components, namely subbasin, reservoir/pond routing, and channel routing. Each of the components includes several subcomponents. For example, the subbasin component consists of eight subcomponents, namely hydrology, weather, sedimentation, soil moisture, crop growth, nutrients, agricultural management, and pesticides. The hydrology subcomponent, in turn, includes surface runoff, lateral subsurface flow, percolation, groundwater flow, snowmelt, evapotranspiration, transmission losses, and ponds. In the Waffle® project, the Soil Conservation Service (SCS) runoff curve number, adjusted according to soil moisture conditions, was used to estimate surface runoff, the Priestley-Taylor method used to estimate potential evapotranspiration, and the Muskingum method used for channel routing. Because SWAT was used to simulate the spring snowmelt flooding in this project, the subcomponents of snowmelt hydrology, pond routing, and the Muskingum channel routing were described as follows.

Snowmelt Hydrology

In SWAT, snowmelt hydrology is realized on an HRU (hydrologic response unit) basis. A watershed is subdivided into a number of subbasins for modeling purposes. Portions of a

subbasin that possess unique land use/management/soil attributes are grouped together and defined as one HRU. Depending on data availability and modeling accuracy, one subbasin may have one or several HRUs defined. When the mean daily air temperature is less than the snowfall temperature, as specified by the variable SFTMP, the precipitation within an HRU is classified as snow and the liquid water equivalent of the snow precipitation is added to the snowpack.

The snowpack increases with additional snowfall, but decreases with snowmelt or sublimation. The mass balance for the snowpack is computed as:

$$SNO_i = SNO_{i-1} + R_{sfi} - E_{subi} - SNO_{mlti} \quad (1)$$

where SNO_i and SNO_{i-1} are the water equivalents of the snowpack on the current day (i) and previous day (i-1), respectively, R_{sfi} is the water equivalent of the snow precipitation on day i, E_{subi} is the water equivalent of the snow sublimation on day i, and SNO_{mlti} is the water equivalent of the snowmelt on day i. All of these variables are reported in terms of the equivalent water depth (mm) over the total HRU area.

The snowpack is rarely uniformly distributed over the total area, resulting in a fraction of the area that is bare of snow. In SWAT, the areal coverage of snow over the total HRU area is defined using an areal depletion curve, which describes the seasonal growth and recession of the snowpack and is defined as:

$$sno_{covi} = \frac{SNO_i}{SNOCOVMX}\left[\frac{SNO_i}{SNOCOVMX} + \exp(cov_1 - cov_2 \cdot \frac{SNO_i}{SNOCOVMX})\right]^{-1} \quad (2)$$

where sno_{covi} is the fraction of the HRU area covered by snow on the current day (i), SNOCOVMX is the minimum snow water content that corresponds to 100% snow cover (mm H_2O), and cov_1 and cov_2 are the coefficients that define the shape of the curve. The values used for cov_1 and cov_2 are determined by solving equation 2 using two known points: (1) 95% coverage at 95% SNOCOVMX, and (2) 50% coverage at a fraction of SNOCOVMX, specified by the variable SNO50COV. For example, assuming that SNO50COV is equal to 0.2, cov_1 and cov_2 will take the values of -1.2399 and 1.8482, respectively.

The value of sno_{covi} is assumed to be equal to 1.0 once the water content of the snowpack exceeds SNOCOVMX, indicating a uniform depth of snow over the HRU area. The areal depletion curve affects snowmelt only when the snowpack water content is between 0.0 and SNOCOVMX. Consequently, a small value for SNOCOVMX will assume a minimal impact of the areal depletion curve on snowmelt, whereas as the value of SNOCOVMX increases, the curve will assume a more important role in approximating the snowmelt process.

In addition to the areal coverage of snow, snowmelt is also controlled by the snowpack temperature and melting rate. The snowpack temperature is a function of the mean daily temperature during the preceding days and varies as a dampened function of air temperature. The influence of the previous day's snowpack temperature on the current day's snowpack temperature is described by a lag factor, specified by the variable TIMP, which implicitly accounts for snowpack density, water content, and exposure. The snowpack temperature is calculated as:

$$T_{spi} = T_{spi-1}(1 - TIMP) + \overline{T}_{ai} \cdot TIMP \quad (3)$$

where T_{spi} and T_{spi-1} are the snowpack temperatures on the current day (i) and the previous day (i-1), respectively, and $\overline{T}_{ai}$ is the mean air temperature on day i. As TIMP approaches

1.0, $\overline{T}_{ai}$ exerts an increasingly greater influence on T_{spi}; conversely, as TIMP moves away from 1.0, T_{spi-1} becomes more important.

The amount of snowmelt on the current day (i), SNO_{mlti}, expressed in terms of the equivalent amount of water in mm, or melting rate, is calculated in SWAT as:

$$SNO_{mlti} = b_{mlti} \cdot sno_{covi} \left(\frac{T_{spi} + T_{maxi}}{2} - SMTMP \right) \tag{4}$$

where T_{maxi} is the maximum air temperature on day i (°C), SMTMP is the base temperature above which snowmelt is allowed (°C), and b_{mlti} is the melt factor on day i (mm H_2O/°C-day), which is calculated as:

$$b_{mlti} = \frac{SMFMX + SMFMN}{2} + \frac{SMFMX - SMFMN}{2} \cdot \sin\left[\frac{2\pi}{365}(i - 81)\right] \tag{5}$$

where SMFMX and SMFMN are the maximum and minimum snowmelt factors, respectively (mm H_2O/°C-day).

Pond Routing

In SWAT, a pond is defined within a subbasin to receive inflow from a fraction of the subbasin area. Thus, ponds can be used to appropriately mimic the hydrologic functions of Waffle® storages. The water balance for a pond is:

$$V = V_{stored} + V_{flowin} - V_{flowout} + V_{pcp} - V_{evap} - V_{seep} \tag{6}$$

where V is the volume of water in the pond at the end of the day (m^3 H_2O), V_{stored} is the volume of water stored in the pond at the beginning of the day (m^3 H_2O), V_{flowin} is the volume of water entering the pond during the day (m^3 H_2O), $V_{flowout}$ is the volume of water flowing out of the pond during the day (m^3 H_2O), V_{pcp} is the volume of precipitation falling in the pond during the day (m^3 H_2O), V_{evap} is the volume of water removed from the pond by evaporation during the day (m^3 H_2O), and V_{seep} is the volume of water lost from the pond by seepage (m^3 H_2O).

To estimate the terms in Equation (6), SWAT updates the surface area in a daily time step using the equation:

$$SA = \beta_{sa} \cdot V^{expsa} \tag{7}$$

where SA is the surface area of the pond (ha), β_{sa} is a coefficient, and expsa is an exponent. expsa and β_{sa} are computed as:

$$expsa = \frac{\log_{10}(SA_{em}) - \log_{10}(SA_{pr})}{\log_{10}(V_{em}) - \log_{10}(V_{pr})} \tag{8}$$

$$\beta_{sa} = \frac{SA_{em}}{V_{em}^{expsa}} \tag{9}$$

where SA_{em} is the surface area of the pond when filled to the emergency spillway (ha), SA_{pr} is the surface area of the pond when filled to the principal spillway (ha), V_{em} is the volume of water held in the pond when filled to the emergency spillway (m^3 H_2O), and V_{pr} is the volume of water held in the pond when filled to the principal spillway (m^3 H_2O).

V_{pcp} is computed as:

$$V_{pcp} = 10 \cdot R_{day} \cdot SA \tag{10}$$

where R_{day} is the amount of precipitation falling during the day (mm H_2O).

V_{flowin} is computed as:

$$V_{flowin} = fr_{imp} \cdot 10 \cdot (Q_{surf} + Q_{gw} + Q_{lat}) \cdot (A_{sub} - SA) \tag{11}$$

where fr_{imp} is the fraction of the subbasin area draining into the pond, Q_{surf} is the surface runoff from the subbasin during the day (mm H_2O), Q_{gw} is the groundwater flow generated in the subbasin during the day (mm H_2O), Q_{lat} is the lateral flow generated in the subbasin during the day (mm H_2O), and A_{sub} is the subbasin area (ha).

V_{evap} is computed as:

$$V_{evap} = 10 \cdot \eta \cdot E_0 \cdot SA \tag{12}$$

where η is an evaporation coefficient with a default value of 0.6, and E_0 is the potential evapotranspiration for the day (mm H_2O).

V_{seep} is computed as:

$$V_{seep} = 240 K_{sat} \cdot SA \tag{13}$$

where K_{sat} is the effective saturated hydraulic conductivity of the pond bottom (mm/hr).

$V_{flowout}$ is computed as:

$$V_{flowout} = \frac{V - V_{targ}}{ND_{targ}} \tag{14}$$

where V_{targ} is the target pond volume for the day (m^3 H_2O), and ND_{targ} is the number of days required for the pond to reach the target volume.

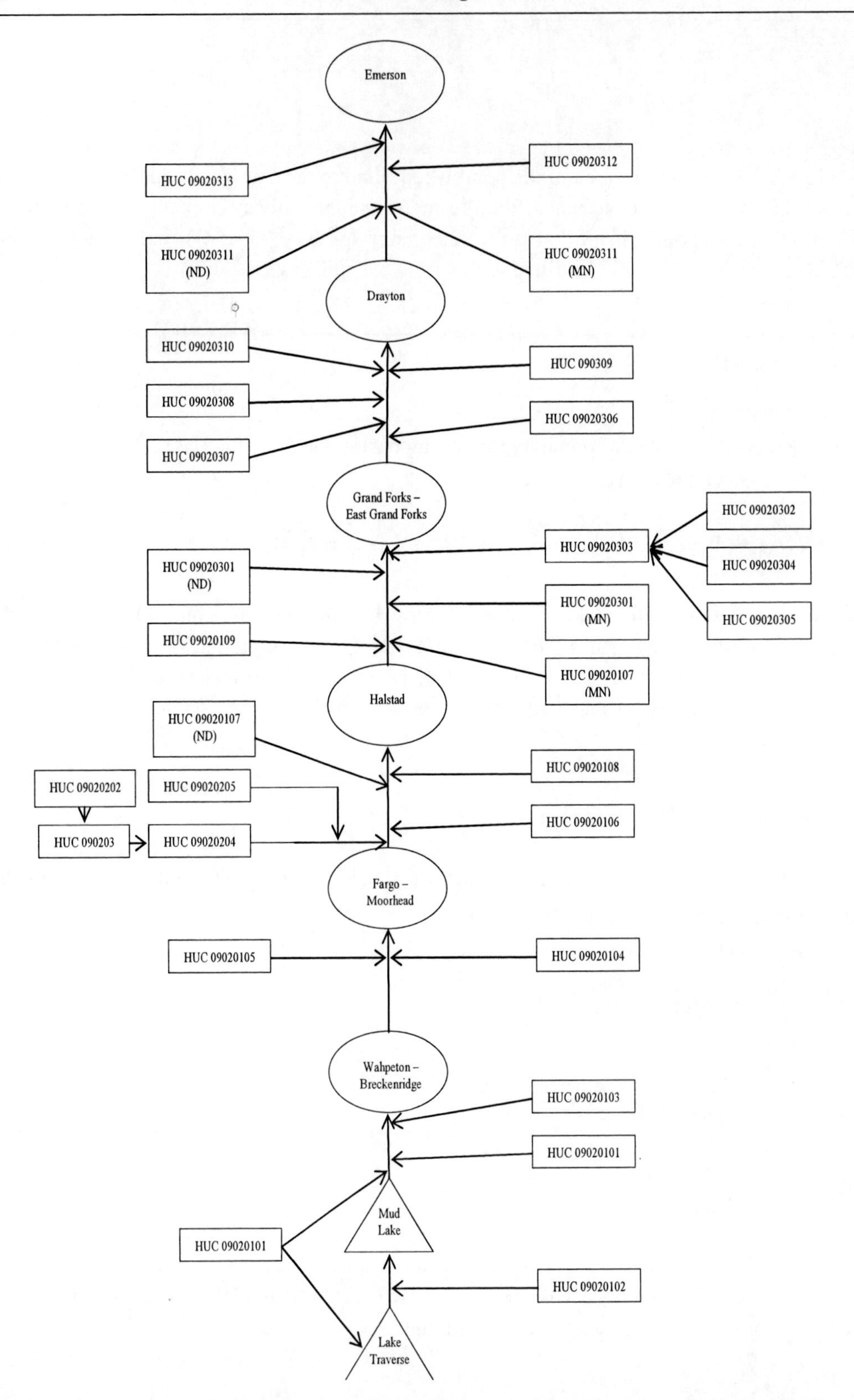

Figure 7-1. Hydraulic connectivity of the hydrologic unit codes (HUCs), which comprises the Red River of the North Basin.

Table 7-1. The hydrologic unit codes (HUCs) comprising the Red River of the North Basin

No.	HUC	Name	Drainage Area (mi^2)	Administration Boundary
1	09020101	Bois de Sioux	1,140	Minnesota, North Dakota
2	09020102	Mustinka	825	Minnesota
3	09020103	Otter Tail	1,980	Minnesota
4	09020104	Upper Red	594	Minnesota, North Dakota
5	09020105	Western Wild Rice	2,380	North Dakota
6	09020106	Buffalo	1,150	Minnesota
7	09020107	Elm–Marsh	1,150	Minnesota, North Dakota
8	09020108	Eastern Wild Rice	1,670	Minnesota
9	09020109	Goose	1,280	North Dakota
10	09020202	Upper Sheyenne	1,940	North Dakota
11	09020203	Middle Sheyenne	2,070	North Dakota
12	09020204	Lower Sheyenne	1,640	North Dakota
13	09020205	Maple	1,620	North Dakota
14	09020301	Sandhill–Wilson	1,130	Minnesota, North Dakota
15	09020302	Red Lakes	2,040	Minnesota
16	09020303	Red Lake	1,450	Minnesota
17	09020304	Thief	994	Minnesota
18	09020305	Clearwater	1,350	Minnesota
19	09020306	Grand Marais–Red	482	Minnesota, North Dakota
20	09020307	Turtle	714	North Dakota
21	09020308	Forest	875	North Dakota
22	09020309	Snake	953	Minnesota
23	09020310	Park	1,080	North Dakota
24	09020311	Lower Red	1,320	Minnesota, North Dakota
25	09020312	Two Rivers	958	Minnesota
26	09020313	Pembina	2,020	North Dakota
27	09020314	Roseau	1,230	Minnesota

To model the storage and release process of Waffle®, i.e., the water is stored for specified days and then released in the following days, the SWAT algorithm for computing V_{targ} was revised as:

$$V_{targ} = \begin{cases} V_{em} & \text{if } jday_{stor,beg} \le jday \le jday_{stor,end} \\ 0 & \text{otherwise} \end{cases} \quad (15)$$

where $jday_{stor,beg}$ is the beginning Julian day of storage, $jday_{stor,end}$ is the end Julian day of storage, and jday is the current simulating Julian day.

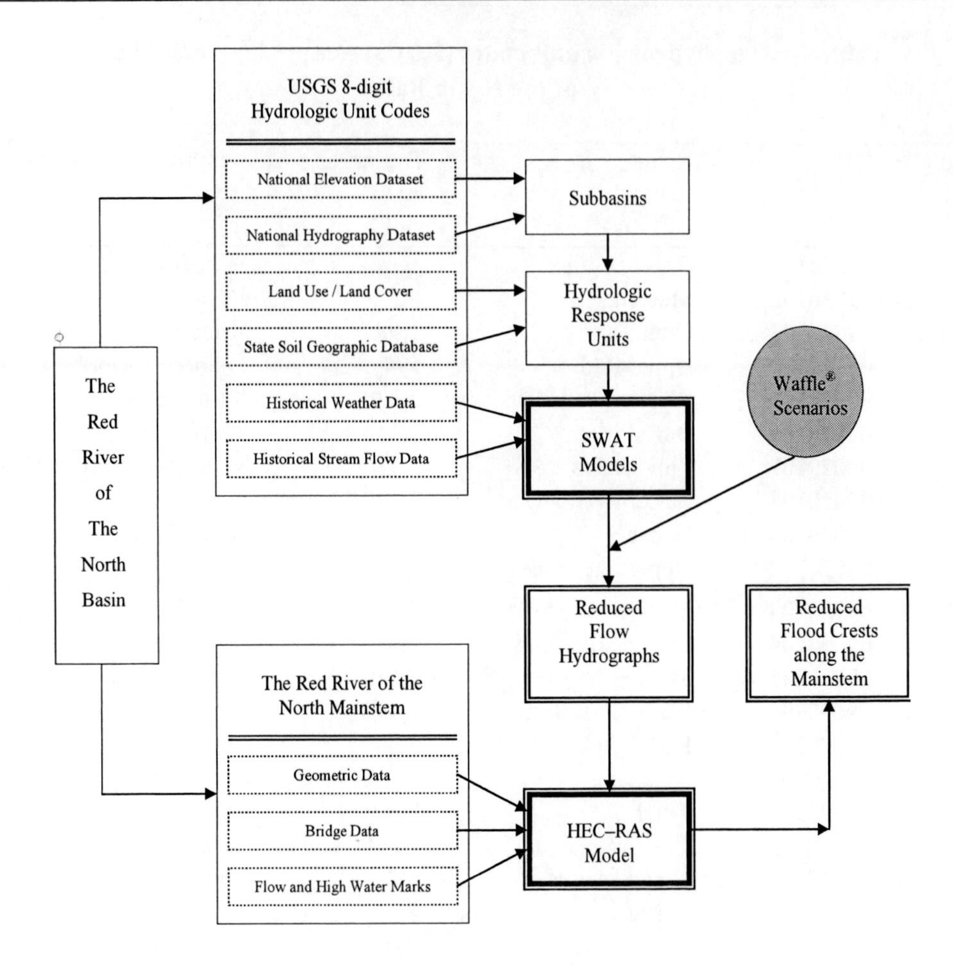

Figure 7-2. The general framework of the integrated modeling approach for evaluating the effects of the Waffle® on flood reduction.

The Muskingum Channel Routing

The Muskingum routing method models the storage volume in a channel length as a combination of wedge and prism storages. When a flood wave advances into a reach segment, inflow exceeds outflow and a wedge of storage is produced. As the flood wave recedes, outflow exceeds inflow in the reach segment and a negative wedge is produced. In addition to the wedge storage, the reach segment contains a prism of storage formed by a volume of constant cross-section along the reach length. For a simulation day, the outflow is computed as:

$$q_{out,i+1} = C_1 \cdot q_{in,i+1} + C_2 \cdot q_{in,i} + C_3 \cdot q_{out,i} \tag{16}$$

where $q_{in,i}$ is the inflow rate at day i (m^3/s), $q_{in,i+1}$ is the inflow rate at day i+1 (m^3/s), $q_{out,i}$ is the outflow rate at day i (m^3/s), $q_{out,i+1}$ is the outflow at day i+1 (m^3/s), and C_1, C_2, C_3 are coefficients with a summation of unity, i.e., $C_1 + C_2 + C_3 = 1$, and are computed as:

$$C_1 = \frac{\Delta t - 2 \cdot K \cdot X}{2 \cdot K \cdot (1-X) + \Delta t} \quad (17)$$

$$C_2 = \frac{\Delta t + 2 \cdot K \cdot X}{2 \cdot K \cdot (1-X) + \Delta t} \quad (18)$$

$$C_3 = \frac{2 \cdot K \cdot (1-X) - \Delta t}{2 \cdot K \cdot (1-X) + \Delta t} \quad (19)$$

where K is the ratio of storage to outflow (s), X is a weighting factor that controls the relative importance of inflow and outflow in determining the storage in a reach, and Δt is the simulation time step (i.e., 1 day). To maintain numerical stability and avoid the comutation of negative outflows, the following condition must be net:

$$2 \cdot K \cdot X \leq \Delta t \leq 2 \cdot K \cdot (1-X) \quad (20)$$

K is computed as:

$$K = coef_1 \cdot K_{bnkfull} + coef_2 \cdot K_{0.1bnkful} \quad (21)$$

where $coef_1$ and $coef_2$ are weighting coefficients, $K_{bnkfull}$ is the storage time constant calculated for the reach segment with the bankfull flow (s), and $K_{0.1bnkfull}$ is the storage time constant calculated for the reach segment with one-tenth of the bankfull flow (s).

$K_{bnkfull}$ is computed as:

$$K_{bnkfull} = \frac{1000 L_{ch,bnkfull}}{c_{k,bnkfull}} \quad (22)$$

where $L_{ch,bnkfull}$ is the channel length at the bankfull flow (km), and $c_{k,bnkful}$ is the celerity at the bankfull flow, i.e., the velocity with which a variation in flow rate travels along the channel.

$c_{k,bnkful}$ is computed as:

$$c_{k,bnkfull} = \frac{5}{3} \cdot v_{bnkful} \quad (23)$$

where v_{bnkful} is the flow velocity at the bankfull flow and computed using the Manning's equation.

$K_{0.1bnkfull}$ is computed as:

$$K_{0.1bnkfull} = \frac{1000\,L_{ch,0.1bnkfull}}{c_{k,0.1bnkfull}} \quad (22)$$

where $L_{ch,0.1bnkfull}$ is the channel length at one-tenth of the bankfull flow (km), and $c_{k,0.1bnkful}$ is one-tenth of the celerity at the bankfull flow, i.e., the velocity with which a variation in flow rate travels along the channel.

$c_{k,0.1bnkful}$ is computed as:

$$c_{k,0.1bnkfull} = \frac{5}{3} \cdot v_{0.1bnkful} \quad (23)$$

where v_{bnkful} is the flow velocity at one-tenth of the bankfull flow and computed using the Manning's equation.

Description of HEC–RAS

HEC–RAS is designed to perform one-dimensional hydraulic calculations for a full network of natural and constructed channels. The current version of HEC–RAS supports steady and unsteady flow water surface profile calculations. The hydraulic calculations for cross sections, bridges, culverts, and other hydraulic structures that were developed for the steady flow component were incorporated into the unsteady flow module. The unsteady flow module has the ability to model storage areas and hydraulic connections between storage areas as well as between stream reaches. However, this module was developed primarily for subcritical flow regime calculations.

Assuming a horizontal water surface at each cross section normal to the direction of flow, i.e., a negligible exchange of momentum between the channel and floodplain, HEC – RAS distributes the discharge according to conveyance and solves a set of one-dimensional equations expressed as:

$$\frac{\partial A}{\partial t} + \frac{\partial(\phi \cdot Q)}{\partial x_c} + \frac{\partial[(1-\phi) \cdot Q]}{\partial x_f} = 0 \quad (24)$$

$$\frac{\partial Q}{\partial t} + \frac{\partial(\phi^2 \cdot Q^2 / A_c)}{\partial x_c} + \frac{\partial[(1-\phi)^2 \cdot Q^2 / A_f]}{\partial x_f} + g \cdot A_c \left[\frac{\partial Z}{\partial x_c} + S_{fc}\right] + g \cdot A_f \left[\frac{\partial Z}{\partial x_f} + S_{ff}\right] = 0 \quad (25)$$

where Q is the total flow, Z is the water surface elevation, the subscripts c and f refer to the channel and floodplain, respectively, and ϕ is the ratio of the channel conveyance to the total conveyance (i.e., the summation of the channel conveyance and floodplain conveyance).

Equations (24) and (25) are solved using the four-point implicit scheme, also known as the box scheme. With this scheme, Equation (24) has a finite difference form expressed as:

$$\Delta Q+\frac{\Delta A_c}{\Delta t}\cdot(\Delta x_c)+\frac{\Delta A_f}{\Delta t}\cdot(\Delta x_f)+\frac{\Delta S}{\Delta t}\cdot(\Delta x_f)-\overline{Q}_L=0 \tag{26}$$

where $\overline{Q}_L$ is the average lateral inflow.

Equation (25) has a finite difference form expressed as:

$$\frac{\Delta(Q_c\cdot\Delta x_c+Q_f\cdot\Delta x_f)}{\Delta t\cdot\Delta x_e}+\frac{\Delta(\beta\cdot V\cdot Q)}{\Delta x_e}+g\cdot\overline{A}(\frac{\Delta z}{\Delta x_e}+\overline{S}_f+\overline{S}_h)=\xi\cdot\frac{Q_L\cdot V_L}{\Delta x_e} \tag{27}$$

where Q_L is the lateral inflow, V_L is the average velocity of the lateral inflow, ξ is the fraction of the momentum entering the receiving stream, $\overline{A}$ is equal to $\overline{A}_c+\overline{A}_f$, $\overline{S}_f$ is the friction slope for the entire section, and Δx_e is the equivalent flow path computed as:

$$\Delta x_e=\frac{\overline{A}_c\cdot\overline{S}_{fc}\cdot\Delta x_c+\overline{A}_f\cdot\overline{S}_{ff}\cdot\Delta x_f}{\overline{A}\cdot\overline{S}_f} \tag{28}$$

The convective term β is defined as:

$$\beta=\frac{V_c\cdot Q_c+V_f\cdot Q_f}{V\cdot Q} \tag{29}$$

$\overline{S}_h$ is the rate of energy loss caused by structures such as bridge piers, navigation dams, and cofferdams, and is computed as:

$$\overline{S}_h=\frac{h_L}{\Delta x_e} \tag{30}$$

where h_L is the head loss. Within HEC–RAS, the steady flow bridge and culvert routines are used to compute a family of rating curves for the structure. During the simulation, for a given flow and tailwater, a resulting headwater elevation is interpolated from the curves. The difference between the headwater and tailwater is set to h_L.

Using the Preissmann technique, for a computation reach from node j to j+1, Equations (26) and (27) are linearized as:

$$CQ1_j\cdot\Delta Q_j+CZ1_j\cdot\Delta z_j+CQ2_j\cdot\Delta Q_{j+1}+CZ2_j\cdot\Delta z_{j+1}=CB_j \tag{31}$$

$$MQ1_j\cdot\Delta Q_j+MZ1_j\cdot\Delta z_j+MQ2_j\cdot\Delta Q_{j+1}+MZ2_j\cdot\Delta z_{j+1}=MB_j \tag{32}$$

where

$$CQ1_j = \frac{-\theta}{\Delta x_{ej}} \tag{33}$$

$$CZ1_j = \frac{0.5}{\Delta t \cdot \Delta x_{ej}} \left[\left(\frac{dA_c}{dz} \right)_j \cdot \Delta x_{cj} + \left(\frac{dA_f}{dz} + \frac{dS}{dz} \right)_j \cdot \Delta x_{fj} \right] \tag{34}$$

$$CQ2_j = \frac{\theta}{\Delta x_{ej}} \tag{35}$$

$$CZ2_j = \frac{0.5}{\Delta t \cdot \Delta x_{ej}} \left[\left(\frac{dA_c}{dz} \right)_{j+1} \cdot \Delta x_{cj} + \left(\frac{dA_f}{dz} + \frac{dS}{dz} \right)_{j+1} \cdot \Delta x_{fj} \right] \tag{36}$$

$$CB_j = -\frac{Q_{j+1} - Q_j}{\Delta x_{ej}} + \frac{Q_L}{\Delta x_{ej}} \tag{37}$$

$$MQ1_j = 0.5 \frac{\Delta x_{cj} \cdot \phi_j + \Delta x_{fj} \cdot (1 - \phi_j)}{\Delta x_{ej} \cdot \Delta t} - \frac{\beta_j \cdot V_j \cdot \theta}{\Delta x_{ej}} + \theta \cdot g \cdot \overline{A} \frac{S_{fj} + S_{hj}}{Q_j} \tag{38}$$

$$MZ1_j = -\frac{g \cdot \overline{A} \cdot \theta}{\Delta x_{ej}} + 0.5 \cdot g \cdot (z_{j+1} - z_j) \left(\frac{dA}{dz} \right)_j \left(\frac{\theta}{\Delta x_{ej}} \right) - g \cdot \theta \cdot \overline{A} \left[\left(\frac{dK}{dz} \right)_j \left(\frac{S_{fj}}{K_j} \right) + \left(\frac{dA}{dz} \right)_j \left(\frac{S_{hj}}{A_j} \right) \right] + 0.5 \cdot \theta \cdot g \left(\frac{dA}{dz} \right)_j \left(\overline{S}_f + \overline{S}_h \right) \tag{39}$$

$$MQ2_j = 0.5 \frac{\Delta x_{cj+1} \cdot \phi_{j+1} + \Delta x_{fj+1} \cdot (1 - \phi_{j+1})}{\Delta x_{ej} \cdot \Delta t} + \frac{\beta_{j+1} \cdot V_{j+1} \cdot \theta}{\Delta x_{ej}} + \theta \cdot g \cdot \overline{A} \frac{S_{fj+1} + S_{hj+1}}{Q_{j+1}} \tag{40}$$

$$MZ2_j = -\frac{g \cdot \overline{A} \cdot \theta}{\Delta x_{ej}} + 0.5 \cdot g \cdot (z_{j+1} - z_j) \left(\frac{dA}{dz} \right)_{j+1} \left(\frac{\theta}{\Delta x_{ej}} \right) - g \cdot \theta \cdot \overline{A} \left[\left(\frac{dK}{dz} \right)_{j+1} \left(\frac{S_{fj+1}}{K_{j+1}} \right) + \left(\frac{dA}{dz} \right)_{j+1} \left(\frac{S_{hj+1}}{A_{j+1}} \right) \right] + 0.5 \cdot \theta \cdot g \left(\frac{dA}{dz} \right)_{j+1} \left(\overline{S}_f + \overline{S}_h \right) \tag{41}$$

$$MB_j = -\left[\left(\beta_{j+1}\cdot V_{j+1}\cdot Q_{j+1} - \beta_j\cdot V_j\cdot Q_j\right)\left(\frac{1}{\Delta x_{ej}}\right) + \left(\frac{g\cdot \bar{A}}{\Delta x_{ej}}\right)\left(z_{j+1} - z_j\right) + g\cdot \bar{A}\left(\bar{S}_f + \bar{S}_h\right)\right] \tag{42}$$

$$\phi_j = \frac{K_{cj}}{K_{cj} + K_{fj}} \tag{43}$$

$$\Delta x_{ej} = \frac{(A_{cj} + A_{cj+1})\Delta x_{cj} + (A_{fj} + A_{fj+1})\Delta x_{fj}}{A_j + A_{j+1}} \tag{44}$$

For a junction, HEC–RAS applies flow continuity to reaches upstream of flow splits and downstream of flow combinations, whereas stage continuity is used for all other reaches. In addition, upstream boundary conditions are required at the upstream end of all reaches that are not connected to other reaches or storage areas. An upstream boundary condition is applied as a flow hydrograph of discharge versus time. On the other hand, downstream boundary conditions are required at the downstream end of all reaches which are not connected to other reaches or storage areas. In this study, a single-valued rating curve and/or normal depth from Manning's equation are specified as the downstream boundary condition.

THE SWAT MODELS

Data Quality and Availability

The basic model inputs included the 30 m USGS National Elevation Dataset (NED), the EPA 1:250,000-scale Land Use Land Cover (LULC), and the USDA-NRCS (Natural Resources Conservation Service) State Soil Geographic database (STATSGO). The NED was developed by merging the highest-resolution, best-quality elevation data available across the U.S. into a seamless raster format. The LULC was developed by combining the data obtained from 1970s and 1980s aerial photography surveys with land use maps and surveys. Because there have been negligible changes in the types of land use of RRB in the past two decades, the LULC was an appropriate choice. Data for the STATSGO are collected at the USGS 1:250,000-scale in 1- by 2-degree topographic quadrangle units, and then merged and distributed as state coverages. The STATSGO has a county-level resolution and can readily be used for river-basin water resource studies. The NED and LULC were downloaded from the USGS website (http://edc.usgs.gove/geodata), and the STATSGO was downloaded from the USDA-NRCS website (http://www.ncgc.nrcs.usda.gov/branch /ssb/products). In addition to these three datasets, the USGS National Hydrography Dataset (NHD) was also used as a model input. The NHD is a comprehensive set of digital spatial data that contains information about surface water features such as lakes, ponds, streams, rivers, springs, and wells. The stream feature provided by NHD was utilized as the reference surface water drainage network to delineate subbasins for each of the USGS 8-digit HUCs for modeling purposes.

The National Weather Service (NWS) National Climate Data Center (NCDC) collects data on daily precipitation and minimum and maximum temperatures at stations across RRB. To minimize modeling uncertainties, the stations that have 30% or more missing values from 1 October to 31 May for the years of 1966, 1969, 1975, 1978, 1979, and 1997 were not selected to set up the models. Because the largest floods occurred as a result of melting snow in spring, the daily flow data for these six years collected at a number of USGS gauging stations were used to calibrate and validate the models. Figure 7-3 shows the locations of the NWS precipitation and temperature stations and USGS flow gauging stations, where the available data were used in this study. Tables 7-2 and 7-3 list the basic information of these stations.

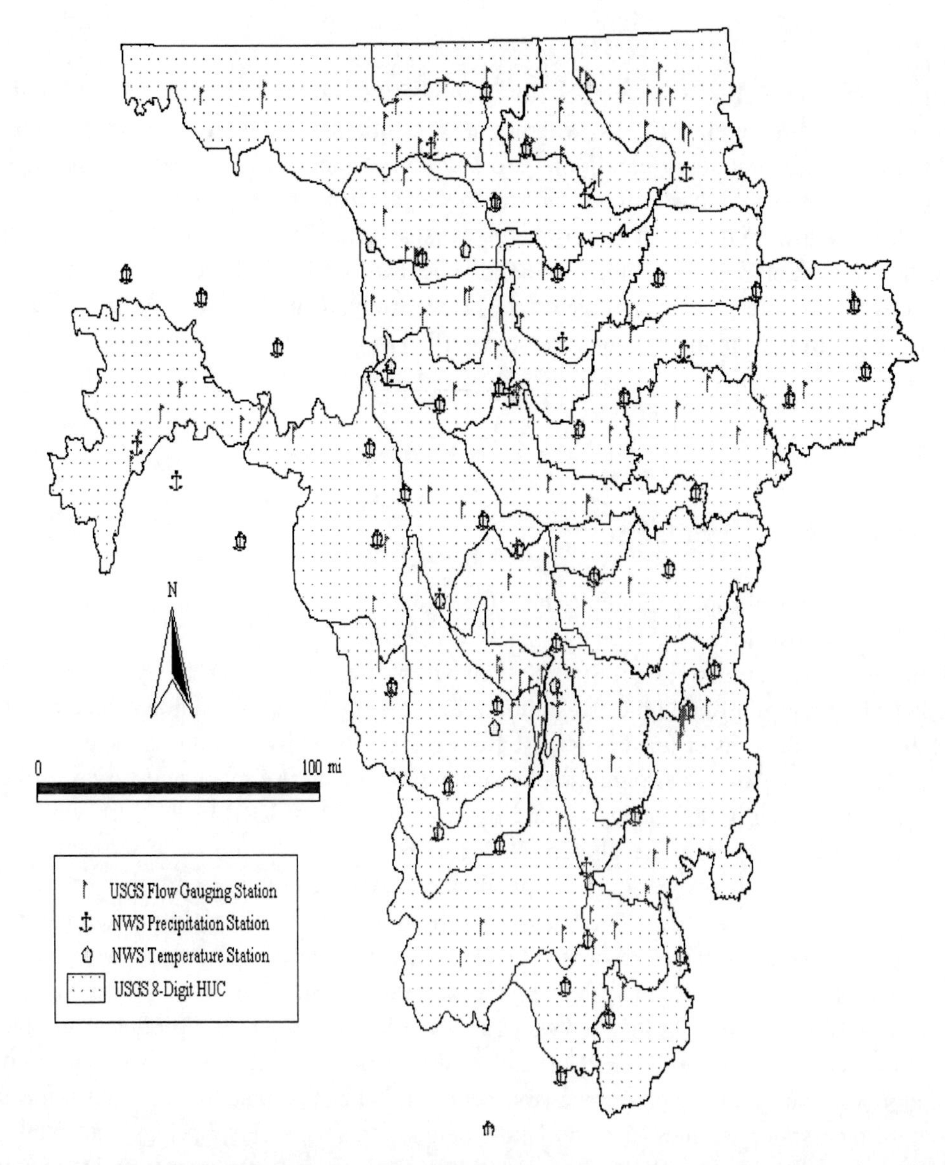

Figure 7-3. Map showing the locations of the weather stations and flow gauging stations, where the data were used in this study.

Table 7-2. Weather stations where data were used in this study

COOP ID[1]	X (m)[2]	Y (m)[2]	Latitude (degree)[3]	Longitude (degree) [3]	Elevation (m)[4]	Para-meter[5]
210018	688243.0000	5241291.0000	47.3000	-96.5100	150.00	P, T
210050	723964.4897	5353834.8508	48.3000	-95.9800	348.10	P, T
210195	671238.3104	5329754.8350	48.3300	-96.7300	264.90	T
210252	668249.3397	5355250.7371	45.6100	-96.8300	265.20	P, T
211063	669208.9776	5052789.9167	48.9600	-96.4500	301.80	P, T
211303	686660.3067	5425924.1108	47.8000	-96.6000	310.90	P
211891	679720.6359	5296641.8282	46.8300	-95.8500	267.70	P, T
212142	740239.0000	5190873.0000	46.0000	-95.9600	151.00	P, T
212476	735397.6073	5098323.7836	47.5600	-95.7500	405.40	P, T
212916	744485.6858	5272299.0043	47.0800	-96.8000	399.30	P, T
213104	667008.6801	5216184.2882	48.7600	-96.9500	269.70	P, T
213455	650661.9192	5402584.8182	47.9300	-94.4500	246.90	P, T
213756	738469.3082	5326593.8255	47.3100	-95.9600	350.50	T
214213	683649.9902	5383542.4572	47.9000	-96.2600	325.20	T
214233	839842.7733	5318328.1180	47.8600	-95.0300	423.70	P, T
215012	729776.0000	5243878.0000	46.4800	-96.2600	118.00	P, T
216787	704784.9849	5308601.3739	46.9600	-95.6500	327.70	P, T
216795	796930.2784	5308154.9370	48.2300	-95.2500	371.90	P, T
217149	710324.8156	5150810.8827	48.1600	-94.5100	376.70	P, T
218191	754873.7493	5205949.3706	45.8000	-96.4800	118.00	P, T
218254	778474.0569	5348446.6926	48.4462	-98.1395	362.70	P, T
218656	739365.7717	5393428.9796	47.4462	-99.1396	362.70	T
218700	833869.8987	5343620.5311	46.8800	-97.2300	365.80	P, T
218907	695835.4783	5074697.1468	46.8000	-97.2600	310.30	P, T
320022	563636.7400	5366031.2300	47.2300	-97.6500	473.70	P
321360	489472.9000	5254538.8500	47.4500	-98.1100	483.00	P, T
321408	634869.0200	5193131.4300	48.1020	-98.8501	285.00	P, T
321435	597902.9100	5403133.5100	48.5800	-97.1800	271.00	T
321477	632780.1400	5184190.7400	46.6100	-97.6000	294.00	P
321686	602196.0000	5231389.0000	46.9352	-96.8169	359.70	P, T
321766	567094.4300	5255338.1400	48.4166	-97.4167	421.00	P, T
322158	511163.0800	5327429.6000	47.9500	-97.1800	446.20	P, T
322312	634236.5658	5382148.1082	47.9100	-97.0800	243.80	P, T
322695	607210.5000	5162559.9000	46.0687	-96.6171	351.00	P, T
322859	666170.6500	5200061.1900	47.4000	-97.0600	274.00	P, T
322949	453899.8700	5277576.6200	47.9000	-97.6300	493.80	T
323594	617156.6800	5363600.9100	48.2756	-99.4345	252.10	P
323616	635896.8999	5312127.0743	46.4519	-97.6837	255.70	P, T
323621	643474.1121	5307862.5778	47.5000	-97.3100	253.00	P, T
323908	684286.0000	5104220.0000	46.4020	-97.2334	326.10	P, T
324013	431200.9000	5290789.4600	47.7600	-98.1600	487.70	T
324203	646388.0000	5251222.0000	48.4000	-97.7500	275.00	P, T
325013	602395.0000	5305875.0000	48.9568	-97.2325	345.00	P, T
325078	467761.6200	5346800.9400	48.0300	-98.0000	466.00	P, T
325220	601095.9700	5144886.3000	48.3517	-100.0067	337.00	P, T
325660	627282.0755	5261895.1789	47.6015	-97.9017	288.30	P, T
325754	635800.0000	5140015.0000	46.9500	-98.0166	327.70	P, T
325764	562952.0000	5289749.0000	46.2600	-96.6000	447.00	P, T
326857	592523.3700	5361295.9500	45.9173	-96.7845	295.70	P, T
326947	629393.3100	5423941.9000	45.4354	-97.3504	241.00	P, T
327027	574554.0000	5319899.0000	47.3000	-96.5100	466.00	P, T
327704	425416.1500	5355663.9400	48.3000	-95.9800	472.00	P, T
327986	582558.9000	5272371.5000	48.3300	-96.7300	464.80	P, T
328937	574835.3300	5199858.6300	45.6100	-96.8300	369.00	P, T
329100	684968.0000	5125517.0000	48.9600	-96.4500	291.40	P, T
398652	671808.1600	5087028.2400	47.8000	-96.6000	329.20	P, T
398980	629029.0000	5032422.0000	46.8300	-95.8500	557.80	P, T

[1] The 6-digit National Weather Service Cooperative Station Identifier (COOP ID) for the station.
[2] NAD 1927 UTM Zone 14N. [3] NAD 1983. [4] NGVD 1929 above sea level.
[5] P signifies precipitation and T signifies minimum and maximum temperatures.

Table 7-3. US Geological Survey (USGS) flow gauging stations which were used to set up the SWAT and HEC-RAS models in this study

USGS ID	Station Name	USGS ID	Station Name
05030000	OTTER TAIL RIVER NEAR DETROIT LAKES, MN	05084500	FOREST RIVER NEAR MINTO, N. DAK.
05030150	OTTERTAIL RIVER NR PERHAM, MINN.	05085000	FOREST RIVER AT MINTO, ND
05030500	OTTER TAIL RIVER NEAR ELIZABETH, MN	05085900	SNAKE RIVER ABOVE ALVARADO, MINN
05033900	PELICAN RIVER AT DETROIT LAKES, MN	05087500	MIDDLE RIVER AT ARGYLE, MN
05034100	PELICAN R AT DETROIT LK OUT NR DETROIT LAKES, MN	05088000	SOUTH BRANCH PARK R NR PARK RIVER, N. DAK.
05035100	LONG LAKE OUTLET NEAR DETROIT LAKES, MN	05088500	HOMME RESERVOIR NR PARK RIVER, ND
05035200	WEST BRANCH COUNTY DIT. #14 NR DETROIT LAKES, MN	05089000	SOUTH BRANCH PARK RIVER BELOW HOMME DAM, ND
05035300	EAST BRANCH COUNTY DIT. #14 NR DETROIT LAKES, MN	05089100	MIDDLE BRANCH PARK RIVER NR UNION, ND
05035500	ST. CLAIR LAKE OUTLET NEAR DETROIT LAKES, MN	05089500	CART CREEK AT MOUNTAIN, ND
05035600	PELICAN R AT MUSKRAT LK OUT NR DETROIT LAKES, MN	05090000	PARK RIVER AT GRAFTON, ND
05037100	PELICAN R AT SALLIE LK OTLT NR DETROIT LAKES, MN	05092000	RED RIVER OF THE NORTH AT DRAYTON, ND
05039100	PELICAN R AT LK MELISSA OUT NR DETROIT LAKES, MN	05092200	PEMBINA COUNTY DRAIN 20 NR GLASSTON, ND
05040000	PELICAN RIVER NEAR DETROIT LAKES, MN	05092500	MIDDLE BRANCH TWO RIVERS NEAR HALLOCK, MN
05040500	PELICAN RIVER NEAR FERGUS FALLS, MN	05093000	SOUTH BRANCH TWO RIVERS AT PELAN, MN
05045950	ORWELL LAKE NEAR FERGUS FALLS, MN	05094000	SOUTH BRANCH TWO RIVERS AT LAKE BRONSON, MN
05046000	OTTER TAIL RIVER BL ORWELL D NR FERGUS FALLS, MN	05095000	TWO RIVERS AT HALLOCK, MN
05047500	MUSTINKA D AB W BR MUSTINKA R NR CHARLESVILLE MN	05095500	TWO RIVERS BELOW HALLOCK, MN

Table 7-3. Continued

USGS ID	Station Name	USGS ID	Station Name
05048000	MUSTINKA D BL W BR MUSTINKA R NR CHARLESVILLE MN	05096000	NORTH BRANCH TWO RIVERS NEAR LANCASTER, MN
05048500	W BR MUSTINKA R BL MUSTINKA D NR CHARLESVILLE MN	05096500	STATE DITCH #85 NEAR LANCASTER, MN
05049000	MUSTINKA RIVER ABOVE WHEATON, MN	05097500	NORTH BRANCH TWO RIVERS NEAR NORTHCOTE, MN
05050000	BOIS DE SIOUX RIVER NEAR WHITE ROCK, SD	05098700	HIDDEN ISLAND COULEE NR HANSBORO, ND
05050500	BOIS DE SIOUX RIVER NEAR FAIRMOUNT, N. DAK.	05098800	CYPRESS CREEK NR SARLES, ND
05051000	RABBIT RIVER AT CAMPBELL, MN	05098820	CYPRESS CREEK ABV INTL BOUNDARY NR SARLES, ND
05051300	BOIS DE SIOUX RIVER NEAR DORAN, MN	05099400	LITTLE SOUTH PEMBINA RIVER NR WALHALLA, ND
05051500	RED RIVER OF THE NORTH AT WAHPETON, ND	05099600	PEMBINA RIVER AT WALHALLA, ND
05051522	RED RIVER OF THE NORTH AT HICKSON, ND	05100000	PEMBINA RIVER AT NECHE, ND
05051600	WILD RICE RIVER NR RUTLAND, ND	05100500	HERZOG CREEK NR CONCRETE, ND
05051700	WILD RICE RIVER NR CAYUGA, ND	05101000	TONGUE RIVER AT AKRA, ND
05052100	RICHLAND COUNTY DRAIN #65 NEAR GREAT BEND, ND	05101500	TONGUE R AT CAVALIER N DAK
05053000	WILD RICE RIVER NR ABERCROMBIE, ND	05102500	RED RIVER OF THE NORTH AT EMERSON, MAN
05054000	RED RIVER OF THE NORTH AT FARGO, ND	05103000	ROSEAU RIVER NEAR MALUNG, MN
05054020	RED RIVER OF THE NORTH BELOW FARGO, ND	05104000	SOUTH FORK ROSEAU RIVER NEAR MALUNG, MN
05054500	SHEYENNE RIVER ABOVE HARVEY, ND	05104500	ROSEAU RIVER BELOW SOUTH FORK NEAR MALUNG, MN
05055000	SHEYENNE RIVER NEAR HARVEY, N. DAK.	05106500	ROSEAU RIVER AT ROSEAU LAKE, MN
05055100	N FK SHEYENNE R NR WELLSBURG N DAK	05107000	PINE CREEK NEAR PINE CREEK, MN
05055200	BIG COULEE NEAR MADDOCK, N. DAK.	05107500	ROSEAU RIVER AT ROSS, MN

Table 7-3. Continued

USGS ID	Station Name	USGS ID	Station Name
05055500	SHEYENNE RIVER AT SHEYENNE, N. DAK.	05108000	ROSEAU RIVER NEAR BADGER, MN
05055520	BIG COULEE NR FT. TOTTEN, N. DAK.	05109000	BADGER CREEK NEAR BADGER, MN
05056000	SHEYENNE RIVER NR WARWICK, ND	05109500	ROSEAU RIVER NEAR HAUG, MN
05057000	SHEYENNE RIVER NR COOPERSTOWN, ND	05112000	ROSEAU RIVER BELOW STATE DITCH 51 NR CARIBOU, MN
05057200	BALDHILL CREEK NR DAZEY, ND	05112500	ROSEAU R AT INTERNATIONAL BOUNDARY NR CARIBOU MN
05058000	SHEYENNE RIVER BELOW BALDHILL DAM, ND	05084500	FOREST RIVER NEAR MINTO, N. DAK.
05058500	SHEYENNE RIVER AT VALLEY CITY, N. DAK.	05085000	FOREST RIVER AT MINTO, ND
05058600	SHEYENNE RIVER NR KATHRYN, N. DAK.	05085900	SNAKE RIVER ABOVE ALVARADO, MINN
05058700	SHEYENNE RIVER AT LISBON, ND	05087500	MIDDLE RIVER AT ARGYLE, MN
05059000	SHEYENNE RIVER NEAR KINDRED, ND	05088000	SOUTH BRANCH PARK R NR PARK RIVER, N. DAK.
05059300	SHEYENNE R AB SHEYENNE R DIVERSION NR HORACE, ND	05088500	HOMME RESERVOIR NR PARK RIVER, ND
05059310	SHEYENNE RIVER DIVERSION NR HORACE, ND	05089000	SOUTH BRANCH PARK RIVER BELOW HOMME DAM, ND
05059400	SHEYENNE RIVER NR HORACE, ND	05089100	MIDDLE BRANCH PARK RIVER NR UNION, ND
05059480	SHEYENNE RIVER DIVERSION AT WEST FARGO, ND	05089500	CART CREEK AT MOUNTAIN, ND
05059500	SHEYENNE RIVER AT WEST FARGO, ND	05090000	PARK RIVER AT GRAFTON, ND
05059600	MAPLE RIVER NR HOPE, ND		

For stations where observed daily stream flows were available, the data were used to calibrate and validate the models.

Table 7-4. Set up of the SWAT models

State	Modeling Domain	Calibration	Validation	Remark
MN	HUC 09020101	No	No	Judgment, a spot 1997 peak of 6000 cfs provided by a consulting engineer
	HUC 09020102	No	No	Judgment, a spot 1997 peak of 8800 cfs at USGS 05049000 gauging station
	HUC 09020103	Yes	Yes	Daily stream flows observed at USGS 05030500 and 05046000
	HUC 09020104	Yes	Yes	Daily stream flows at two USGS gauging stations, validation for 1979 and 1978 only
	HUC 0920106	Yes	Yes	Daily stream flows at USGS 05061500, a spot 1997 peak at USGS 05061200
	HUC 09020107	Yes	Yes	Daily stream flows at USGS 05067500
	HUC 09020108	Yes	Yes	Daily stream flows at USGS 05062500 and 05064000
	HUC 09020301	Yes	Yes	Daily stream flows at USGS 05069000
	HUC 09020302	No	No	Judgment
	HUC09020303	Yes	Yes	Daily stream flows at USGS 05075000 and 05079000
	HUC 09020304	Yes	Yes	Daily stream flows at USGS 05076000
	HUC 09020305	Yes	Yes	Daily stream flows at USGS 05078000, 05078230, and 05078500
	HUC 09020306	No	No	Judgment
	HUC 09020309	Yes	Yes	Daily stream flows at USGS 05087500
	HUC 09020311	No	No	Judgment
	HUC 09020312	Yes	Yes	Daily stream flows at USGS 05094000
	HUC 09020314	Yes	Yes	Daily stream flows at USGS 05112000
ND	HUC 09020101	Yes	No	Daily stream flows at outlet of the Big Slough river obtained from the Corps
	HUC 09020105	Yes	No	Daily stream flows at USGS 05051600 and 05053000
	HUC 09020107	No	Yes	Daily stream flows at USGS 05062200 for the 1978, 1975, and 1969 floods
	HUC 09020109	Yes	Yes	Daily stream flows at USGS 05064900 and 05066500
	HUC 09020202	Yes	No	Daily stream flows at USGS 05054500
	HUC 09020203	Yes	No	Daily stream flows at USGS 05056000 and 05057000
	HUC 09020204	Yes	No	Daily stream flows at USGS 05058700, 05059000 and 05059500
	HUC 09020205	Yes	No	Daily stream flows at USGS 05059700 and 05060100
	HUC 09020301	No	No	Judgment
	HUC 09020307	Yes	No	Daily stream flows at USGS 05082625
	HUC 09020308	Yes	No	Daily stream flows at USGS 05084000 and 05085000
	HUC 09020310	Yes	No	Daily stream flows at USGS 05090000
	HUC 09020311	No	No	Judgment
	HUC 09020313	Yes	No	Daily stream flows at USGS 05100000 and 05101000

Calibration and Validation Strategy

For watersheds where data were available, the daily flows observed from 1 January to 31 May 1997 were used to calibrate the models, whereas, the flows observed from 1 January to 31 May of years 1966, 1969, 1975, 1978, and 1979 were used to validate the models.

The calibration of the models was implemented to adjust three snowmelt-sensitive parameters (variables SMFMX, TIMP, and SMTMP), three additional watershed-level parameters, namely the surface runoff lag coefficient (variable SURLAG) and the Muskingum translation coefficients for normal flow (variable MSK_CO1) and for low flow (variable MSK_CO2), and three HRU-level parameters, namely the SCS curve number for soil moisture condition II (variable CN2), the threshold depth of water in the shallow aquifer required for return flow to occur (variable GWQMN), and the soil evaporation compensation factor (variable ESCO). In addition, for some watersheds, three main channel related parameters, namely average slope (variable CH_S2), length (variable CH_L2), and effective hydraulic conductivity (variable CH_K2), and one ground water related parameter, namely baseflow alpha factor (variable ALPHA_BF), were also adjusted. However, for watersheds where data were unavailable, the SWAT models were set up based on scientific judgment, spot values observed by local engineers, and/or calibrated model parameters in the adjacent watersheds. Table 7-4 summarizes the set up of the SWAT models.

Measures of Model Performance

A SWAT model is said to have a good performance when the simulated flow hydrograph at a given location within a watershed is comparable with the corresponding observed hydrograph in terms of silhouette, volume, and peak. Besides visualization plots showing simulated versus observed values, three statistics, namely Nash-Sutcliffe coefficient, volume deviation, and error function, were also used to model performance in this study. These statistics can be applied for daily, monthly, seasonal, and annual evaluation time steps. The Nash-Sutcliffe coefficient measures the overall fit to the silhouette of an observed flow hydrograph, but it may be an inappropriate measure for use in simulating the volume, which is computed by integrating the flow hydrograph over the evaluation period, and for predicting the peak(s) of the hydrograph. , in addition to the Nash-Sutcliffe coefficient, two extra statistics, namely deviation of volume and error function, are generally employed to test whether the volume and peak(s) of an observed hydrograph are appropriately predicted. Therefore, in addition to the Nash-Sutcliffe coefficient, the deviation of volume was employed to test whether the volume of an observed hydrograph are appropriately predicted.

The Nash-Sutcliffe coefficient (E_j^2) is computed as:

$$E_j^2 = 1 - \frac{\sum_{i=1}^{n_j} (Q_{obsi}^j - Q_{simi}^j)^2}{\sum_{i=1}^{n_j} (Q_{obsi}^j - Q_{mean}^j)^2} \quad (45)$$

where Q_{simi}^j and Q_{obsi}^j are the simulated and observed stream flows, respectively, on the *i*th time step for station *j*, and Q_{mean}^j is the average of Q_{obsi}^j across the n_j evaluation time steps.

The deviation of volume (D_{vj}) is computed as:

$$D_{vj} = \frac{\sum_{i=1}^{n_j} Q^j_{simi} - \sum_{i=1}^{n_j} Q^j_{obsi}}{\sum_{i=1}^{n_j} Q^j_{obsi}} \times 100\% \qquad (46)$$

The peak-flow-weighted error function (E_{RRj}) is computed as:

$$E_{RRj} = \frac{\sum_{k=1}^{m_j} Q^{jkp}_{obs} \left[\left(\frac{Q^{jkp}_{obs} - Q^{jkp}_{sim}}{Q^{jkp}_{obs}} \right)^2 + \left(\frac{T^{jkp}_{obs} - T^{jkp}_{sim}}{T_c} \right)^2 \right]^{\frac{1}{2}}}{\sum_{k=1}^{m_j} Q^{jkp}_{obs}} \times 100\% \qquad (47)$$

where m_j is the number of evaluation years at station j, Q^{jkp}_{sim} and Q^{jkp}_{obs} are the simulated and observed peak discharges, respectively, for evaluation year k at station j, T^{jkp}_{sim} and T^{jkp}_{obs} are the timings of the simulated and observed peaks, respectively, for evaluation year k at station j, and T_c is the SWAT-estimated time of concentration for the watershed.

The value of E_j^2 can range from $-\infty$ to 1.0, with higher values indicating a better overall fit and 1.0 indicating a perfect fit. A negative E_j^2 indicates that for station j the simulated stream flows are less reliable than if one had used the average of the observed stream flows, while a positive value indicates that they are more reliable than using this average. The value of D_{vj} can range from very small negative to very large positive values, with values close to zero indicating a better simulation and zero indicating an exact prediction of the observed volume. In contrast with E_j^2, E_{RRj} can range from 0.0 to $+\infty$, with lower values indicating a better simulation of the observed peak and 0.0 indicating that both the magnitude and timing of the observed peak can be exactly predicted by the model.

SWAT Models for the Minnesota Watersheds

As shown in Table 7-4, the Minnesota jurisdiction in the RRB was divided into 17 modeling domains, leading to 17 SWAT models. These domains coincide with the USGS 8-digit HUCs that are solely located in the jurisdiction (e.g., HUC 09020106 and 09020108) but consist of portions of the HUCs that located both in the Minnesota and North Dakota jurisdictions (e.g., HUC 09020101 and 09020107). Table 7-5 summarizes the measuring statistics of the model performances.

For the calibration year of 1997, the E_j^2 values for most evaluation stations are greater than 0.36, indicating that the SWAT models were calibrated to have a satisfactory simulation performance. Because these values are comparable with, or greater than, that reported in the literature, these models are considered to be sufficiently calibrated, i.e., given the available data, the further improvement of the model performance would be very limited. The poor model performance at station USGS 05078230 might be attributed to that there was insufficient data on the large amount of marshes/wetlands (e.g., storage volumes and geographic locations) located in HUC 09020305 (the Clearwater River watershed). The E_j^2 value at station USGS 05067500 is slightly lower than 0.36 because the data on the water diverted from HUC 09020108 (the Eastern Wild Rice River watershed) to HUC 09020107 (the Marsh River watershed) was unavailable and assumed to be a constant computed based on the geometry of the diversion channel, leading to 18.8% overestimation of the total stream flow for the calibration period. In addition, the models did a good job on reproducing the total runoff volumes observed at most of the stations. The large prediction errors for HUC 09020103 (the Otter Tail River watershed), HUC 09020104 (the Upper Red River watershed), HUC 09020305, and HUC 09020314 (The Roseau River watershed) might be caused by the inaccurate data on marshes/wetlands, whereas, the large prediction error at station USGS 05067500 might be because there was no data on the water diverted from HUC 09020108. Further, the low E_{RRj} values indicate that the models can accurately predict both the magnitudes and timings of the peaks observed at most of the stations. Again, the poor prediction of the peak at station USGS 05112000 was probably resulted from the insufficient data on marshes/wetlands.

For the validation periods, the three statistics exhibit large variations (Table 7-5) across the evaluation stations. This indicates that the models are more robust for some historical flood events than the others. Generally, as expected, the models have a better simulation performance for the flood events occurred in 1970s than the ones occurred in 1960s. Compared with 1970s, the watershed conditions in 1960s were likely more different from the ones used to set up and calibrate the models. Nevertheless, for most stations, the values of the statistics are comparable with that reported in the literature. Hence, the models are considered to be reliable for predicting stream flows and peaks of the typical historical flood events.

To further evaluate the model performance, Figures 7-4 to 7-8 give the plots showing the model predicted versus observed flow hydrographs at selected stations for the calibration year of 1997. These stations were selected because the models had noticeably different simulation performances as indicated by the E_j^2 values ranging from 0.27 to 0.86. The visual inspection revealed that for the stations with a high E_j^2 value, the models successfully reproduced both the peaks and total stream flow volumes. For the stations with a low E_j^2 value, the models successfully reproduced the peaks but tended to have a large prediction error of the total stream flow volumes. A close examination of the flow hydrographs at the other evaluation stations (not shown to be concise) indicated a similar prediction pattern. Again, the insufficiency of the data for characterizing the marshes and wetlands might be one reason for the inaccurate predictions of the total stream volumes. Another reason might be that the generated data to fill the missing values for daily precipitation and minimum and maximum temperatures could not accurately represent the weather conditions actually occurred. Nevertheless, the models were judged to be accurate enough for evaluating the effects of the Waffle® on reducing the 1997-type flood, which is the focus of the study. For an "interest" location within a given watershed, the effect is defined as the change of the peak before and after the Waffle®.

Table 7-5. Nash-Sutcliffe coefficient (E_j^2), deviation of volume (D_{vj}), and error function (E_{RRj}) of the Minnesota SWAT models

Modeling Domain	Calibration Parameters	Calibration			Validation		
		E_j^2	D_{vj} (%)	E_{RRj} (%)	E_j^2	D_{vj} (%)	E_{RRj} (%)
HUC 09020101	SMFMX (6.9), TIMP (0.3), SMTMP (1.5), SURLAG (5), MSK_CO1 (0.6), MSK_CO2 (0.6), ESCO (0.95), GWQMN (0), CN2 (-5%)	–	–	–	–	–	–
HUC 09020102	SMFMX (6.9), TIMP (0.3), SMTMP (1.5), SURLAG (5), MSK_CO1 (1.7), MSK_CO2 (1.7), ESCO (0.95), GWQMN (0), CN2 (-5%)	–	–	–	–	–	–
HUC 09020103	SMFMX (5.43), TIMP (0.88), SMTMP (0.278), SURLAG (24), MSK_CO1 (15.046), MSK_CO2 (15.161), ESCO (0.95), GWQMN (0), CN2 (-10%)	0.47 ~ 0.67	23 ~ 28	13.2 ~ 20.3	-0.56 ~ 0.73	3 ~ 57	22.5 ~ 53.1
HUC 09020104	SMFMX (7.5), TIMP (0.6), SMTMP (1.5), SURLAG (2.5), MSK_CO1 (1.2), MSK_CO2 (1.2), ESCO (0.95), GWQMN (0), CN2 (-10%)	0.47 ~ 0.61	28 ~ 37	10.4 ~ 12.5	0.61 ~ 0.92	0.8 ~ 53	1.6 ~ 45.9
HUC 0920106[1]	SMFMX (6.9), TIMP (0.1 ~ 0.15), SMTMP (1.0 ~ 1.5), SURLAG (0.3 ~ 0.82), MSK_CO1 (0.35 ~ 1.2), MSK_CO2 (0.35~1.2), ESCO (0.95), GWQMN (0), CN2 (-6.4 ~ +7.0)	0.76	-1.9	1.1	-1.23 ~ 0.90	-179.0 ~ 0.4	1.1 ~ 114.4
HUC 09020107	SMFMX (6.9), TIMP (0.1), SMTMP (1.5), SURLAG (3.5), MSK_CO1 (1.2), MSK_CO2 (1.2), ESCO (0.95), GWQMN (0), CN2 (+8.0)	0.27	18.8	7.7	0.54 ~ 0.89	-9.8 ~ 2.5	2.9 ~ 15.4
HUC 09020108	SMFMX (10), TIMP (0.6), SMTMP (3.5), SURLAG (1.5), MSK_CO1 (1.2), MSK_CO2 (1.2), ESCO (0.95), GWQMN (0), CN2 (+3.0)	0.61 ~ 0.86	-0.1 ~ 2.3	2.2 ~ 7.0	0.11 ~ 0.79	-21.4 ~ 16.6	3.3 ~ 76.4
HUC 09020301	SMFMX (7), TIMP (0.9), SMTMP (0.5), SURLAG (15), MSK_CO1 (2.8), MSK_CO2 (2.8), ESCO (0.95), GWQMN (0), CN2 (default)	0.54	4.8	2.9	0.59 ~ 0.90	-21.5 ~ - 1.1	1.0 ~ 7.7
HUC 09020302	SMFMX (6.9), TIMP (0.3), SMTMP (1.2), SURLAG (12), MSK_CO1 (1.2), MSK_CO2 (1.2), ESCO (0.95), GWQMN (0), CN2 (default)	–	–	–	–	–	–
HUC09020303	SMFMX (7.9), TIMP (0.3), SMTMP (2.5), SURLAG (1), MSK_CO1 (3.9), MSK_CO2 (3.9), ESCO (0.95), GWQMN (0), CN2 (default), CH_S2 (+0.006), CH_L2 (-20%)	0.76	-5.7 ~ - 1.7	4.6 ~ 10.6	-0.14 ~ 0.89	-22.2 ~ 23.0	1.4 ~ 53.9
HUC 09020304	SMFMX (7.9), TIMP (0.3), SMTMP (2.5), SURLAG (1), MSK_CO1 (3.9), MSK_CO2 (3.9), ESCO (0.95), GWQMN (0), CN2 (default), CH_S2 (+0.004), CH_L2 (-70%)	0.86	-3.5	15.4	0.54 ~ 0.89	-3.9 ~ 38.5	5.3 ~ 68.6
HUC 09020305	SMFMX (7.9), TIMP (0.3), SMTMP (2.5), SURLAG (1), MSK_CO1 (3.9), MSK_CO2 (3.9), ESCO (0.95), GWQMN (0), CN2 (default), CH_S2 (+0.002), CH_K2 (+0.02), ALPHA_BF (-0.02)	-0.04 ~ 0.66	-41.7 ~ - 12.6	18.3 ~ 26.1	-1.66 ~ 0.86	-48.3 ~ 28.7	0.1 ~ 49.7

Table 7-5. Continued

Modeling Domain	Calibration Parameters	*Calibration*			*Validation*		
		E_j^2	D_{vj} (%)	E_{RRj} (%)	E_j^2	D_{vj} (%)	E_{RRj} (%)
HUC 09020306	SMFMX (7.0), TIMP (0.4), SMTMP (4.0), SURLAG (1), MSK_CO1 (3.9), MSK_CO2 (3.9), ESCO (0.95), GWQMN (0), CN2 (default)	–	–	–	–	–	–
HUC 09020309	SMFMX (7.0), TIMP (0.15), SMTMP (3.0), SURLAG (2), MSK_CO1 (1.2), MSK_CO2 (1.2), ESCO (0.95), GWQMN (0), CN2 (default)	0.57	0.2	2.8	0.27 ~ 0.88	-5.1 ~ 23.4	1.9 ~ 23.4
HUC 09020311	SMFMX (7.0), TIMP (0.15), SMTMP (3.0), SURLAG (2), MSK_CO1 (1.2), MSK_CO2 (1.2), ESCO (0.95), GWQMN (0), CN2 (default)	–	–	–	–	–	–
HUC 09020312	SMFMX (7.5), TIMP (0.35), SMTMP (1.0), SURLAG (4), MSK_CO1 (3.5), MSK_CO2 (3.5), ESCO (0.95), GWQMN (0), CN2 (default)	0.59	16.9	5.6	0.40 ~ 0.86	1.8 ~ 45.1	1.1 ~ 60.4
HUC 09020314	SMFMX (6.9), TIMP (0.3), SMTMP (1.5), SURLAG (1.4), MSK_CO1 (1.2), MSK_CO2 (1.2), ESCO (0.95), GWQMN (0), CN2 (default)	0.77	-22.4	60.7	-0.72 ~ 0.75	-68.6 ~ 43.1	52.6 ~ 119.3

[1] The watershed was modeled using two separate SWAT models: one for the drainage area upstream of Sabin (USGS 05061500) and another for the remaining drainage area.

Table 7-6. Nash-Sutcliffe coefficient (E_j^2), deviation of volume (D_{vj}), and error function (E_{RRj}) of the North Dakota SWAT models

Modeling Domain	Calibration Parameters	*Calibration*			*Validation*		
		E_j^2	D_{vj} (%)	E_{RRj} (%)	E_j^2	D_{vj} (%)	E_{RRj} (%)
HUC 09020101[1]	SMFMX (8), TIMP (0.4 ~ 0.95), SMTMP (0), SURLAG (1.3 ~ 4), MSK_CO1 (0.8 ~ 1.2), MSK_CO2 (1.2 ~ 3.0), ESCO (0.95), GWQMN (0), CN2 (+5%)	0.77	1.8	17.2	–	–	–
HUC 09020105	SMFMX (6.0), TIMP (0.15), SMTMP (3), SURLAG (2), MSK_CO1 (1), MSK_CO2 (1), ESCO (0.95), GWQMN (0), CN2 (default)	0.55 ~ 0.64	-5.7 ~ 25.9	7.1 ~ 37.4	–	–	–
HUC 09020107	SMFMX (10.0), TIMP (0.4), SMTMP (3.5), SURLAG (1.5), MSK_CO1 (0.2), MSK_CO2 (0.2), ESCO (0.95), GWQMN (0), CN2 (default)	–	–	–	0.50 ~ 0.73	-14.6 ~ 12.2	7.1 ~ 32.6
HUC 09020109	SMFMX (7.0), TIMP (0.3), SMTMP (2.5), SURLAG (15.0), MSK_CO1 (1.2), MSK_CO2 (1.2), ESCO (0.95), GWQMN (0), CN2 (+8), ALPHA_BF (+0.3)	0.55 ~ 0.65	-3.6 ~ -2.1	7.1 ~ 27.8	-4.96 ~ 0.76	-18.0 ~ 270.6	4.7 ~ 60.6

Table 7-6. Continued

Modeling Domain	Calibration Parameters	*Calibration*			*Validation*		
		E_j^2	D_{vj} (%)	E_{RRj} (%)	E_j^2	D_{vj} (%)	E_{RRj} (%)
HUC 09020202	SMFMX (10.0), TIMP (0.9), SMTMP (2.5), SURLAG (1.0), MSK_CO1 (0.6), MSK_CO2 (0.6), ESCO (0.95), GWQMN (0), CN2 (+10%)	0.18	5.1	12.9	–	–	–
HUC 09020203	SMFMX (10.0), TIMP (0.5), SMTMP (4.0), SURLAG (1.0), MSK_CO1 (1.8), MSK_CO2 (1.8), ESCO (0.95), GWQMN (0), CN2 (+10%)	-0.01 ~ 0.63	-88.5 ~ -44.2	33.8 ~ 97.5	–	–	–
HUC 09020204	SMFMX (8), TIMP (0.172), SMTMP (2.5), SURLAG (1), MSK_CO1 (1.8), MSK_CO2 (1.8), ESCO (0.95), GWQMN (0), CN2 (+3.0)	0.87 ~ 0.91	-9.2 ~ -5.6	7.4 ~ 21.2	–	–	–
HUC 09020205	SMFMX (8), TIMP (0.05), SMTMP (1.2), SURLAG (0.85), MSK_CO1 (0.65), MSK_CO2 (0.65), ESCO (0.95), GWQMN (0), CN2 (default)	0.65 ~ 0.74	-28.9 ~ -2.7	26.7 ~ 39.0	–	–	–
HUC 09020301	SMFMX (8), TIMP (0.2), SMTMP (3.5), SURLAG (4), MSK_CO1 (1), MSK_CO2 (3), ESCO (0.95), GWQMN (0), CN2 (default)	–	–	–	–	–	–
HUC 09020307	SMFMX (6.5), TIMP (0.25), SMTMP (1.5), SURLAG (1), MSK_CO1 (3.5), MSK_CO2 (3.5), ESCO (0.95), GWQMN (0), CN2 (+3.0)	0.90	-9.9	10.9	–	–	–
HUC 09020308	SMFMX (6.0), TIMP (0.9), SMTMP (2.8), SURLAG (1), MSK_CO1 (0.5), MSK_CO2 (3.5), ESCO (0.95), GWQMN (0), CN2 (default)	0.67 ~ 0.69	-7.6 ~ 15.2	14.6 ~ 18.3	–	–	–
HUC 09020310	SMFMX (8), TIMP (0.5), SMTMP (4), SURLAG (2), MSK_CO1 (3), MSK_CO2 (1.3), ESCO (0.95), GWQMN (0), CN2 (default)	0.77	-20.3	31.9	–	–	–
HUC 09020311	SMFMX (6.5), TIMP (0.15), SMTMP (1.5), SURLAG (1), MSK_CO1 (1.2), MSK_CO2 (1.2), ESCO (0.95), GWQMN (0), CN2 (default)	–	–	–	–	–	–
HUC 09020313	SMFMX (6.5), TIMP (0.15), SMTMP (1.5), SURLAG (1), MSK_CO1 (1.2), MSK_CO2 (1.2), ESCO (0.95), GWQMN (0), CN2 (default)	0.75 ~ 0.97	-12.9 ~ 16.4	5.2 ~ 14.7	–	–	–

SWAT Models for the North Dakota Watersheds

As shown in Table 7-4, the North Dakota jurisdiction in the RRB was divided into 14 modeling domains, leading to 14 SWAT models. These domains coincide with the USGS 8-digit HUCs that are solely located in the jurisdiction (e.g., HUC 09020109 and 09020308) but consist of portions of the HUCs that located both in the Minnesota and North Dakota jurisdictions (e.g., HUC 09020101 and 09020107). Table 7-6 summarizes the measuring statistics of the model performances.

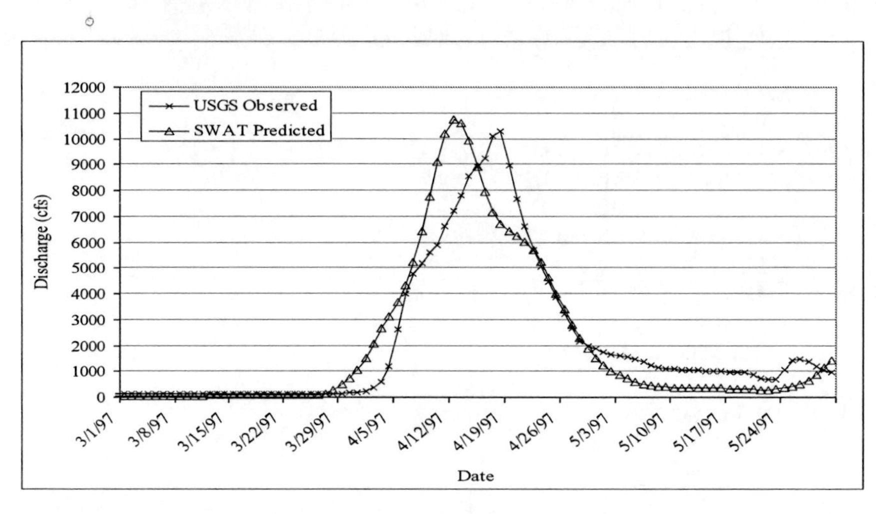

Figure 7-4. The model predicted versus observed stream flow hydrographs at Hendrum (USGS 05064000) in the Eastern Wild Rice River watershed (HUC 09020108). The Nash-Sutcliffe coefficient $E_j^2 = 0.86$.

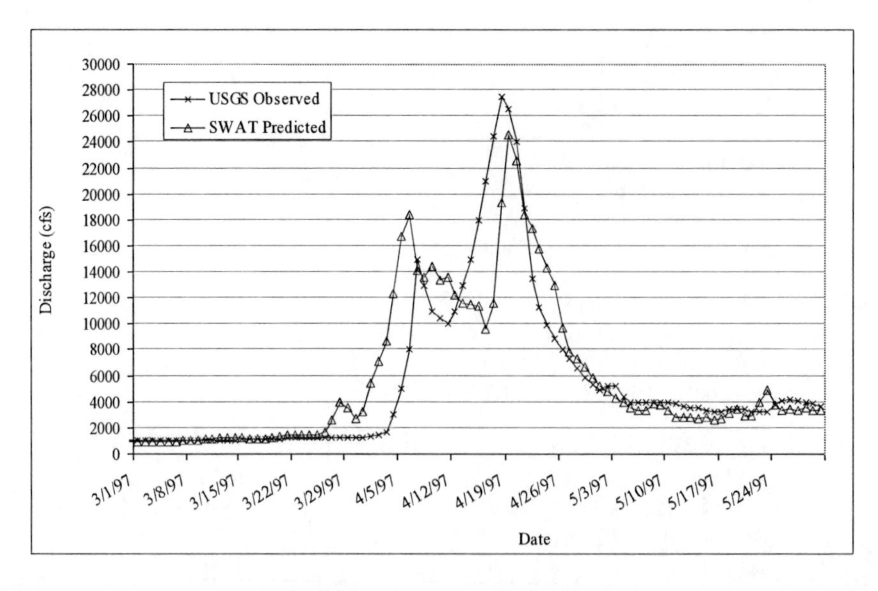

Figure 7-5. The model predicted versus observed stream flow hydrographs at Crookston (USGS 05079000) in the Red Lake River watershed (HUC 09020303). The Nash-Sutcliffe coefficient $E_j^2 = 0.76$.

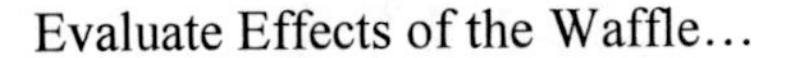

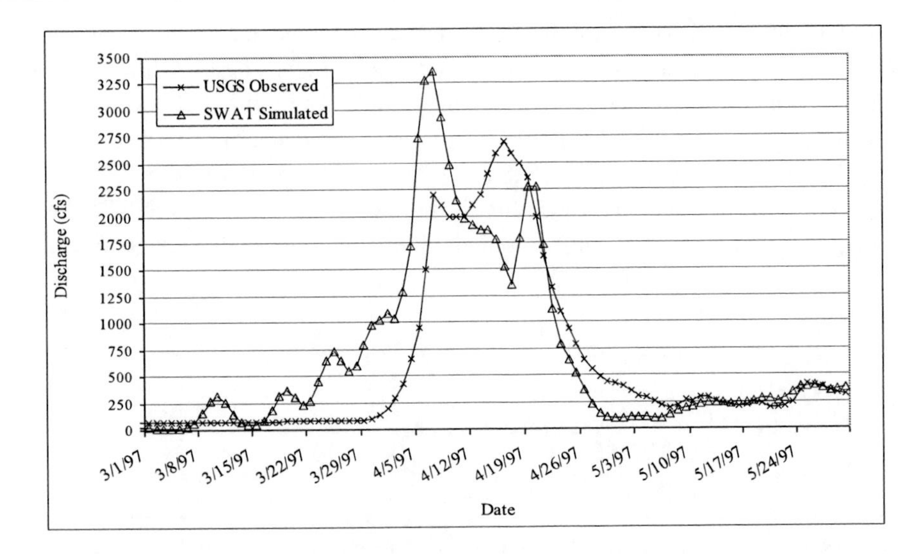

Figure 7-6. The model predicted versus observed stream flow hydrographs at Plummer (USGS 05078000) in the Clearwater River watershed (HUC 09020305). The Nash-Sutcliffe coefficient E_j^2 = 0.66.

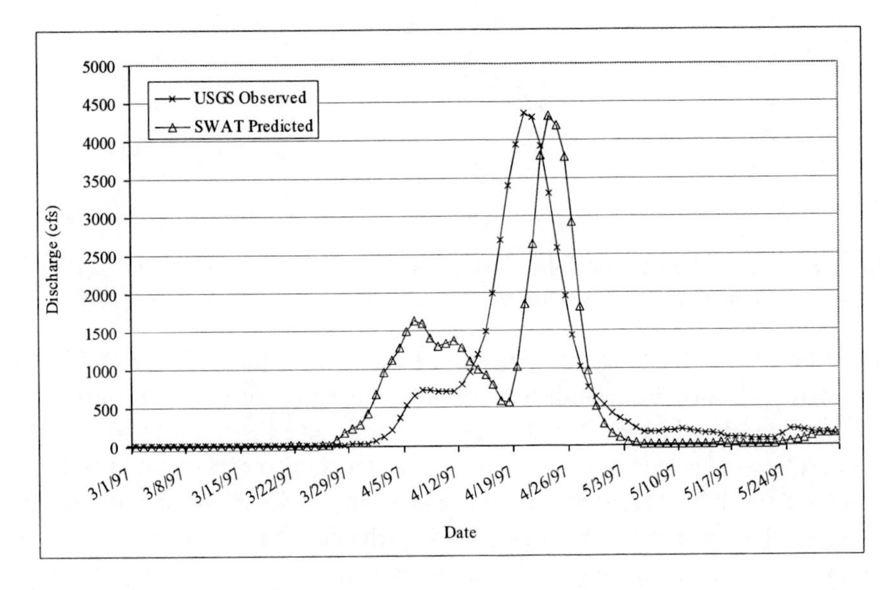

Figure 7-7. The model predicted versus observed stream flow hydrographs at Climax (USGS 05069000) in the Sandhill River watershed (HUC 09020301). The Nash-Sutcliffe coefficient E_j^2 = 0.54.

For the calibration year of 1997, the E_j^2 values for most evaluation stations are greater than 0.55, indicating that the SWAT models were calibrated to have a satisfactory simulation performance. Because these values are comparable with, or greater than, that reported in the literature, these models are considered to be sufficiently calibrated, i.e., given the available data, the further improvement of the model performance would be very limited. The poor model performance at station USGS 05056000 can be attributed to that the SWAT model for the modeling domain HUC 09020203 might be attributed to that there was insufficient data on the large amount of marshes/wetlands (e.g., storage volumes and geographic locations) located in HUC 09020203 (the Middle Sheyenne River watershed). The model noticeably underestimated the total stream flows for this modeling domain. The low E_{RRj} values indicate that the models can accurately predict both the magnitudes and timings of the peaks observed at most of the stations. The poor prediction of peaks at some stations was probably resulted from the insufficient data on marshes/wetlands (e.g., USGS 05056000 and 05057000 in the Middle Sheyenne River watershed) and due to the SWAT's inability to handle the ice jam occurred in channels (e.g., USGS 05059700 and 05060100 in the Maple River watershed).

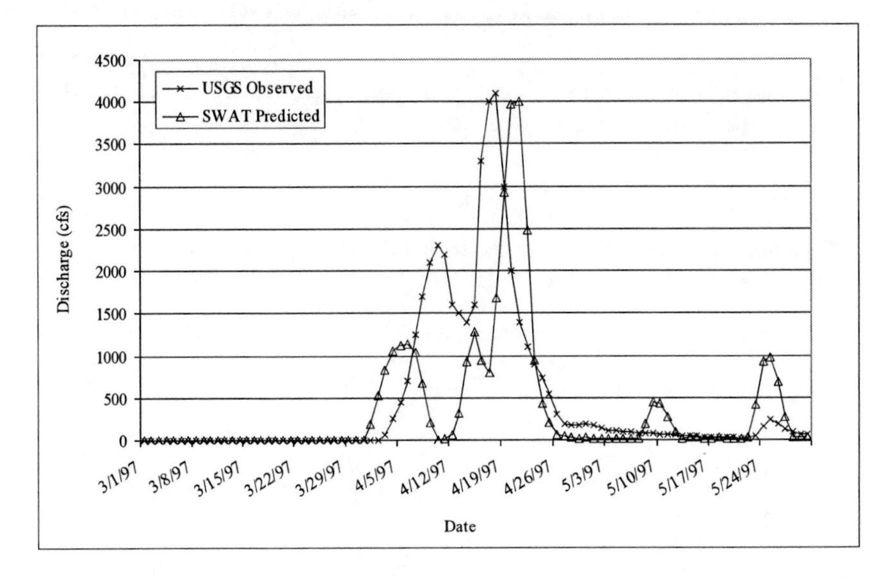

Figure 7-8. The model predicted versus observed stream flow hydrographs at Shelly (USGS 05067500) in the Marsh River watershed (HUC 09020107). The Nash-Sutcliffe coefficient $E_j^2 = 0.27$.

Limited by time and resources, only the SWAT models for HUC 09020107 and HUC 09020109 were validated in accordance with the other historical floods (Table 7-6). The validation indicates that the models are more robust for some historical flood events than the others. Generally, as expected, the models have a better simulation performance for the flood events occurred in 1970s than the ones occurred in 1960s. Compared with 1970s, the watershed conditions in 1960s were likely more different from the ones used to set up and calibrate the models. Nevertheless, for most stations, the values of the statistics are comparable with that reported in the literature. Hence, the models are considered to be reliable for predicting stream flows and peaks of the typical historical flood events.

To further evaluate the model performance, Figures 7-9 to 7-12 give the plots showing the model predicted versus observed flow hydrographs at selected stations for the calibration year of 1997. These stations were selected because the models had noticeably different simulation performances as indicated by the E_j^2 values ranging from -0.01 to 0.91. The visual inspection revealed that for the stations with a high E_j^2 value, the models successfully reproduced both the peaks and total stream flow volumes. For the stations with a low E_j^2 value, the models successfully reproduced the peaks but tended to have a large prediction error of the total stream flow volumes. A close examination of the flow hydrographs at the other evaluation stations (not shown to be concise) indicated a similar prediction pattern.

Again, the insufficiency of the data for characterizing the marshes and wetlands might be one reason for the inaccurate predictions of the total stream volumes. Another reason might be that the generated data to fill the missing values for daily precipitation and minimum and maximum temperatures could not accurately represent the weather conditions actually occurred. Nevertheless, the models were judged to be accurate enough for evaluating the effects of the Waffle® on reducing the 1997-type flood, which is the focus of the study. For an "interest" location within a given watershed, the effect is defined as the change of the peak before and after the Waffle®.

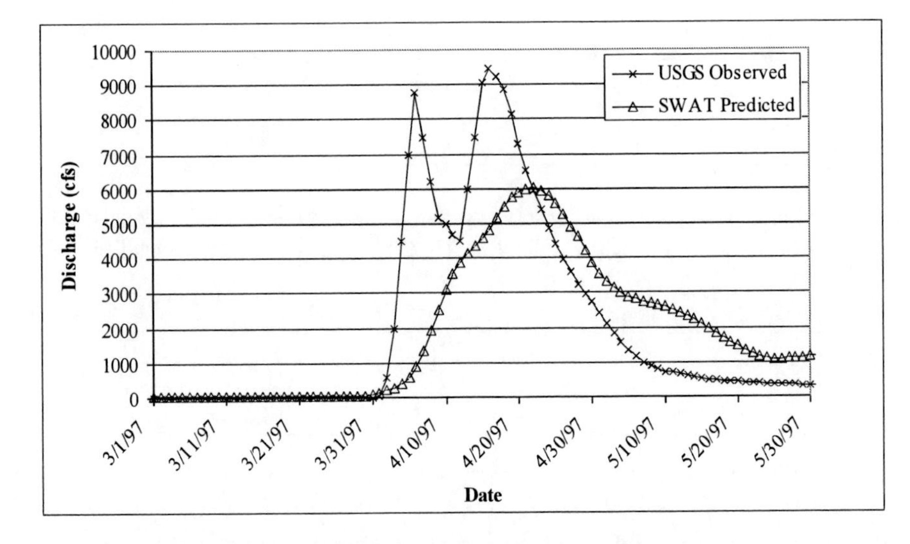

Figure 7-9. The model predicted versus observed stream flow hydrographs at Ambercrombie (USGS 05053000) in the Western Wild Rice River watershed (HUC 09020105). The Nash-Sutcliffe coefficient $E_j^2 = 0.64$.

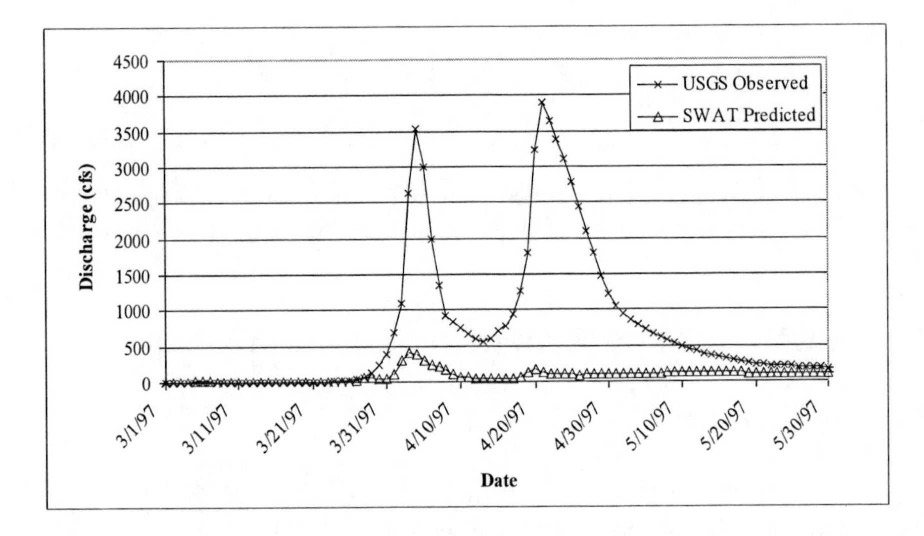

Figure 7-10. The model predicted versus observed stream flow hydrographs at Warrick (USGS 05056000) in the Middle Sheyenne River watershed (HUC 09020203). The Nash-Sutcliffe coefficient $E_j^2 = -0.01$.

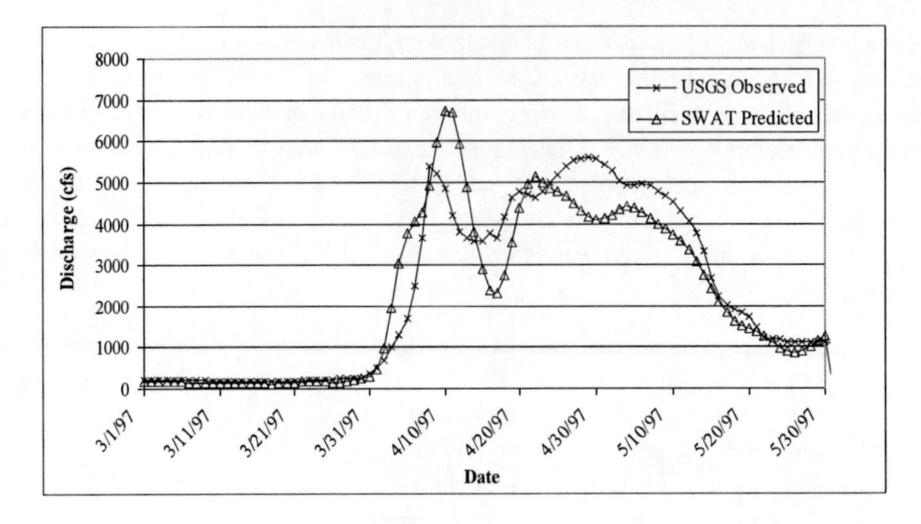

Figure 7-11. The model predicted versus observed stream flow hydrographs at Kindred (USGS 05059000) in the Lower Sheyenne River watershed (HUC 09020204). The Nash-Sutcliffe coefficient $E_j^2 = 0.91$.

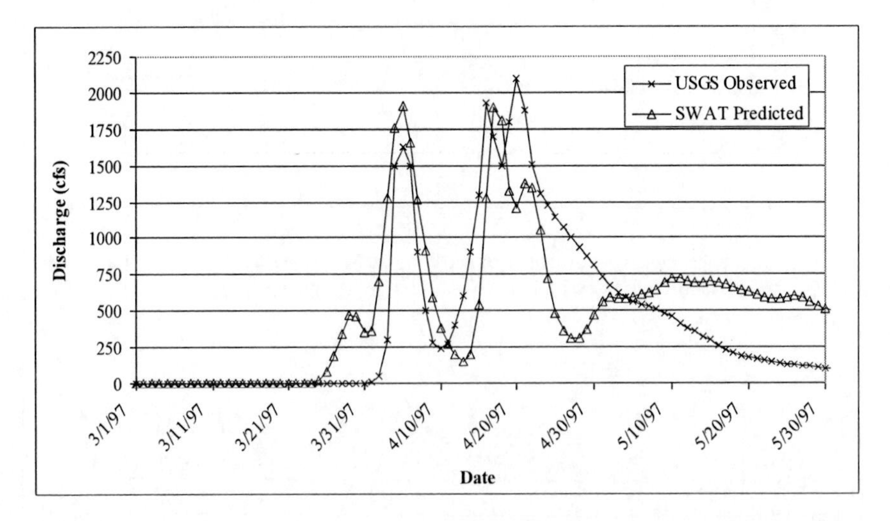

Figure 7-12. The model predicted versus observed stream flow hydrographs at Minto (USGS 05085000) in the Forest River watershed (HUC 09020308). The Nash-Sutcliffe coefficient $E_j^2 = 0.69$.

Modeled Flow Reductions in the Watersheds

Waffle® Scenarios

For each modeling domain (Table 7-4), three Waffle® scenarios were generated and evaluated. Scenario I (S-I) considers 100% of the identified storage, whereas, Scenario II (S-II) and Scenario III (S-III) evaluate 75% and 50% of the identified storage, respectively. Table 7-7 lists the Waffle® storages of the three scenarios for the modeling domains, except for HUC 09020202, HUC 09020203, HUC 09020302, and HUC 09020314. The runoff generated in HUC 09020202 and HUC 09020203 is regulated by the Baldhill Dam, which could offset effects of Waffle® storages in these two watersheds on the flood reduction of RR, whereas, the runoff generated in HUC 09020302 is regulated by the Red Lake Dam and HUC 09020314 does not directly contribute runoff to the RR reach in US. Thus, the Waffle® storages for these four modeling domains were not modeled and are not shown in Table 7-7. In addition, the Waffle® storages for the modeling domain HUC 09020303 (Table 7-7) are the corresponding total volumes for the modeling domains HUC 09020303, HUC 09020304, and HUC 09020305 listed in Tables 7-4 and 7-5. Details on the identification of the potential Waffle® storages across RRB are documented in a separate report.

Readers are reminded that the storages presented in that report are somewhat discrepant from the corresponding values listed in Table 7-7. This discrepancy is caused by three main reasons. Firstly, in that report, the storages are summarized in terms of the USGS 8-digit HUCs provided by NHD, whereas, the corresponding values in Table 7-7 were reported in terms of the modeling domains delineated by SWAT using the 30-m NED data. Although efforts were made to make the delineated boundaries closely match the corresponding ones provided by NHD, a close examination indicated that these two types of boundaries could be offset by as much as 10% in the resulted drainage areas. This small offset is considered acceptable given the course resolution of, and inherent errors in, the NED data. Secondly, the USGS 8-digit HUCs that cover both Minnesota and North Dakota, including Bios de Sioux (09020101), Upper Red (09020104), Elm-Marsh (09020107), Sandhill-Wilson (09020301), and Lower Red (092020311), were split into two modeling domains, which lost 5 to 10% of the derange areas adjacent to RR because of the coarse NED resolution. Finally, the GIS procedure used to clip the RRB Waffle® storage map (reported in the aforementioned separate report) might inappropriately include and/or exclude the sections that intersect with the boundaries of the delineated modeling domains. As a result, the modeled Waffle® storages are less than the corresponding identified values, which would make the analyzed, and presented hereinafter, Waffle® effects on flood reduction more conservative.

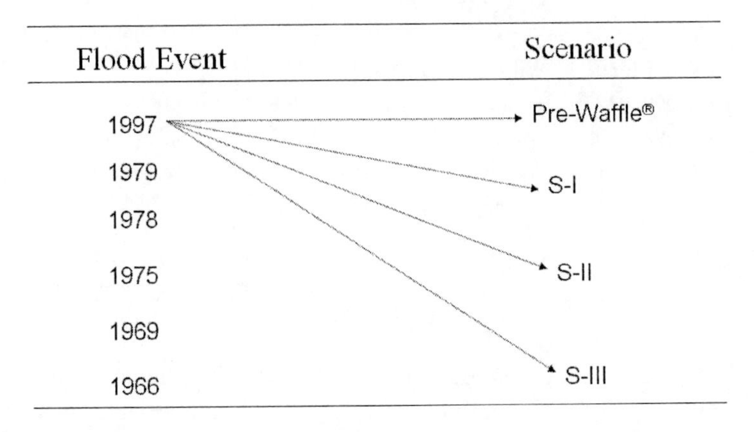

Figure 7-13. Possible 24 model runs for a modeling domain to analyze the three scenarios: Scenario I (S-I), Scenario II (S-II), and Scenario III (S-III). S-I considers 100% of the identified storage, whereas, S-II and S-III evaluate 75% and 50% of the identified storage, respectively.

Table 7-7. The Waffle® storages of the three analyzed scenarios: Scenario I (S-I), Scenario II (S-II), and Scenario III (S-III). S-I considers 100% of the identified storage, whereas, S-II and S-III evaluate 75% and 50% of the identified storage, respectively

State	Modeling Domain	Water-shed	HUC	S-I (ac-ft)	S-II (ac-ft)	S-III (ac-ft)
MN	HUC 09020101	Rabbit	9020101	22,783.87	17,188.09	13,272.48
	HUC 09020102	Mustinka	9020102	6,503.75	5,190.26	3,196.04
	HUC 09020103	Otter Tail	9020103	2,365.02	1,743.68	878.04
	HUC 09020104	Upper Red	9020104	38,934.07	29,434.28	16,684.88
	HUC 09020106	Buffalo	9020106	21,495.07	16,289.70	10,300.54
	HUC 09020107	Marsh	9020107	35,039.74	27,269.58	16,111.29
	HUC 09020108	Wild Rice MN	9020108	20,277.10	15,064.42	10,278.15
	HUC 09020301	Sandhill	9020301	16,272.37	12,775.44	9,482.69
	HUC 09020303	Red Lake	9020303	60,665.30	46,918.60	31,577.28
	HUC 09020306	Grand Marais	9020306	25,161.26	18,821.44	12,444.75
	HUC 09020309	Snake	9020309	12,515.09	9,154.24	5,657.30
	HUC 09020311	Lower Red	9020311	36,057.96	27,359.15	16,092.12
	HUC 09020312	Two Rivers	9020312	18,463.99	14,618.23	8,819.16
ND	HUC 09020101	Bois de Sioux	9020101	3,313.70	2,800.10	1,762.00
	HUC 09020105	Wild Rice	9020105	26,985.40	20,966.30	13,054.00
	HUC 09020107	Elm	9020107	32,695.70	24,725.50	16,591.50
	HUC 09020109	Goose	9020109	20,359.06	14,629.40	11,295.31
	HUC 09020204	Lower Sheyenne	9020204	27,237.50	19,280.10	12,856.40
	HUC 09020205	Mapple	9020205	14,180.70	10,366.20	7,036.60
	HUC 09020301	Wilson	9020301	19,671.80	14,682.00	9,832.00
	HUC 09020307	Turtle	9020307	5,328.50	4,076.60	3,129.20
	HUC 09020308	Forest	9020308	5,643.40	4,596.30	2,804.80
	HUC 09020310	Park	9020310	26,116.40	20,367.30	12,388.30
	HUC 09020311	Lower Red	9020311	16,038.30	12,554.30	7,845.70
	HUC 09020313	Pembina	9020313	9,193.30	7,392.10	5,073.80
Total				523,298.35	398,263.31	258,464.33

The Waffle® effects were measured as reductions of peaks simulated for the post-Waffle® conditions from the pre-Waffle® condition at outlet of, and other selected evaluation locations within, a modeling domain, that is,

$$\text{Effect} = \frac{(\text{Post-Waffle}^{®}\ \text{Peak}) - (\text{Pre-Waffle}^{®}\ \text{Peak})}{(\text{Pre-Waffle}^{®}\ \text{Peak})} \times 100\% \tag{48}$$

To quantify the reductions, for a modeling domain, 24 model runs might be implemented (Figure 7-13). Again, limited by time and resources, the evaluations for the North Dakota modeling domains were only conducted for the 1997 flood event. In addition, the simulated Waffle® effects for these modeling domains were only examined at the corresponding outlets. However, the evaluations for the Minnesota modeling domains were conducted for the six historical flood events and examined at both outlets of, and other selected "interest" points within, the domains.

Representation of Waffle® Storages in the SWAT Models

As mentioned above, the Waffle® was modeled using the pond function in SWAT. With this regard, the identified Waffle® storage areas within a modeling domain were allocated to each of the subbasins, resulting from the subdivision of the modeling domain for modeling purposes. Subsequently, for each subbasin, the allocated storage areas were lumped into one "synthetic" pond. The allocation was implemented by overlaying the SWAT delineated subbasin layer with the storage areas layer (Figure 7-14), whereas, the allocated storage areas were lumped based on an algorithm developed through the project. Assuming that for a 1-mi by 1-mi section, the total storage volume can be proportionally partitioned into the subbasins covering the section, that is, a subbasin covering a larger area of the section can be assigned a greater portion of the storage volume (Figure 7-15), the algorithm can be expressed as:

$$SCV_i^j = \frac{SCA_i^j}{259} \times TWV_j \qquad (49)$$

where SCA_i^j is the area of section j covered by subbasin i (ha); TWV_j is the identified total Waffle® storage volume in section j (ac-ft); and SCV_i^j is the storage volume in section j that is allocated to subbasin i (ac-ft).

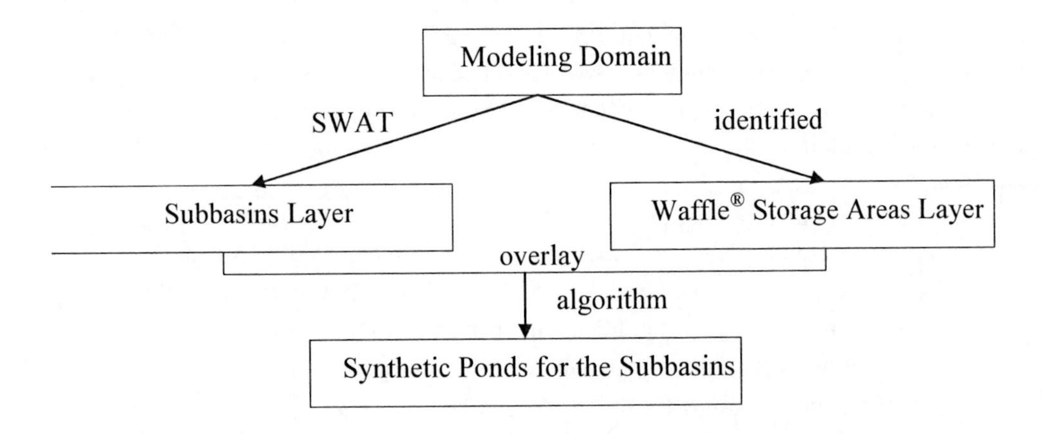

Figure 7-14. Flowchart showing the approach for defining "synthetic" ponds.

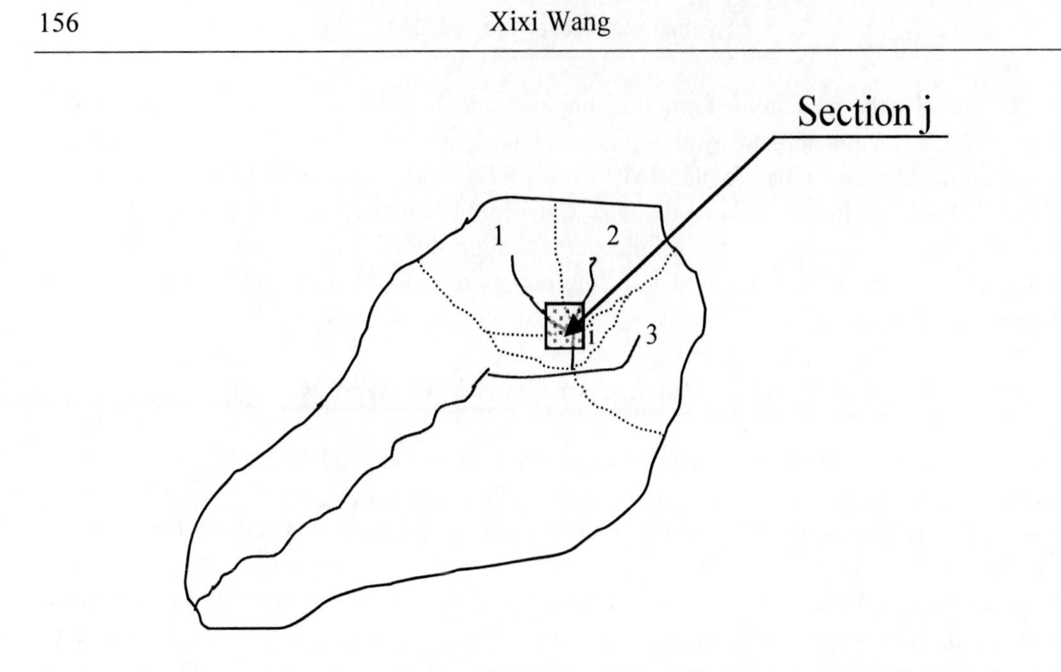

Figure 7-15. Schematic showing section j, which is covered by subbasins 1, 2, …, i with areas of $SCA_1^j, SCA_2^j, \ldots, SCA_i^j$, respectively. $\sum_{m=1}^{i} SCA_m^j = 1\,mi^2$.

The total storage volume in subbasin i, $SynPV_i$, is defined as the summation of the storages of the sections that are both completely and partially included in the subbasin. Thus, the synthetic pond for subbasin i is assumed to have a maximum storage equal to $SynPV_i$. Its corresponding maximum area, $SynPA_i$, is computed as:

$$SynPA_i = \frac{SynPV_i}{\bar{h}} \tag{50}$$

where $\bar{h}$ is the average depth of the Waffle storage areas in the watershed, within which the modeling domain is located. The values for $\bar{h}$ were determined based on the sample sections used to identify the Waffle® storages across RRB and are shown in Table 7-8.

To define a synthetic pond, seven parameters in Equations (7) – (14) need to be determined, including:

- The surface area of the pond when filled to the emergency spillway (SA_{em}),
- The surface area of the pond when filled to the principal spillway (SA_{pr}),
- The volume of water held in the pond when filled to the emergency spillway (V_{em}),
- The volume of water held in the pond when filled to the principal spillway (V_{pr}),
- The fraction of the subbasin area draining into the pond (fr_{imp}),
- The target pond volume for the day (V_{targ}), and
- The number of days required for the pond to reach the target volume (ND_{targ})

Table 7-8. Average depths of the Waffle® storage areas in the watersheds across RRB. Note that a watershed could include two modeling domains listed in Table 7-4

Name	USGS 8-digit Hydrologic Unit Code	Average Depth (m)	Average Depth (ft)
Bois de Sioux	09020101	0.44	1.44
Buffalo	09020106	0.81	2.66
Clearwater	09020305	0.52	1.71
Eastern Wild Rice	09020108	0.69	2.26
Elm-Marsh	09020107	0.63	2.07
Forest	09020308	0.43	1.41
Goose	09020109	0.62	2.03
Grand Marais-Red	09020306	0.36	1.18
Lower Red	09020311	0.42	1.38
Lower Sheyenne	09020204	0.95	3.12
Maple	09020205	0.63	2.07
Middle Sheyenne	09020203	0.60	1.97
Mustinka	09020102	0.35	1.15
Otter Tail	09020103	0.57	1.87
Park	09020310	0.52	1.71
Pembina	09020313	0.59	1.94
Red Lake	09020303	0.37	1.21
Roseau	09020314	0.35	1.15
Sandhill-Wilson	09020301	0.36	1.18
Snake	09020309	0.34	1.12
Tamarac[1]	09020311	0.37	1.21
Thief	09020304	0.48	1.57
Turtle	09020307	0.64	2.10
Two Rivers	09020312	0.34	1.12
Upper Red	09020104	0.45	1.48
Upper Sheyenne	09020202	0.73	2.40
Western Wild Rice	09020105	0.71	2.33
Average across RRB	–	0.53	1.73

[1] This river drains partial area of the Lower Red watershed. The depth should be used for the Waffle® storage areas located in the area drained by the Tamarac River.

Installed to have a top elevation around 1 ft (0.31m) below the lowest point along the roads surrounding a section, the standpipe proposed for the Waffle® functions as an emergency spillway. In addition, the storage areas within the section were identified to have a 1 ft freeboard. Thus, for subbasin i, SA_{em} and V_{em} can be determined as:

$$SA_{em} = SynPA_i \tag{51}$$

$$V_{em} = SynPV_i \tag{52}$$

Further, the lowest inlet elevation of the culvert(s) that would be modified to control water in the section is supposed to function as a principal spillway. Because at this inlet elevation, the surface area and volume of water held in the pond would be negligible, in this study, SA_{pr} and V_{pr} were assumed to have small constant values as:

$$SA_{pr} = 0.1 \text{ ha} \quad (53)$$

$$V_{pr} = 0.81 \text{ ac-ft} = 0.1\times10^4 \text{ m}^3 \quad (54)$$

fr_{imp} was computed as:

$$fr_{imp} = \frac{5\times SynPA_i}{DA_i} \quad (55)$$

where DA_i is the drainage area of subbasin i (ha).

An examination of the historical flow hydrographs indicated that it would make a noticeable difference for the Waffle® to hold water for two to three weeks, which was verified by the field trials conducted in the springs of 2004 and 2005. Also, this storage period is likely to be acceptable for landowner and would cause an unimportant delay to the subsequent planting activity. Hence, ND_{targ} was assumed to be 20 days.

V_{targ} was set as:

$$V_{targ} = \begin{cases} SynPV_i & \text{During the storage period of } ND_{targ} \text{ days} \\ 0.0 & \text{Otherwise} \end{cases} \quad (56)$$

Results and Discussion

For the 1997-type flood, S-I was predicted to result in a reduction of the peaks at outlets of the modeling domains by 0.3 to 59.2%, whereas, S-II and S-III would reduce the peaks by 0.3 to 45.2% and 0.0 to 27.2%, respectively (Table 7-9 and Figures 7-16 and 7-17). The percentage reductions would be overall larger for the watersheds with a greater south-north width than for that with a greater east-west length. For example, the Upper Red River watershed (modeling domain HUC 09020104) has a south-north width much greater than its east-west length and was predicted to have a reduction of 59.2%. The Lower Sheyenne River watershed (modeling domain HUC 09020204), on the other hand, has a south-north width much smaller than its east-west length and was predicted to have a reduction of only 1.4%. One explanation might be that the dominant drainage area of a watershed with a larger width-to-length ratio is adjacent to the watershed outlet. As a result, the effect of the Waffle® storages can be noticed without much dissipation. In contrast, the Waffle® effect for a watershed with a smaller width-to-length ratio tends to dissipate much before the effect can be noticed at the watershed outlet. Another explanation is that a watershed with a greater width-to-length ratio tends to have the overland process more dominant than the channel process, that is, a water drop has a longer travel time on the land than along the associated streams. Thus, the overland runoff could have a higher chance to be intercepted and regulated by the Waffle® storage areas before the runoff would become the concentrated stream flows. Because the Waffle® storage areas are scattered across the watershed and an individual Waffle® (i.e., a section) usually has limited storage capacity, the effect of the storage areas on handling the concentrated stream flows is much lower than that on regulating the corresponding overland runoff. This indicates that it is more important to exploit the cumulative effects of the Waffle® storage areas.

In addition, a watershed with a smaller drainage area tends to have a larger percentage reduction (Tables 7-1 and 7-9). This is also true for the magnitude reduction of the peak. For example, S-I was predicted to reduce the peak at the outlet of the 370 mi^2 Marsh River watershed (modeling domain HUC 09020107) by 2,370 cfs (from 7,910 cfs to 5,540 cfs), which is one of the smallest modeling domains in RRB. Again, this can be attributed to that for the watersheds with a smaller drainage area the overland runoff process might be more dominant than the channel process.

Further, as expected, more Waffle® storage areas always correspond to larger reductions. For all watersheds, S-I was predicted to have more effects than S-II, which in turn was predicted to have more effects than S-III (Table 7-9). The average difference of the effects between two consecutive scenarios (i.e., S-I versus S-II and S-II versus S-III) was predicted to be around 3.2%. The watersheds with a smaller drainage area and/or a greater width-to-length ratio are more sensitive to the change of the Waffle® storage areas. For example, for the Marsh River watershed, the reduction difference between the consecutive scenarios was predicted to be about 10.5%, whereas, for the modeling domain HUC 09020303, which has a much larger drainage area of 3,533 mi^2, the reduction difference would be only 1%. Again, this is an indication that the Waffle® may be more effective on controlling the overland runoff than the concentrated stream flows. Compared with conventional reservoirs, which are usually situated on drainage channels and intercept all upstream stream flows, the Waffle® may reduce flood peaks as a result of the cumulative effects of individual, small storage areas.

Besides, the spatial distribution of the Waffle® storage areas within a watershed (modeling domain) is also important for flood reduction. For S-I, the Rabbit and Buffalo River watersheds were identified to have near-equivalent Waffle® storage volumes (22,783.87 ac-ft versus 21,495.07 ac-ft; Table 7-7). However, the spatial locations of the storage areas within the inclusive watersheds are distinctly different (Figure 7-18). In the Buffalo River watershed, the Waffle® storage areas are primarily located in the lower portion, where the hydrologic processes were dominated by concentrated stream flows. As a result, the Waffle® storages would have a very limited effect, as indicated by the small percentage reduction of 1.4% for the peak at the watershed outlet. In contrast, the Waffle® storage areas in the Rabbit River watershed cover most of the upland areas that have hydrologic processed primarily dominated by overland runoff. This spatial distribution is ideal for achieving flood reduction using the Waffle® concept, as indicated by the large percentage reduction of 19.2% for the peak at the watershed outlet.

The importance of the spatial distribution of the Waffle® storage areas on flood reduction for a watershed can be further verified by examining the percentage reductions at "interest" points within the watershed. Tables 7-10 to 7-24 present the predicted percentage reductions of the peaks as a result of S-I at the selected evaluation points within the Minnesota modeling domains and the Elm River watershed (modeling domain HUC 09020107) in North Dakota. The locations of these evaluation points are depicted in Figures 7-19 to 7-33. For the 1997-

type flood, the predicted flood reductions within the Rabbit River watershed vary from 4.1% at one location (Loc4) to 19.2% at another (Loc2; Figure 7-19 and Table 7-10). For the Red Lake River watershed (modeling domain HUC 09020303), while the S-I was predicted to reduce the peak at the watershed outlet by only 4.9%, the percentage reduction at Loc3 (Figure 7-27) could be as much as 9.7% (Table 7-18). The similar spatial variations can be noticed by examining the results for the other watersheds or modeling domains. Different from the percentage reduction of the peak at the outlet of a watershed, the spatial variation of the flood reductions seems to be irrelevant to the watershed shape as measured by the width-to-length ratio. Instead, the spatial variation is closely related to the spatial distribution of the Waffle® storage areas within the watershed. For a watershed with the storage areas scattered across the drainage area, the corresponding flood reductions tend to exhibit a larger spatial variation. For example, compared with that within the Rabbit River watershed, the predicted flood reductions within the Buffalo River watershed have a spatial variation of less than 0.7% (Figure 7-23 and Table 7-14).

The flood reductions would also vary from one flood event to another (Tables 7-10 to 7-24). For a watershed, the flood reductions tend to be larger for a flood event with a smaller magnitude of peak. For example, at the outlet of the Rabbit River River watershed, the 1969 flood peak (1590 cfs) was much higher than the peaks occurred in 1979 (690 cfs), 1978 (445 cfs), 1975 (575 cfs), and 1966 (520 cfs; Figure 7-19 and Table 7-10). The predicted reduction as a result of the S-I for the 1969-type flood is about 57% of that for the other three historical flood events. In addition, the shape of the flow hydrographs is also a determinative factor for the flood reductions, particularly at evaluation points where the channel process is dominant. For the Rabbit River watershed, although the flood peak in 1969 was larger than that in 1997, the predicted reductions for the 1969-type flood are higher than the corresponding values for 1997 (Table 7-10). The reduction effect of the Waffle® storages is likely to become smaller for flood events with a prolonged rising limb because the water prior to the peak tends to fill the storages, reducing the storage volume available for regulating the peak.

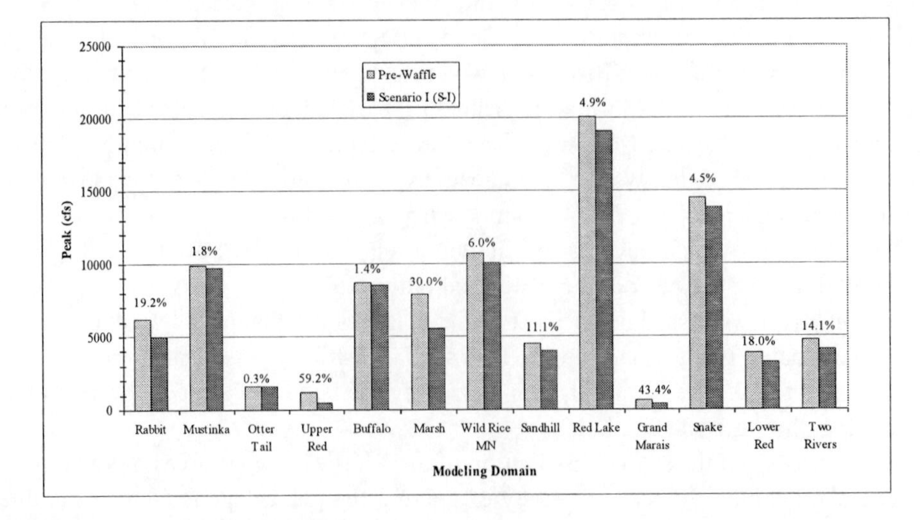

Figure 7-16. Plot showing the predicted reductions of the 1997-type flood peaks at outlets of the modeling domains in Minnesota, resulting from Scenario I.

Table 7-9. Effects of on reducing 1997-type peaks as measured at outlets of the modeling domains for the three Waffle® scenarios

State	Modeling Domain	Watershed	Pre-Waffle®	Scenario I (S-I)		Scenario II (S-II)		Scenario III (S-III)	
			Peak (cfs)	Peak (cfs)	Effect (%)	Peak (cfs)	Effect (%)	Peak (cfs)	Effect (%)
MN	HUC 09020101	Rabbit	6185	5000	19.2	5320	14.0	5458	11.8
	HUC 09020102	Mustinka	9915	9735	1.8	9780	1.4	9830	0.9
	HUC 09020103	Otter Tail	1615	1610	0.3	1610	0.3	1615	0.0
	HUC 09020104	Upper Red	1250	510	59.2	685	45.2	910	27.2
	HUC 09020106	Buffalo	8700	8575	1.4	8610	1.0	8640	0.7
	HUC 09020107	Marsh	7910	5540	30.0	6385	19.3	7215	8.8
	HUC 09020108	Wild Rice MN	10735	10095	6.0	10255	4.5	10405	3.1
	HUC 09020301	Sandhill	4515	4015	11.1	4100	9.2	4250	5.9
	HUC 09020303	Red Lake	20070	19090	4.9	19270	4.0	19540	2.6
	HUC 09020306	Grand Marais	680	385	43.4	450	33.8	500	26.5
	HUC 09020309	Snake	14480	13835	4.5	13995	3.3	14175	2.1
	HUC 09020311	Lower Red	3890	3190	18.0	3360	13.6	3480	10.5
	HUC 09020312	Two Rivers	4775	4100	14.1	4230	11.4	4445	6.9
ND	HUC 09020101	Bois de Sioux	2428	2080	14.3	2084	14.2	2090	13.9
	HUC 09020105	Wild Rice	8529	8084	5.2	8264	3.1	8296	2.7
	HUC 09020107	Elm	4885	3460	29.2	3760	23.0	4120	15.7
	HUC 09020109	Goose	7695	7430	3.4	7508	2.4	7554	1.8
	HUC 09020204	Lower Sheyenne	4775	4708	1.4	4729	1.0	4747	0.6
	HUC 09020205	Mapple	6586	6488	1.5	6516	1.1	6537	0.7
	HUC 09020301	Wilson	5745	4780	16.8	5135	10.6	5477	4.7
	HUC 09020307	Turtle	2265	2168	4.3	2188	3.4	2207	2.6
	HUC 09020308	Forest	2956	2768	6.4	2826	4.4	2906	1.7
	HUC 09020310	Park	7374	6286	14.7	6724	8.8	7335	0.5
	HUC 09020311	Lower Red	3456	2770	19.8	2878	16.7	2999	13.2
	HUC 09020313	Pembina	19205	18680	2.7	18774	2.2	18929	1.4
Ave-rage			6825	6215	13.3	6377	10.1	6546	6.7

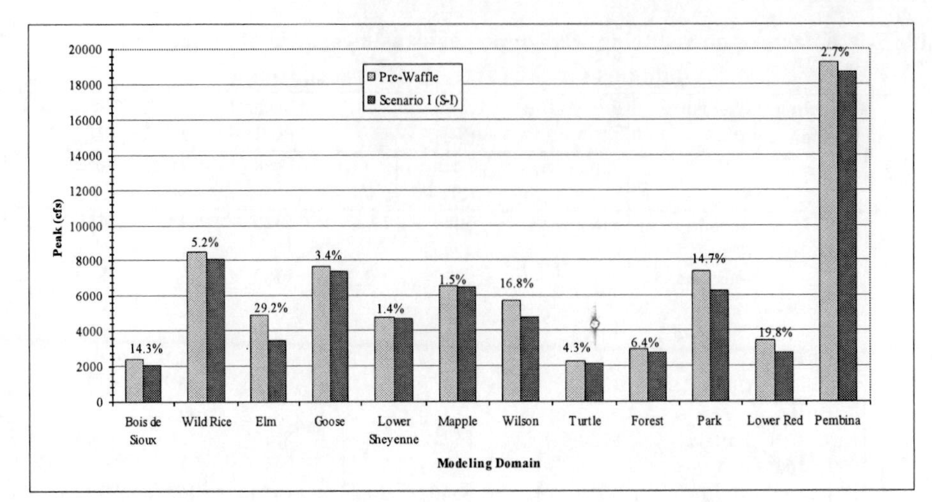

Figure 7-17. Plot showing the predicted reductions of the 1997-type flood peaks at outlets of the modeling domains in North Dakota, resulting from Scenario I.

Table 7-10. SWAT predicted reductions of peaks for the historical floods, resulting from Scenario I (S-I), at the selected "interest" points in the Minnesota modeling domain of HUC 09020101 shown in Figure 7-19

Flood Event	Loc1			Loc2			Loc3			Loc4		
	Pre-Waffle® (cfs)	S-I (cfs)	Redu-tion (%)	Pre-Waffle® (cfs)	S-I (cfs)	Redu-tion (%)	Pre-Waffle® (cfs)	S-I (cfs)	Redu-tion (%)	Pre-Waffle® (cfs)	S-I (cfs)	Redu-tion (%)
1997	6125	5000	-18.4	6185	5000	-19.2	325	295	-9.2	740	710	-4.1
1979	4410	3180	-27.9	4535	3270	-27.9	285	215	-24.6	690	490	-29.0
1978	3235	2480	-23.3	3215	2460	-23.5	175	130	-25.7	445	340	-23.6
1975	3725	2745	-26.3	3635	2675	-26.4	250	180	-28.0	575	425	-26.1
1969	11320	9510	-16.0	11435	9515	-16.8	635	510	-19.7	1590	1360	-14.5
1966	3830	2870	-25.1	3875	2900	-25.2	215	160	-25.6	520	400	-23.1

Table 7-11. SWAT predicted reductions of peaks for the historical floods, resulting from Scenario I (S-I), at the selected "interest" points in the Minnesota modeling domain of HUC 09020102 shown in Figure 7-20

Flood Event	Loc1			Loc2		
	Pre-Waffle® (cfs)	S-I (cfs)	Redu-tion (%)	Pre-Waffle® (cfs)	S-I (cfs)	Redu-tion (%)
1997	8830	8685	-1.6	9915	9735	-1.8
1979	9350	9240	-1.2	9585	9450	-1.4
1978	4490	4390	-2.2	4590	4470	-2.6
1975	7830	7565	-3.4	8115	7870	-3.0
1969	15170	14835	-2.2	15735	15330	-2.6
1966	5635	5415	-3.9	5830	5620	-3.6

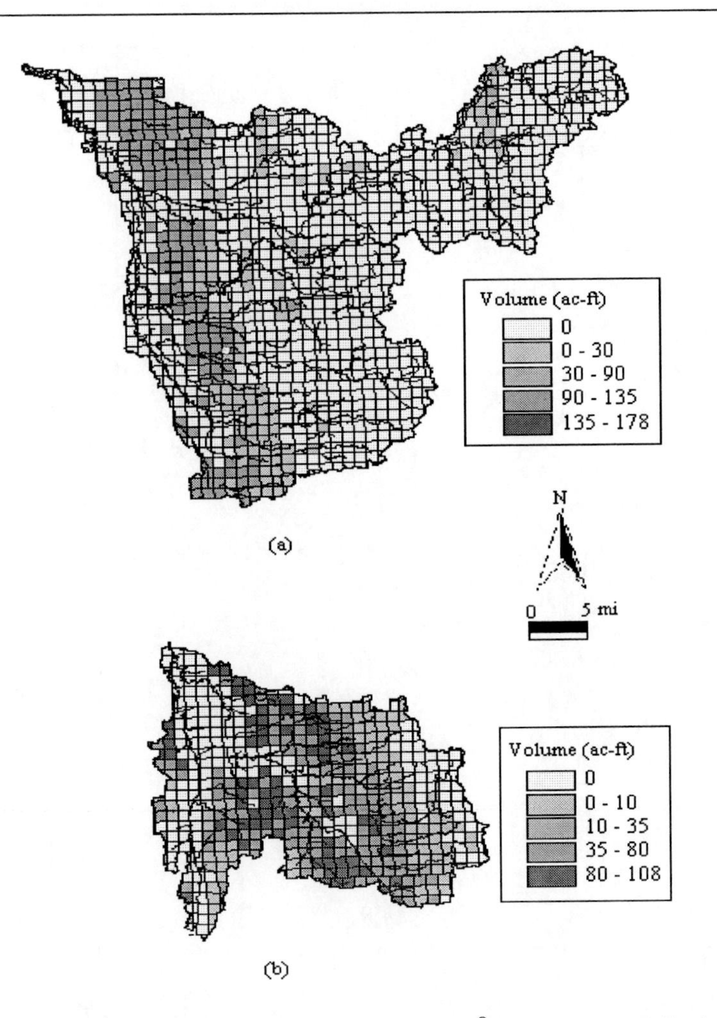

Figure 7-18. Map showing the spatial distributions of the Waffle® storage areas within the (a) Buffalo River watershed (modeling domain HUC 09020106) and (b) Rabbit River watershed (modeling domain HUC 09020101).

Summary

This study evaluated three Waffle® scenarios. Scenario I (S-I) considers 100% of the identified storage, whereas, Scenario II (S-II) and Scenario III (S-III) represent 75% and 50% of the identified storage, respectively. The evaluation was conducted on the basis of the 25 modeling domains or watersheds. For each of the 13 Minnesota watersheds, the evaluation was implemented for six historical floods occurred in 1966, 1969, 1975, 1978, 1979, and 1997, respectively. The predicted Waffle® effects were measured as reductions of peaks at the watershed outlets. In addition, the effects on reducing the 1997-type flood were assessed at "interest" points (i.e., selected evaluation locations) within the watersheds. However, limited by time and resources, the similar detailed assessments could not be implemented for the North Dakota modeling domains except for the Elm River watershed (modeling domain HUC 09020107). For the other 11 watersheds, the evaluation was conducted only to examine reductions of the 1997-type flood peaks at the corresponding watershed outlets.

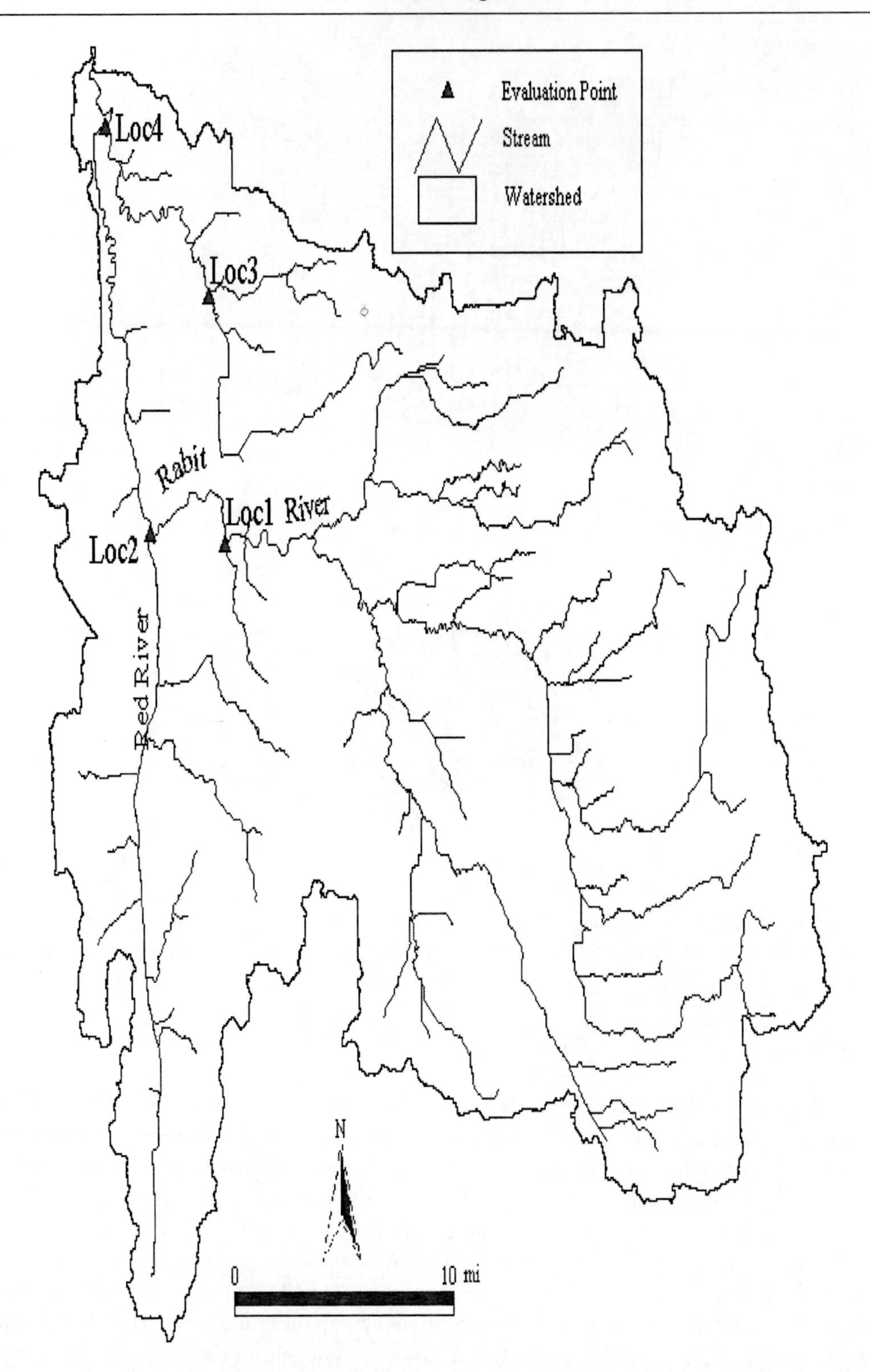

Figure 7-19. Map showing the selected "interest" points in the Minnesota modeling domain of HUC 09020101 (Table 7-4), where the predicted reductions of peaks for the historical floods are presented in Table 7-10, resulting from Scenario I.

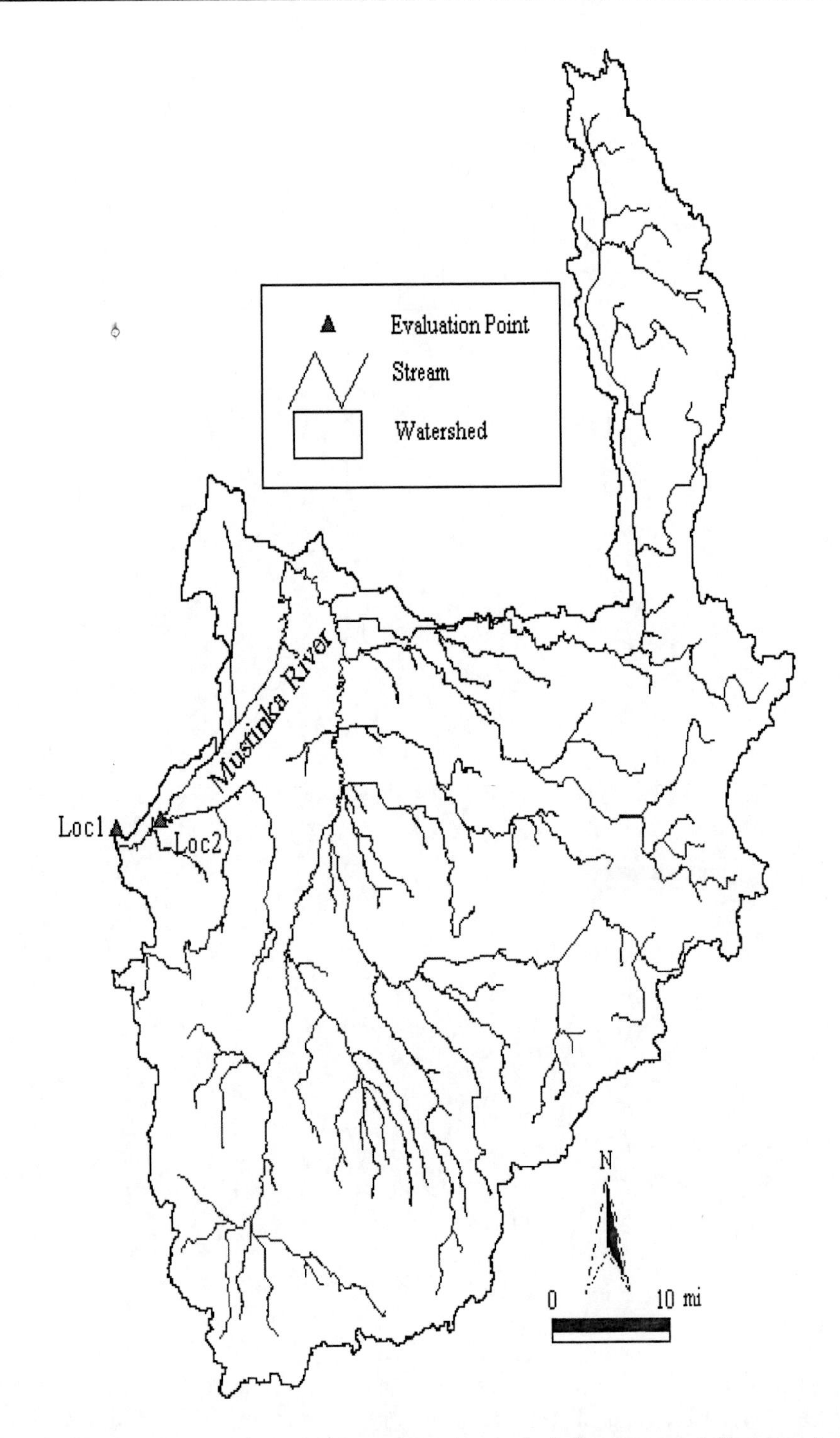

Figure 7-20. Map showing the selected "interest" points in the Minnesota modeling domain of HUC 09020102 (Table 7-4), where the predicted reductions of peaks for the historical floods are presented in Table 7-11, resulting from Scenario I.

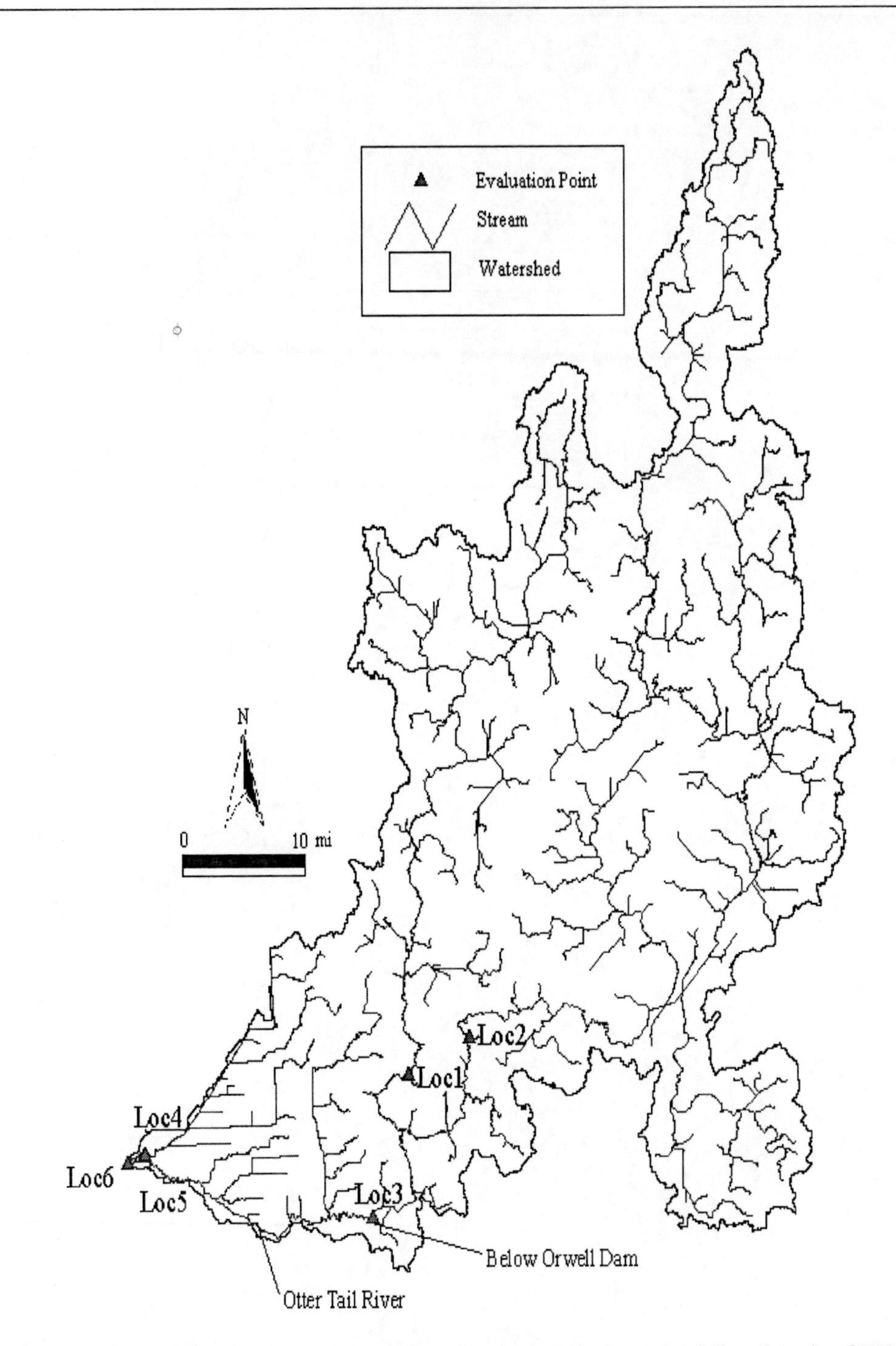

Figure 7-21. Map showing the selected "interest" points in the Minnesota modeling domain of HUC 09020103 (Table 7-4), where the predicted reductions of peaks for the historical floods are presented in Table 7-12, resulting from Scenario I.

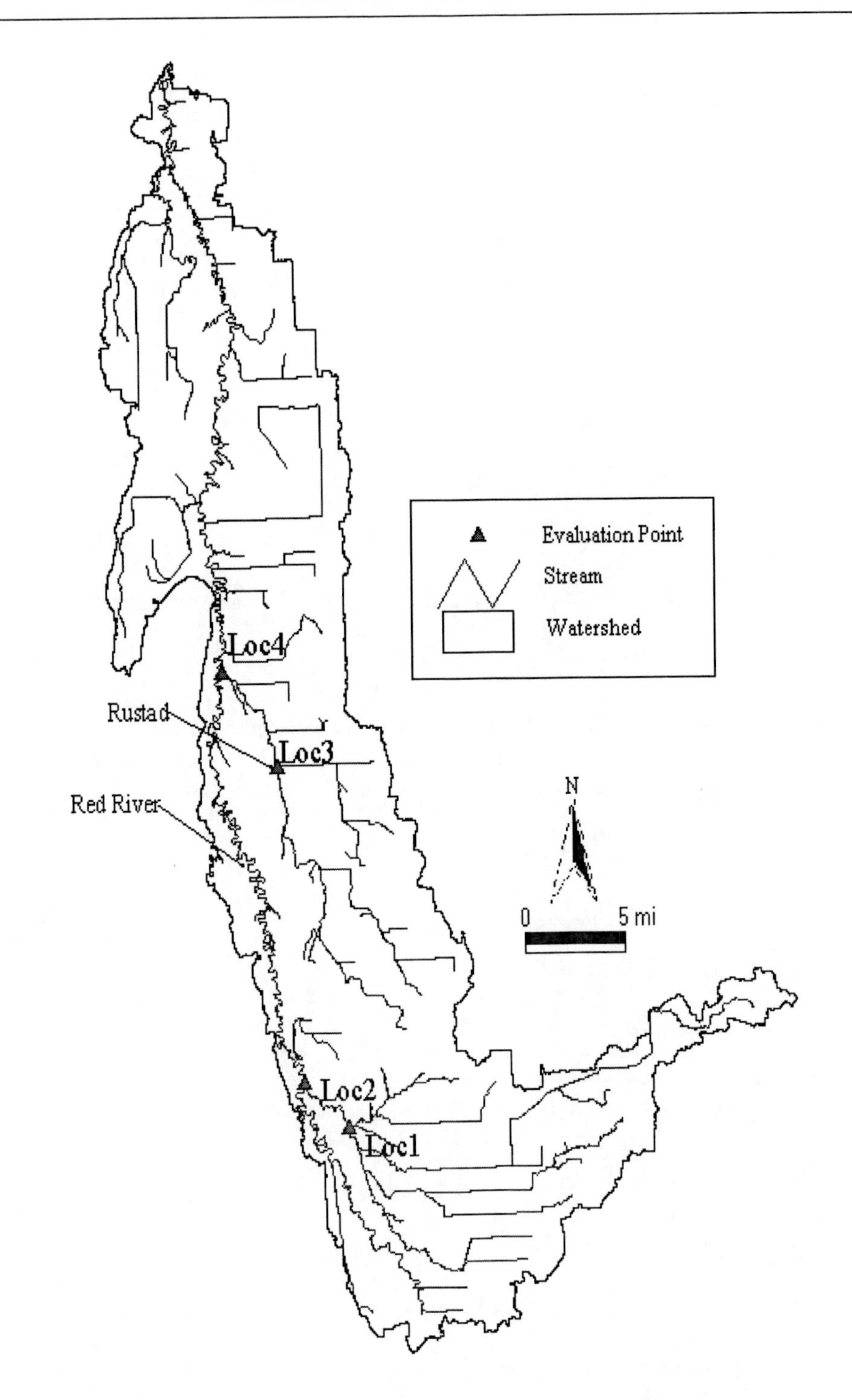

Figure 7-22. Map showing the selected "interest" points in the Minnesota modeling domain of HUC 09020104 (Table 7-4), where the predicted reductions of peaks for the historical floods are presented in Table 7-13, resulting from Scenario I.

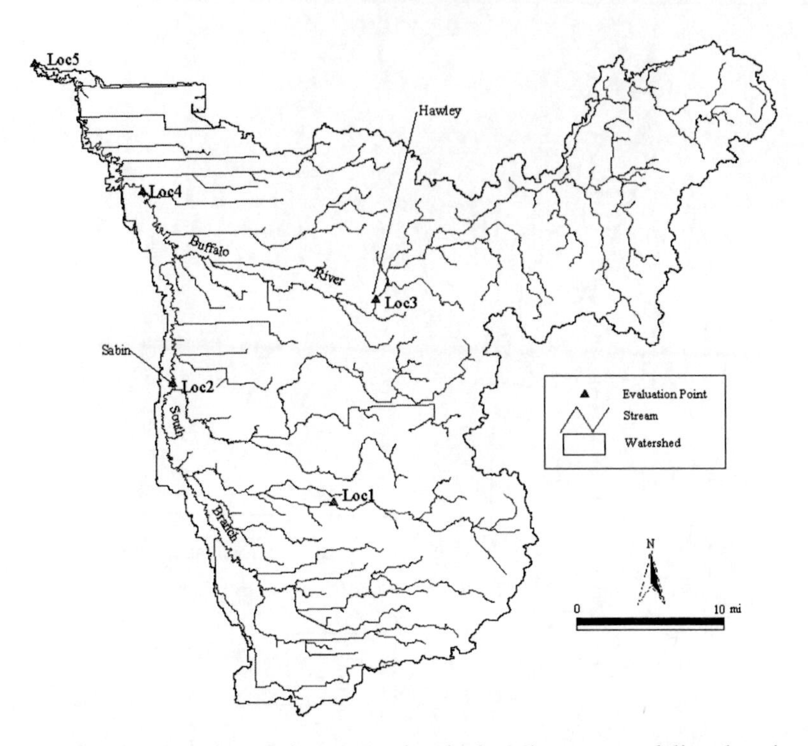

Figure 7-23. Map showing the selected "interest" points in the Minnesota modeling domain of HUC 09020106 (Table 7-4), where the predicted reductions of peaks for the historical floods are presented in Table 7-14, resulting from Scenario I.

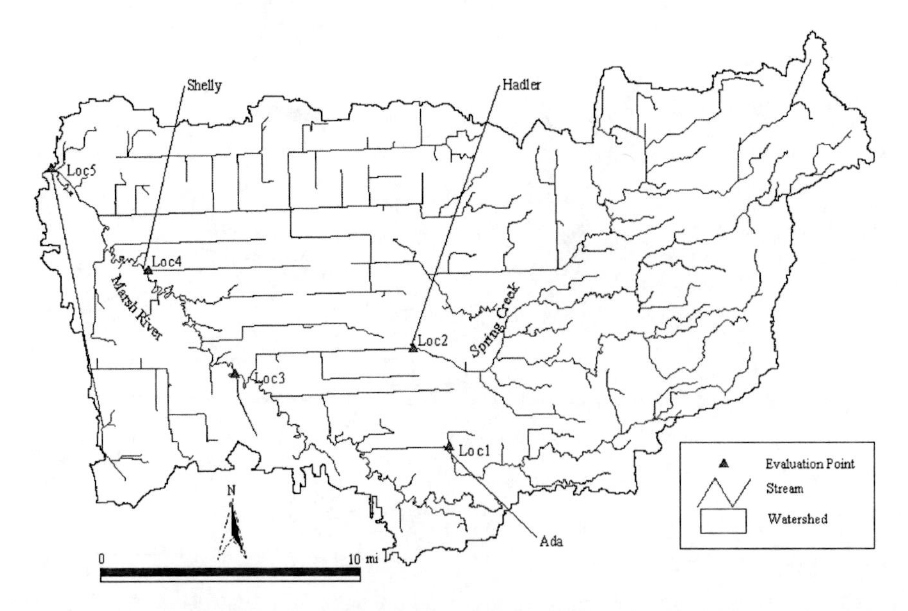

Figure 7-24. Map showing the selected "interest" points in the Minnesota modeling domain of HUC 09020107 (Table 7-4), where the predicted reductions of peaks for the historical floods are presented in Table 7-15, resulting from Scenario I.

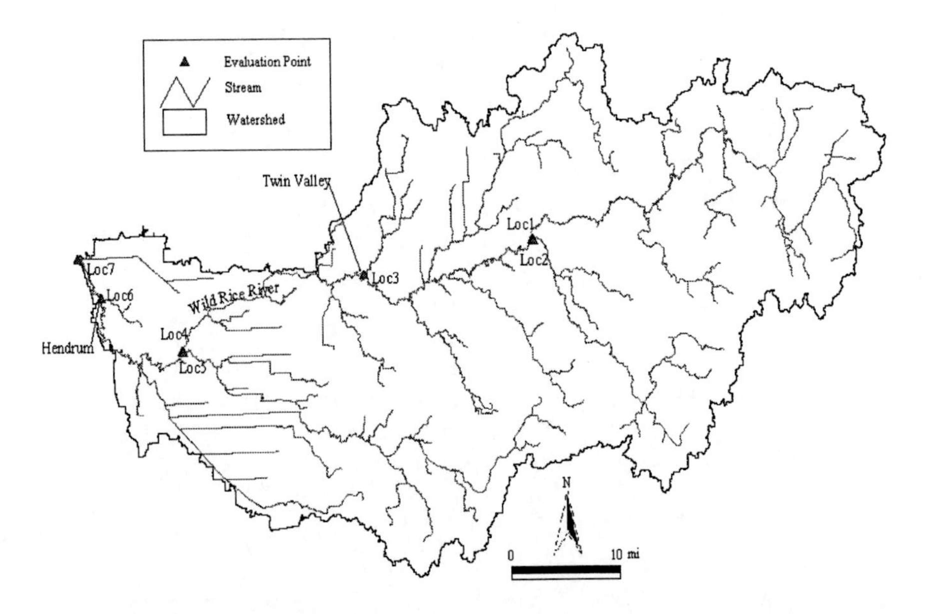

Figure 7-25. Map showing the selected "interest" points in the Minnesota modeling domain of HUC 09020108 (Table 7-4), where the predicted reductions of peaks for the historical floods are presented in Table 7-16, resulting from Scenario I.

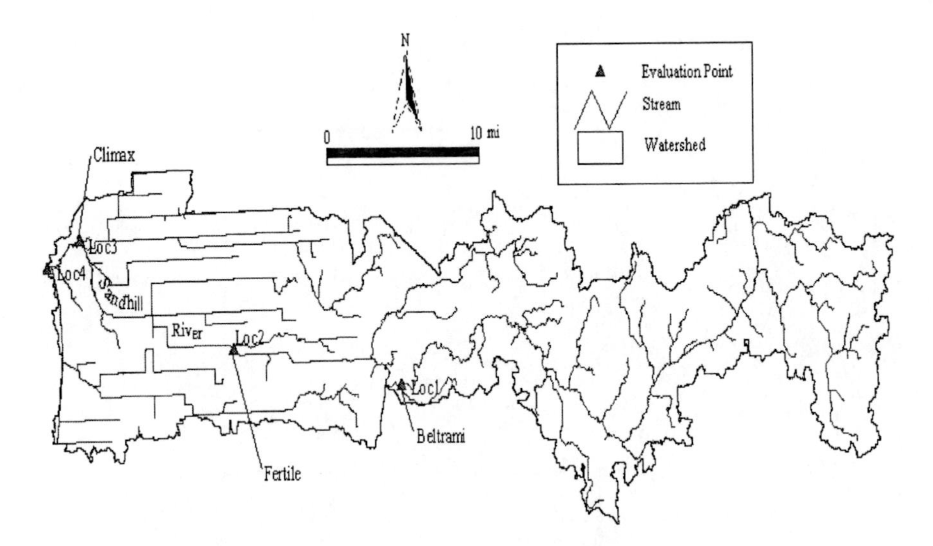

Figure 7-26. Map showing the selected "interest" points in the Minnesota modeling domain of HUC 09020301 (Table 7-4), where the predicted reductions of peaks for the historical floods are presented in Table 7-17, resulting from Scenario I.

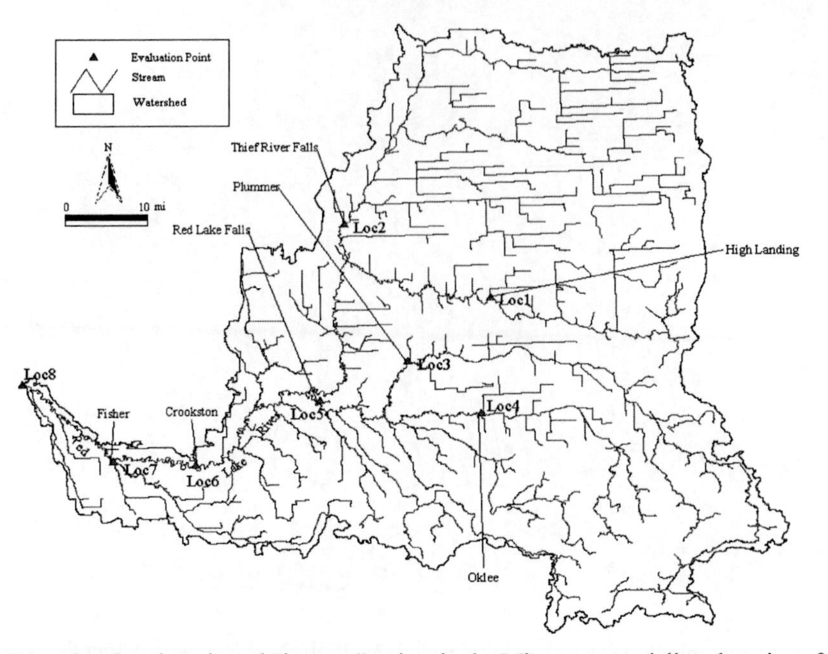

Figure 7-27. Map showing the selected "interest" points in the Minnesota modeling domains of HUC 09020303, HUC 09020304, and HUC 09020305 (Table 7-4), where the predicted reductions of peaks for the historical floods are presented in Table 7-18, resulting from Scenario I.

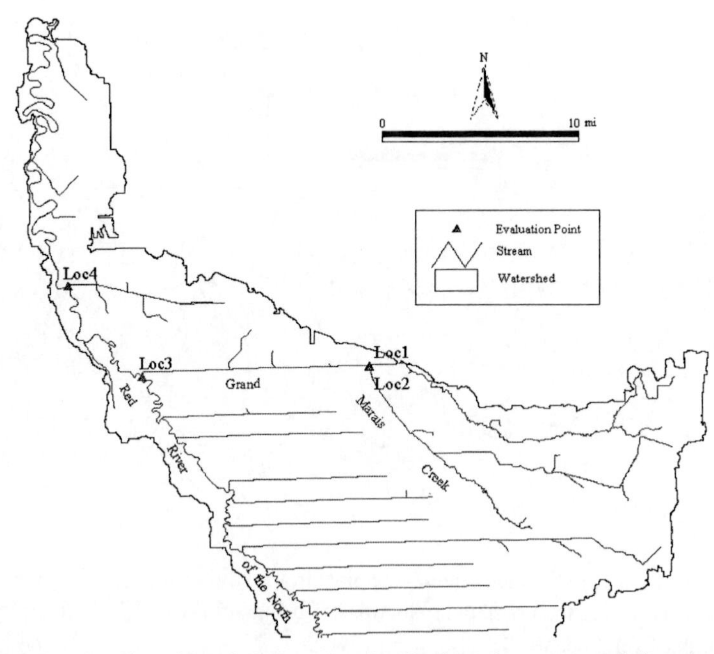

Figure 7-28. Map showing the selected "interest" points in the Minnesota modeling domain of HUC 09020306 (Table 7-4), where the predicted reductions of peaks for the historical floods are presented in Table 7-19, resulting from Scenario I.

Table 7-12. SWAT predicted reductions of peaks for the historical floods, resulting from Scenario I (S-I), at the selected "interest" points in the Minnesota modeling domain of HUC 09020103 shown in Figure 7-21

Flood Event	Loc1			Loc2			Loc3		
	Pre-Waffle® (cfs)	S-I (cfs)	Reduction (%)	Pre-Waffle® (cfs)	S-I (cfs)	Reduction (%)	Pre-Waffle® (cfs)	S-I (cfs)	Reduction (%)
1997	790	790	0.0	940	940	0.0	1500	1500	0.0
1979	450	450	0.0	770	770	0.0	1130	1130	0.0
1978	405	405	0.0	755	755	0.0	1110	1110	0.0
1975	550	550	0.0	895	895	0.0	1390	1390	0.0
1969	635	635	0.0	785	785	0.0	1115	1115	0.0
1966	295	295	0.0	540	540	0.0	815	815	0.0
Flood Event	Loc4			Loc5			Loc6		
	Pre-Waffle® (cfs)	S-I (cfs)	Reduction (%)	Pre-Waffle® (cfs)	S-I (cfs)	Reduction (%)	Pre-Waffle® (cfs)	S-I (cfs)	Reduction (%)
1997	1465	1465	0.0	215	210	-2.3	1615	1610	-0.3
1979	1000	1000	0.0	120	120	0.0	1035	1035	0.0
1978	1000	1000	0.0	75	75	0.0	920	920	0.0
1975	1285	1285	0.0	100	100	0.0	1330	1330	0.0
1969	1070	1070	0.0	160	155	-3.1	1165	1165	0.0
1966	735	735	0.0	70	70	0.0	775	770	-0.6

Table 7-13. SWAT predicted reductions of peaks for the historical floods, resulting from Scenario I (S-I), at the selected "interest" points in the Minnesota modeling domain of HUC 09020104 shown in Figure 7-22

Flood Event	Loc1			Loc2			Loc3			Loc4		
	Pre-Waffle® (cfs)	S-I (cfs)	Reduction (%)	Pre-Waffle® (cfs)	S-I (cfs)	Reduction (%)	Pre-Waffle® (cfs)	S-I (cfs)	Reduction (%)	Pre-Waffle® (cfs)	S-I (cfs)	Reduction (%)
1997	995	550	-44.7	990	545	-44.9	650	310	-52.3	1250	510	-59.2
1979	815	370	-54.6	800	360	-55.0	430	235	-45.3	995	675	-32.2
1978	675	335	-50.4	675	335	-50.4	385	230	-40.3	820	445	-45.7
1975	1035	510	-50.7	1025	510	-50.2	500	210	-58.0	810	335	-58.6
1969	2025	1200	-40.7	2000	1185	-40.8	985	590	-40.1	1855	925	-50.1
1966	560	400	-28.6	555	390	-29.7	500	235	-53.0	980	450	-54.1

Table 7-14. SWAT predicted reductions of peaks for the historical floods, resulting from Scenario I (S-I), at the selected "interest" points in the Minnesota modeling domain of HUC 09020106 shown in Figure 7-23

Flood Event	Loc1			Loc2			Loc3		
	Pre-Waffle® (cfs)	S-I (cfs)	Reduction (%)	Pre-Waffle® (cfs)	S-I (cfs)	Reduction (%)	Pre-Waffle® (cfs)	S-I (cfs)	Reduction (%)
1997	355	355	0.0	5840	5750	-1.5	2310	2300	-0.4
1979	180	180	0.0	2790	2740	-1.8	930	925	-0.5
1978	390	390	0.0	3395	3345	-1.5	1950	1940	-0.5
1975	260	260	0.0	3025	2975	-1.7	1075	1070	-0.5
1969	545	545	0.0	6295	6200	-1.5	1655	1640	-0.9
1966	215	215	0.0	3055	3015	-1.3	1535	1530	-0.3

Flood Event	Loc4			Loc5		
	Pre-Waffle® (cfs)	S-I (cfs)	Reduction (%)	Pre-Waffle® (cfs)	S-I (cfs)	Reduction (%)
1997	8290	8220	-0.8	8700	8575	-1.4
1979	4375	4155	-5.0	5740	5335	-7.1
1978	5190	5135	-1.1	5525	5425	-1.8
1975	2855	2820	-1.2	3120	3055	-2.1
1969	9880	9780	-1.0	10550	10350	-1.9
1966	4940	4830	-2.2	5660	5455	-3.6

Table 7-15. SWAT predicted reductions of peaks for the historical floods, resulting from Scenario I (S-I), at the selected "interest" points in the Minnesota modeling domain of HUC 09020107 shown in Figure 7-24

Flood Event	Loc1			Loc2			Loc3		
	Pre-Waffle® (cfs)	S-I (cfs)	Reduction (%)	Pre-Waffle® (cfs)	S-I (cfs)	Reduction (%)	Pre-Waffle® (cfs)	S-I (cfs)	Reduction (%)
1997	815	680	-16.6	1405	1240	-11.7	3255	2610	-19.8
1979	975	880	-9.7	2110	1755	-16.8	3925	3085	-21.4
1978	465	295	-36.6	950	635	-33.2	1915	1155	-39.7
1975	535	380	-29.0	1290	900	-30.2	2145	1345	-37.3
1969	700	550	-21.4	1555	1195	-23.2	3230	2320	-28.2
1966	285	210	-26.3	765	540	-29.4	1130	830	-26.5

Table 7-15. Continued

Flood Event	Loc4			Loc5		
	Pre-Waffle® (cfs)	S-I (cfs)	Reduction (%)	Pre-Waffle® (cfs)	S-I (cfs)	Reduction (%)
1997	4010	3240	-19.2	7910	5540	-30.0
1979	4700	3520	-25.1	9065	6640	-26.8
1978	2285	1280	-44.0	4395	2120	-51.8
1975	2330	1430	-38.6	4625	2540	-45.1
1969	3810	2540	-33.3	7470	4455	-40.4
1966	1370	910	-33.6	2655	1670	-37.1

Table 7-16. SWAT predicted reductions of peaks for the historical floods, resulting from Scenario I (S-I), at the selected "interest" points in the Minnesota modeling domain of HUC 09020108 shown in Figure 7-25

Flood Event	Loc1			Loc2			Loc3			Loc4		
	Pre-Waffle® (cfs)	S-I (cfs)	Reduc-tion (%)	Pre-Waffle® (cfs)	S-I (cfs)	Reduc-tion (%)	Pre-Waffle® (cfs)	S-I (cfs)	Reduc-ton (%)	Pre-Waffle® (cfs)	S-I (cfs)	Reduc-tion (%)
1997	1460	1430	-2.1	845	835	-1.2	6135	5920	-3.5	8160	7875	-3.5
1979	1020	1010	-1.0	1070	1045	-2.3	4920	4730	-3.9	6055	5825	-3.8
1978	1185	1175	-0.8	830	810	-2.4	5040	4875	-3.3	5785	5600	-3.2
1975	850	845	-0.6	550	545	-0.9	2970	2880	-3.0	3705	3580	-3.4
1969	710	705	-0.7	645	645	0.0	4235	4090	-3.4	5680	5440	-4.2
1966	395	390	-1.3	325	325	0.0	2730	2625	-3.8	3400	3240	-4.7

Flood Event	Loc5			Loc6			Loc7		
	Pre-Waffle® (cfs)	S-I (cfs)	Reduction (%)	Pre-Waffle® (cfs)	S-I (cfs)	Reduc-tion (%)	Pre-Waffle® (cfs)	S-I (cfs)	Reduc-ion (%)
1997	945	875	-7.4	10470	9830	-6.1	10735	10095	-6.0
1979	720	670	-6.9	7365	6950	-5.6	7560	7130	-5.7
1978	750	700	-6.7	7055	6690	-5.2	7145	6785	-5.0
1975	565	525	-7.1	4790	4465	-6.8	4950	4635	-6.4
1969	765	705	-7.8	7355	6810	-7.4	7615	7085	-7.0
1966	370	340	-8.1	4000	3770	-5.8	4140	3920	-5.3

Table 7-17. SWAT predicted reductions of peaks for the historical floods, resulting from Scenario I (S-I), at the selected "interest" points in the Minnesota modeling domain of HUC 09020301 shown in Figure 7-26

Flood Event	Loc1			Loc2			Loc3			Loc4		
	Pre-Waffle® (cfs)	S-I (cfs)	Reduction (%)	Pre-Waffle® (cfs)	S-I (cfs)	Reduction (%)	Pre-Waffle® (cfs)	S-I (cfs)	Reduction (%)	Pre-Waffle® (cfs)	S-I (cfs)	Reduction (%)
1997	1480	1445	-2.4	285	280	-1.8	4215	3645	-13.5	4515	4015	-11.1
1979	1010	980	-3.0	270	255	-5.6	3395	2660	-21.6	3730	2940	-21.2
1978	1670	1645	-1.5	130	120	-7.7	2995	2430	-18.9	3060	2490	-18.6
1975	1070	1055	-1.4	100	95	-5.0	2285	1820	-20.4	2450	2020	-17.6
1969	1745	1690	-3.2	220	200	-9.1	3645	2780	-23.7	3725	2935	-21.2
1966	2630	2575	-2.1	180	165	-8.3	4095	3640	-11.1	4405	3905	-11.4

Table 7-18. SWAT predicted reductions of peaks for the historical floods, resulting from Scenario I (S-I), at the selected "interest" points in the Minnesota modeling domains of HUC 09020303, HUC 09020304, and HUC 09020305 shown in Figure 7-27

	Loc1			Loc2			Loc3			Loc4		
Flood Event	Pre-Waffle® (cfs)	S-I (cfs)	Reduction (%)	Pre-Waffle® (cfs)	S-I (cfs)	Reduction (%)	Pre-Waffle® (cfs)	S-I (cfs)	Reduction (%)	Pre-Waffle® (cfs)	S-I (cfs)	Reduction (%)
1997	2290	2210	-3.5	4680	4330	-7.5	3365	3040	-9.7	2250	2180	-3.1
1979	3660	3605	-1.5	3230	2550	-21.1	3840	3720	-3.1	2060	2005	-2.7
1978	2225	2160	-2.9	2700	2460	-8.9	3070	2915	-5.0	2910	2825	-2.9
1975	2030	1950	-3.9	2740	2495	-8.9	2400	2190	-8.8	2095	2040	-2.6
1969	2195	2045	-6.8	3235	2890	-10.7	3520	3320	-5.7	3080	3010	-2.3
1966	3200	3135	-2.0	5260	4670	-11.2	1980	1720	-13.1	2670	2575	-3.6

	Loc5			Loc6			Loc7			Loc8		
Flood Event	Pre-Waffle® (cfs)	S-I (cfs)	Reduction (%)	Pre-Waffle® (cfs)	S-I (cfs)	Reduction (%)	Pre-Waffle® (cfs)	S-I (cfs)	Reduction (%)	Pre-Waffle® (cfs)	S-I (cfs)	Reduction (%)
1997	9340	8950	-4.2	24600	23670	-3.8	25190	24235	-3.8	20070	19090	-4.9
1979	9930	9575	-3.6	23020	21620	-6.1	25680	24090	-6.2	19650	18875	-3.9
1978	9960	9125	-8.4	15640	14650	-6.3	16080	15085	-6.2	13000	12140	-6.6
1975	7330	7090	-3.3	14020	12985	-7.4	16210	14935	-7.9	13135	12080	-8.0
1969	9000	8425	-6.4	27275	25570	-6.3	27910	26155	-6.3	20590	19185	-6.8
1966	8840	8395	-5.0	19765	18655	-5.6	21650	20425	-5.7	17830	16785	-5.9

Table 7-19. SWAT predicted reductions of peaks for the historical floods, resulting from Scenario I (S-I), at the selected "interest" points in the Minnesota modeling domain of HUC 09020306 shown in Figure 7-28

Flood Event	Loc1			Loc2			Loc3			Loc4		
	Pre-Waffle® (cfs)	S-I (cfs)	Reduction (%)	Pre-Waffle® (cfs)	S-I (cfs)	Reduction (%)	Pre-Waffle® (cfs)	S-I (cfs)	Reduction (%)	Pre-Waffle® (cfs)	S-I (cfs)	Reduction (%)
1997	195	125	-35.9	540	295	-45.4	680	385	-43.4	95	95	0.0
1979	110	70	-36.4	290	125	-56.9	360	185	-48.6	30	30	0.0
1978	25	15	-40.0	85	45	-47.1	95	55	-42.1	10	10	0.0
1975	45	30	-33.3	100	50	-50.0	130	75	-42.3	40	40	0.0
1969	90	60	-33.3	265	140	-47.2	310	175	-43.5	30	30	0.0
1966	15	15	0.0	30	15	-50.0	75	70	-6.7	115	115	0.0

Table 7-20. SWAT predicted reductions of peaks for the historical floods, resulting from Scenario I (S-I), at the selected "interest" points in the Minnesota modeling domain of HUC 09020309 shown in Figure 7-29

Flood Event	Loc1			Loc2			Loc3		
	Pre-Waffle® (cfs)	S-I (cfs)	Reduction (%)	Pre-Waffle® (cfs)	S-I (cfs)	Reduction (%)	Pre-Waffle® (cfs)	S-I (cfs)	Reduction (%)
1997	2425	2275	-6.2	3795	3590	-5.4	2755	2520	-8.5
1979	1070	950	-11.2	1970	1790	-9.1	2695	2600	-3.5
1978	1050	950	-9.5	1345	1230	-8.6	1175	920	-21.7
1975	615	555	-9.8	935	860	-8.0	775	620	-20.0
1969	1630	1470	-9.8	2320	2140	-7.8	1890	1490	-21.2
1966	970	875	-9.8	1305	1190	-8.8	1055	850	-19.4
1997	4060	3990	-1.7	500	485	-3.0	14480	13835	-4.5
1979	3140	3115	-0.8	575	565	-1.7	11995	11705	-2.4
1978	1450	1400	-3.4	215	205	-4.7	4925	4450	-9.6
1975	965	930	-3.6	140	135	-3.6	3420	3090	-9.6
1969	2535	2445	-3.6	340	330	-2.9	8655	7840	-9.4
1966	1320	1270	-3.8	190	185	-2.6	4765	4295	-9.9
1997	4060	3990	-1.7	500	485	-3.0	14480	13835	-4.5

Table 7-21. SWAT predicted reductions of peaks for the historical floods, resulting from Scenario I (S-I), at the selected "interest" points in the Minnesota modeling domain of HUC 09020311 shown in Figure 7-30

Flood Event	Loc1 Pre-Waffle® (cfs)	Loc1 S-I (cfs)	Loc1 Reduc-tion (%)	Loc2 Pre-Waffle® (cfs)	Loc2 S-I (cfs)	Loc2 Reduc-tion (%)	Loc3 Pre-Waffle® (cfs)	Loc3 S-I (cfs)	Loc3 Reduc-tion (%)	Loc4 Pre-Waffle® (cfs)	Loc4 S-I (cfs)	Loc4 Reduc-tion (%)
1997	4210	3485	-17.2	4700	3840	-18.3	5920	4955	-16.3	6085	5100	-16.2
1979	3600	3070	-14.7	4090	3405	-16.7	5110	4200	-17.8	5190	4260	-17.9
1978	2750	2150	-21.8	3120	2340	-25.0	3700	2815	-23.9	3785	2895	-23.5
1975	2320	1790	-22.8	2615	1950	-25.4	2945	2230	-24.3	3035	2315	-23.7
1969	2805	2130	-24.1	3140	2320	-26.1	4095	3115	-23.9	4155	3195	-23.1
1966	1945	1505	-22.6	2200	1650	-25.0	2400	1830	-23.8	2475	1905	-23.0
1997	470	380	-19.1	220	130	-40.9	2145	1620	-24.5	970	840	-13.4
1979	430	350	-18.6	195	110	-43.6	1925	1470	-23.6	895	745	-16.8
1978	300	230	-23.3	135	75	-44.4	1500	1005	-33.0	705	540	-23.4
1975	255	195	-23.5	115	70	-39.1	1315	890	-32.3	610	475	-22.1
1969	315	245	-22.2	145	85	-41.4	1590	1090	-31.4	740	575	-22.3
1966	215	165	-23.3	95	60	-36.8	1140	780	-31.6	535	415	-22.4
1997	2120	1695	-20.0	3345	2735	-18.2	35	25	-28.6	10	10	0.0
1979	1970	1510	-23.4	2970	2450	-17.5	150	110	-26.7	35	35	0.0
1978	1505	1105	-26.6	2345	1725	-26.4	40	30	-25.0	10	10	0.0
1975	1290	950	-26.4	2060	1520	-26.2	45	35	-22.2	10	10	0.0
1969	1510	1115	-26.2	2490	1855	-25.5	55	40	-27.3	15	15	0.0
1966	1110	825	-25.7	1795	1330	-25.9	60	45	-25.0	15	15	0.0
1997	265	185	-30.2	375	290	-22.7	130	125	-3.8	545	455	-16.5
1979	980	715	-27.0	1395	1105	-20.8	545	515	-5.5	2085	1765	-15.3
1978	245	175	-28.6	345	270	-21.7	130	125	-3.8	515	430	-16.5
1975	305	215	-29.5	425	335	-21.2	155	150	-3.2	620	525	-15.3
1969	385	270	-29.9	540	420	-22.2	190	180	-5.3	775	655	-15.5
1966	425	295	-30.6	600	465	-22.5	205	195	-4.9	860	720	-16.3

Table 7-22. SWAT predicted reductions of peaks for the historical floods, resulting from Scenario I (S-I), at the selected "interest" points in the Minnesota modeling domain of HUC 09020312 shown in Figure 7-31

Flood Event	Loc1			Loc2			Loc3			Loc4		
	Pre-Waffle® (cfs)	S-I (cfs)	Reduction (%)	Pre-Waffle® (cfs)	S-I (cfs)	Reduction (%)	Pre-Waffle® (cfs)	S-I (cfs)	Reduction (%)	Pre-Waffle® (cfs)	S-I (cfs)	Reduction (%)
1997	4145	3695	-10.9	3840	3400	-11.5	510	420	-17.6	4790	4105	-14.3
1979	3145	2760	-12.2	3015	2630	-12.8	1165	1110	-4.7	6550	6090	-7.0
1978	2605	2175	-16.5	2420	2020	-16.5	665	640	-3.8	4095	3780	-7.7
1975	1875	1570	-16.3	1720	1430	-16.9	435	420	-3.4	2175	2035	-6.4
1969	2100	1760	-16.2	1920	1600	-16.7	620	605	-2.4	2980	2820	-5.4
1966	2095	1795	-14.3	1935	1655	-14.5	610	580	-4.9	3230	3060	-5.3

Flood Event	Loc5			Loc6			Loc7			Loc8		
	Pre-Waffle® (cfs)	S-I (cfs)	Reduction (%)	Pre-Waffle® (cfs)	S-I (cfs)	Reduction (%)	Pre-Waffle® (cfs)	S-I (cfs)	Reduction (%)	Pre-Waffle® (cfs)	S-I (cfs)	Reduction (%)
1997	1520	1230	-19.1	295	290	-1.7	60	55	-8.3	4775	4100	-14.1
1979	2255	2125	-5.8	1020	1005	-1.5	195	195	0.0	6595	6175	-6.4
1978	1395	1285	-7.9	610	595	-2.5	110	110	0.0	4180	3895	-6.8
1975	850	810	-4.7	430	420	-2.3	85	85	0.0	2210	2070	-6.3
1969	1180	1135	-3.8	605	590	-2.5	120	120	0.0	3005	2855	-5.0
1966	1090	1045	-4.1	535	520	-2.8	100	95	-5.0	3255	3090	-5.1

Table 7-23. SWAT predicted reductions of peaks for the historical floods, resulting from Scenario I (S-I), at the selected "interest" points in the Minnesota modeling domain of HUC 09020314 shown in Figure 7-32

Flood Event	Loc1		
	Pre-Waffle® (cfs)	S-I (cfs)	Reduction (%)
1997	4315	4120	-4.5
1979	5245	4875	-7.1
1978	2570	2230	-13.2
1975	3560	3285	-7.7
1969	4645	4230	-8.9
1966	4365	3875	-11.2

Table 7-24. SWAT predicted reductions of peaks for the historical floods, resulting from Scenario I (S-I), at the selected "interest" points in the North Dakota modeling domain of HUC 09020107 shown in Figure 7-33

Flood Event	Loc1			Loc2			Loc3		
	Pre-Waffle® (cfs)	S-I (cfs)	Reduction (%)	Pre-Waffle® (cfs)	S-I (cfs)	Reduction (%)	Pre-Waffle® (cfs)	S-I (cfs)	Reduction (%)
1997	660	600	-9.1	735	670	-8.8	825	710	-13.9
1979	215	205	-4.7	570	515	-9.6	1135	955	-15.9
1978	535	495	-7.5	230	210	-8.7	1025	870	-15.1
1975	235	220	-6.4	355	320	-9.9	745	640	-14.1
1969	840	785	-6.5	550	495	-10.0	1330	1135	-14.7
1966	140	130	-7.1	210	185	-11.9	595	505	-15.1
1997	2215	1875	-15.3	1675	1125	-32.8	2140	1805	-15.7
1979	1225	915	-25.3	1335	910	-31.8	1175	885	-24.7
1978	1080	935	-13.4	1340	910	-32.1	1095	960	-12.3
1975	680	595	-12.5	1335	910	-31.8	680	565	-16.9
1969	2185	1835	-16.0	1335	910	-31.8	2100	1785	-15.0
1966	670	525	-21.6	1335	910	-31.8	640	510	-20.3

Flood Event	Loc4			Loc5			Loc6		
	Pre-Waffle® (cfs)	S-I (cfs)	Reduction (%)	Pre-Waffle® (cfs)	S-I (cfs)	Reduction (%)	Pre-Waffle® (cfs)	S-I (cfs)	Reduction (%)
1997	930	610	-34.4	3985	2925	-26.6	4885	3460	-29.2
1979	945	660	-30.2	2490	1675	-32.7	3460	2245	-35.1
1978	880	655	-25.6	1425	1140	-20.0	1955	1495	-23.5
1975	540	405	-25.0	1505	1065	-29.2	1675	1120	-33.1
1969	1335	965	-27.7	3730	2620	-29.8	5115	3445	-32.6
1966	520	395	-24.0	1390	960	-30.9	1570	1015	-35.4

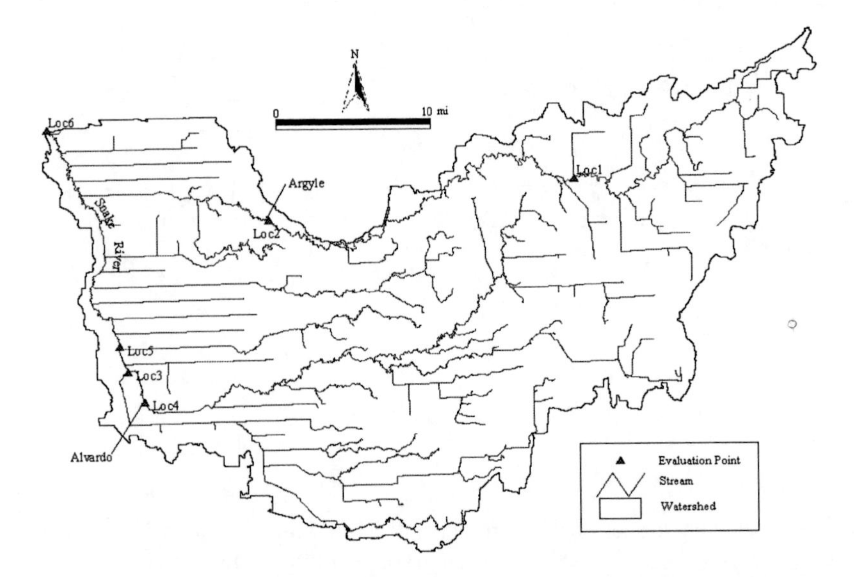

Figure 7-29. Map showing the selected "interest" points in the Minnesota modeling domain of HUC 09020309 (Table 7-4), where the predicted reductions of peaks for the historical floods are presented in Table 7-20, resulting from Scenario I.

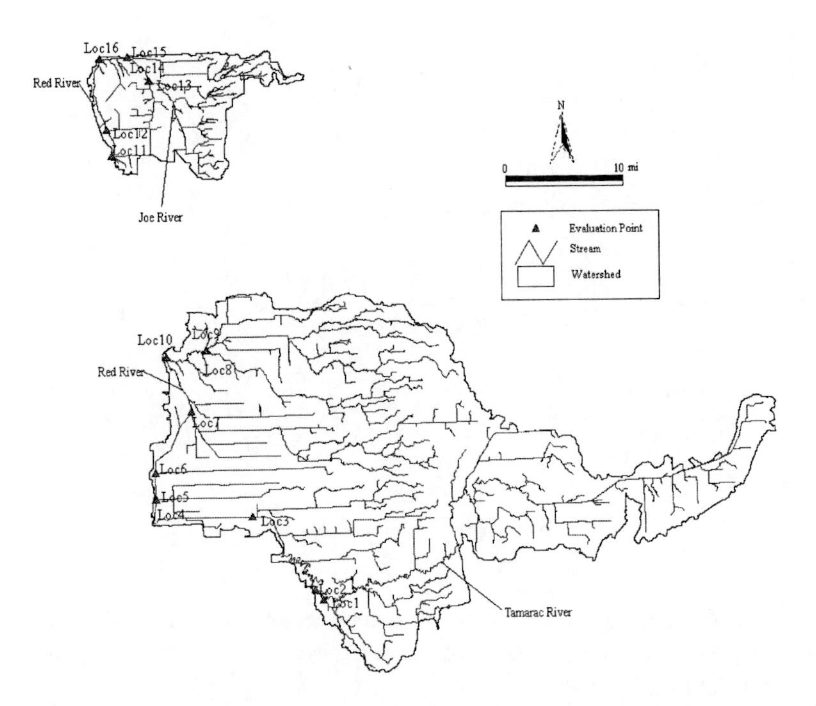

Figure 7-30. Map showing the selected "interest" points in the Minnesota modeling domain of HUC 09020311 (Table 7-4), where the predicted reductions of peaks for the historical floods are presented in Table 7-21, resulting from Scenario I.

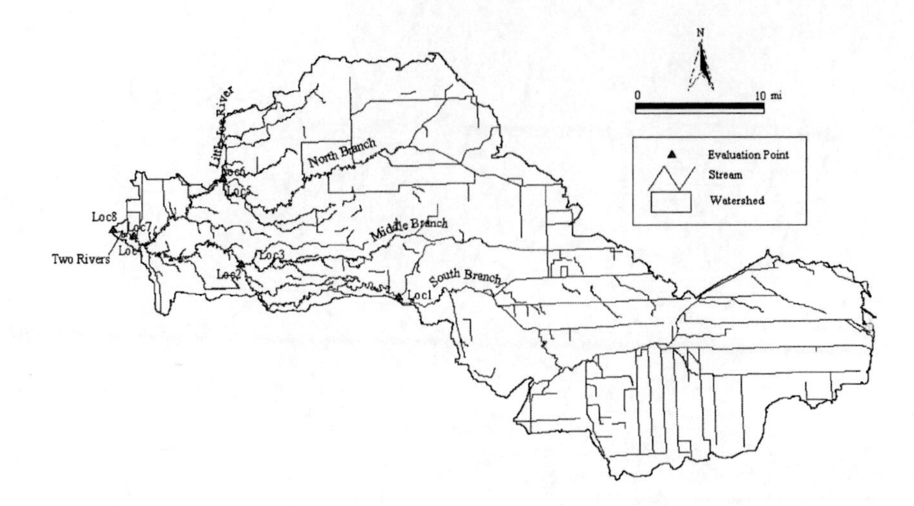

Figure 7-31. Map showing the selected "interest" points in the Minnesota modeling domain of HUC 09020312 (Table 7-4), where the predicted reductions of peaks for the historical floods are presented in Table 7-22, resulting from Scenario I.

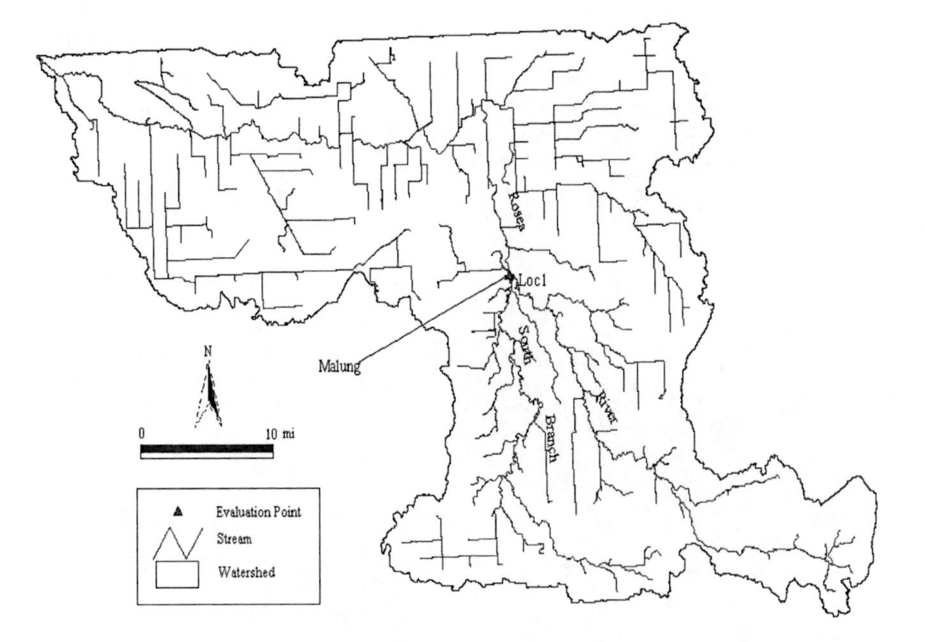

Figure 7-32. Map showing the selected "interest" points in the Minnesota modeling domain of HUC 09020314 (Table 7-4), where the predicted reductions of peaks for the historical floods are presented in Table 7-23, resulting from Scenario I.

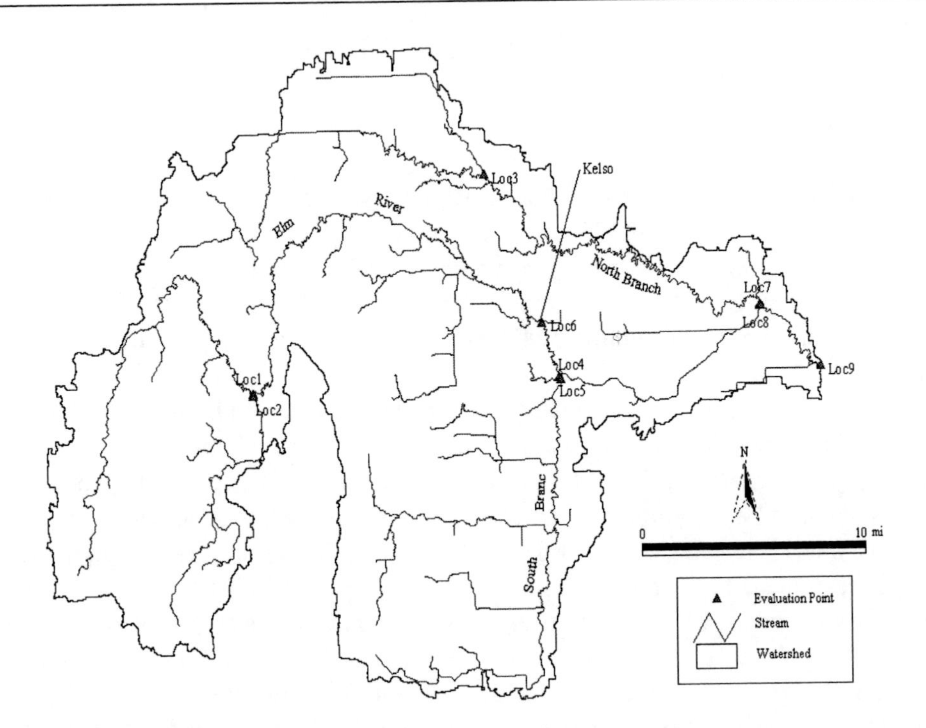

Figure 7-33. Map showing the selected "interest" points in the North Dakota modeling domain of HUC 09020107 (Table 7-4), where the predicted reductions of peaks for the historical floods are presented in Table 7-24, resulting from Scenario I.

In the SWAT models, the Waffle® storages were modeled as "synthetic" ponds. With this regard, this study developed a new algorithm for defining a "synthetic" pond to have a hydrologic function equivalent to the storages to be represented by the pond. The algorithm was implemented for each of the subbasins, resulting from subdividing the inclusive watershed for modeling purposes. As a result, a subbasin may include zero or one "synthetic" pond.

The results indicated that for a watershed, the effects of the Waffle® storages would depend upon both the watershed size and shape as measured by the width-to-length ratio. The percentage reductions would be overall larger for watersheds with a greater width-to-length ratio, but tend to be smaller for watersheds with a larger size. In addition, for a watershed, more Waffle® storage areas always correspond to larger reductions. However, the spatial distribution of the storage areas within the watershed is also a control factor of the reduction effect. When two watersheds have near-equivalent storages, the Waffle® would be more effective for the watershed with the storage areas controlling the upland runoff than that with the storage areas intercepting the concentrated stream flows. Because of the spatial variability, the flood reduction effects at different locations within a watershed could be distinctly different. Further, the reduction effects would be smaller for a flood event with a large peak and/or a prolonged rising limb.

The HEC–RAS Model

Data and Model Set up

In this study, two HEC-RAS hydrodynamic or unsteady-state models were used to predict reductions of the 1997 flood crests along the RR mainstem. The first model, developed by Mr. Stuart Dobberpuhl, a hydraulic engineer from the U.S. Army Corps of Engineers (USACE) St. Paul District, covers the reach from White Rock to Halstad (Figure 7-34). The second model, developed by EERC, covers the reach from Halstad to Emerson (Figure 7-35). The outputs from the first model were taken as the inputs into the second model, enabling a seamless prediction along the mainstem from White Rock to Emerson. For description purposes, hereinafter, the first model is designated "ACE-M," whereas, the second model designated "EERC-M." The common features of these two models are that: 1) the flows simulated by the aforementioned SWAT models were used to define the boundary conditions when the USGS observed data were unavailable; 2) the flows from the ungaged drainage areas, the areas that are uncontrolled by any USGS gauging station, were simulated by the SWAT models; 3) the major tributaries were explicitly modeled; 4) all bridges and major breakout flows, such as that occurred along the Maple and Sheyenne Rivers and at Thompson Bridge, were considered; 5) all available data on cross sections for the RR mainstem were used; and 6) the models incorporated the best knowledge of the engineers, including Mr. Scott Jutila, Mr. Randy Gjestvang, Mr. James Fay, Mr. Stuart Dobberpuhl, and Mr. Michael Lesher, to just name a few.

The geometric data for the cross sections and bridges along the RR mainstem were extracted from the HEC-RAS steady-state model, distributed along with the USACE's "Regional Red River Flood Assessment Report," dated January 2003. In addition, the data on cross sections for the tributaries that were modeled in ACE-M were generated using the USGS 1:24,000 quadrangles maps or extracted from a HEC-RAS unsteady-state model developed by the Pacific International Engineering for the "Maple River and Overflow Area Flood Insurance Study" project. Details on ACE-M can be found in the final report for the USACE's "Fargo-Moorhead Upstream Feasibility Study" project, entitled "Hydrology and Hydraulics Analysis."

Table 7-25 presents the approach of modeling the tributaries in EERC-M. The geometric data for the Red Lake River and Heartsville Coulee were extracted from the HEC-RAS steady-state models provided by Mr. Michael Lesher, a hydraulic engineer from USACE. These data reflect the current ground truth, the topography with the levees along the Red Lake River and the Heartsville Diversion channel constructed. The new bridge crossing the Hartsville Diversion channel was also included in the model. The cross sections for the other tributaries modeled as a branch were generated using the topographic information provided by the NED data. The flows for the upper and lateral boundary conditions and at the inflow points were simulated by the corresponding SWAT models. The RR mainstem was modeled as ten subreaches to account for the hydraulic connections between RR and the modeled tributaries. These subreaches were described by 248 cross sections, extracted the HEC-RAS steady-state model, distributed along with the USACE's "Regional Red River Flood Assessment Report," dated January 2003. The cross sections for the subreach affected by the Grand Forks/East Grand Forks Dike Project were taken from a steady-state HEC-RAS model, developed and used by USACE to update the Grand Forks/East Grand Forks Flood Insurance Rating Map, to reflect the current ground truth. In addition, EERC-M included 19 bridges crossing the RR mainstem. One lateral weir was added at Thompson Bridge to model the overflow from the RR into the Heartsville Coulee. The data used to define the lateral weir were provided by Mr. Michael Lesher and are presented in Table 7-26. The upper boundary condition of EERC-M was specified as the flow hydrograph at Halstad (USGS 05064500), whereas, the lower boundary condition was defined as a normal depth with a friction slope of 0.000065, determined based on the elevation information provided in the geometric data for the cross sections at, and adjacent to, Emerson (USGS 05102500).

As with ACE-M, EERC-M was also calibrated in accordance with the 1997 flood but was not validated for any other historical flood due to limited time and resources. The calibration was implemented to make the model simulated daily stream flow hydrographs closely match the corresponding observed hydrographs at Drayton, Pembina, and Emerson. The observed flow hydrograph at Grand Forks was not used for the model calibration because the Grand Forks/East Grand Forks Dike Project has noticeably changed the topography and geomorphology of the subreach located in the city limits. Given the changes, a hydrologic condition that is identical to the one of 1997 would result in a distinctly different flow hydrograph. In order to evaluate effects of proposed projects (e.g., the Waffle®) on reducing flood crests, which is the main purpose of developing this model, the geometric data for the current ground truth (i.e., the topography with the dikes and diversions constructed) rather than for the 1997 geomorphology should be used to set up the model. One may argue that the model should first be set up using the geometric data for the 1997 geomorphology and calibrated using the observed flow hydrograph. After that, the cross sections for the subreach affected by the construction project should be revised to represent the geometry of the current ground truth. This approach is logic for assessing affects of the Dike Project, but less useful for evaluating effects of other projects because the comparison should be made between the one with those projects and the current rather than the 1997 situation. Thus, instead of this approach, the Manning's n values used to design the dikes and diversions were adopted for the cross sections of the subreach affected by the construction project. The model simulated flow hydrograph was assumed to be the one corresponding to the 1997 hydrologic condition but with the current topography and geomorphology. This flow hydrograph was used as the comparison base to evaluate effects of the Waffle®. However, the observed flow hydrographs at stations downstream of Grand Forks were used to calibrate the model because USACE has shown that the changes in the Grand Forks/East Grand Forks have a negligible influence on the flow regimes at locations one-mile away from the northern boundary of the dike project. The calibration was realized by manually adjusting the Manning's n values for the cross sections. Moreover, the calibration was also implemented to have a close match between the model simulated water surface elevation hydrographs and the corresponding observed ones at Halstad, Drayton, Pembina, and Emerson. Nevertheless, because it is infeasible to have best matches both for flow and elevation, the calibration gave the first priority to having an accurate simulation of flows.

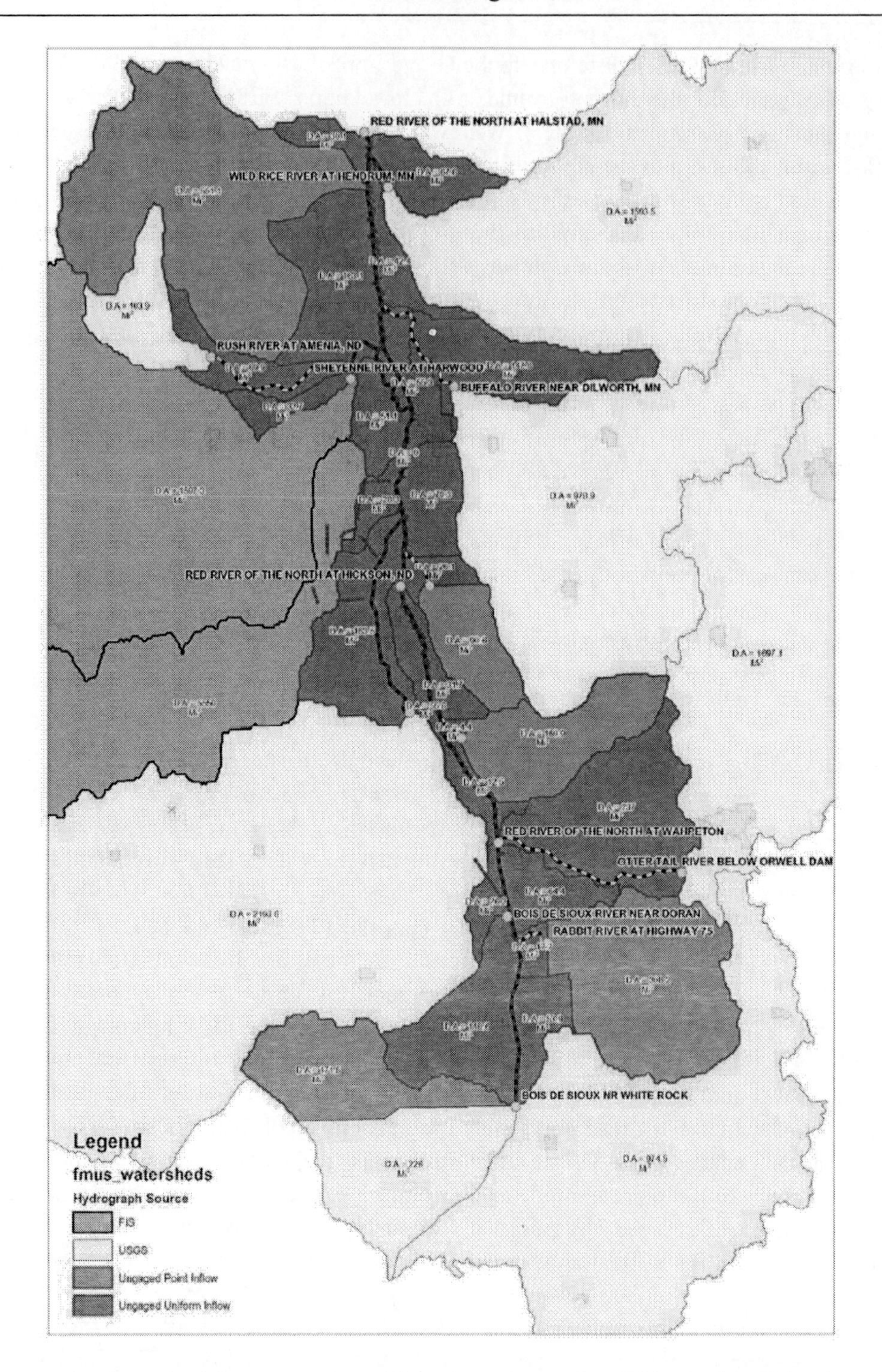

Figure 7-34. Schematic of the HEC-RAS model for the mainstem fromWhite Rock to Halstad.

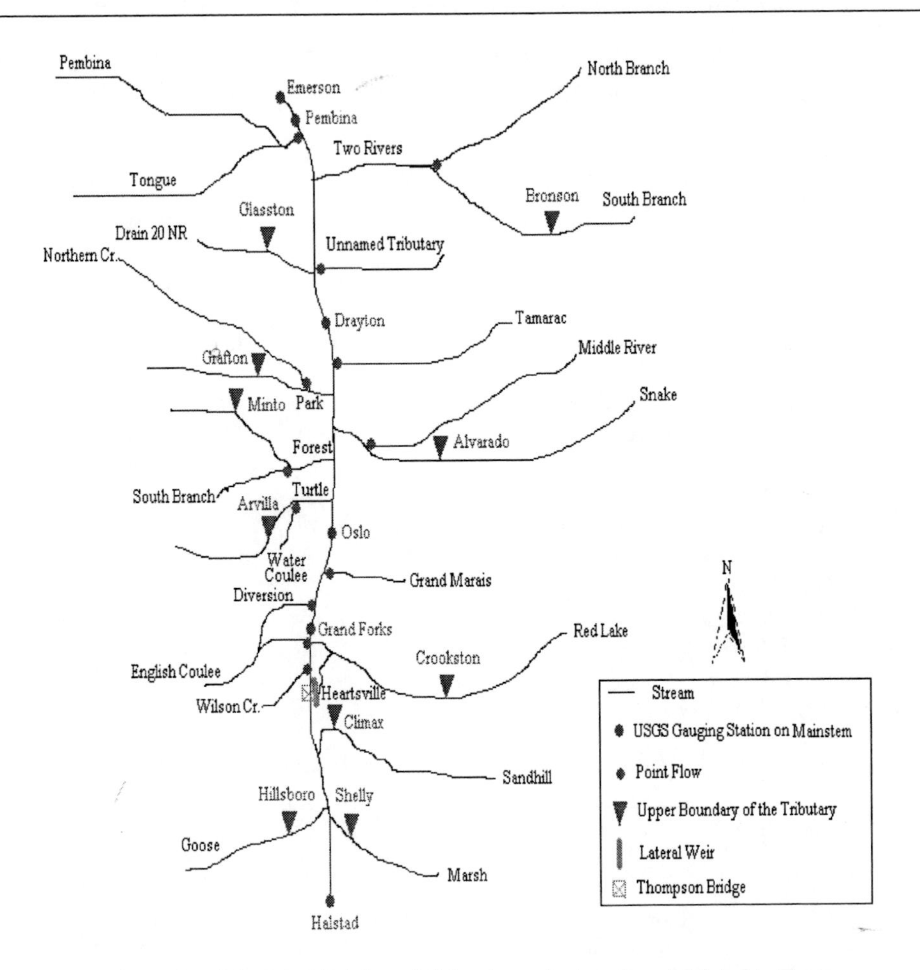

Figure 7-35. Schematic of the HEC-RAS model for the mainstem from Halstad to Emerson.

EERC-M did a good job on predicting both peaks and volumes (Table 7-27). As expected, Halstad is the model upper boundary and thus the predicted and observed values at this station are identical. The model successfully reproduced the peak discharges and timings at the three stations downstream of Grand Forks. The maximum prediction error is only 1.72% or 1 day off. In addition, the prediction error for volumes is less than 1%. The results for Grand Forks are presented for information purposes only because this station was not used for model calibration. Nevertheless, the model performance is acceptable for Grand Forks as well. Further, the model predicted the daily discharges with a sufficient accuracy, as indicated by the high coefficients of determination (R^2) of greater than 0.6 and a near-one slope (Figures 7-36 to 7-39). Also, the predicted stage hydrographs match well with the corresponding observed ones (Figures 7-40 to 7-44). The observed stages were obtained from USACE. Again, the results for Grand Forks are shown for information purposes only because this station was not used for model calibration. The model indicated that a peak discharge of 10,425 cfs might overtop the Thompson Bridge and flow into the Heartsville Coulee in 1997 (Figure 7-45). Considering the flows generated in the area drained by the coulee, the discharge of 12,000 cfs used by USACE to design the Heartsville Diversion is very reasonable.

Table 7-25. Tributaries included in the HEC-RAS unsteady-state model for the RR mainstem from Halstad to Emerson

Tributary	Modeling Approach	Boundary Conditions			Number of Cross Section
		Upper	Lower	Middle	
Marsh	Branch	Flow at Shelly (USGS 05067500)	Junction with RR	Lateral Flow	5
Sandhill	Branch	Flow at Climax (USGS 05069000)	Junction with RR	Lateral Flow	4
Red Lake	Branch	Flow at Crookston (USGS 05079000)	Junction with RR	Lateral Flow	35
Heartsville Coulee	Branch	Lateral Weir at Thompson Bridge	Junction with Red Lake	Uniform Lateral Flow	69
Grand Marais	Point Flow into RR	–	–	–	–
Snake	Branch	Flow at Alvarado (USGS 05085900)	Junction with RR	Lateral Flow	7
Middle River	Point Flow into Snake	–	–	–	–
Tamarac	Point Flow into RR	–	–	–	–
Unnamed Tributary	Point Flow into RR	–	–	–	–
Two Rivers	Branch	Flow at Bronson (USGS 05094000)	Junction with RR	Lateral Flow	7
North Branch	Point Flow into Two Rivers	–	–	–	
Goose	Branch	Flow at Hillsboro (USGS 05066500)	Junction with RR	Lateral Flow	7
English Coulee	Point Flow into RR	–	–	–	–
Diversion	Point Flow into RR	–	–	–	–
Turtle	Branch	Flow at Arvilla (USGS 05082625)	Junction with RR	Lateral Flow	8
Water Coulee	Point Flow into Turtle	–	–	–	–
Forest	Branch	Flow at Minto (USGS 05085000)	Junction with RR	Lateral Flow	8
South Branch	Point Flow into Forest	–	–	–	–
Park	Branch	Flow at Grafton (USGS 05090000)	Junction with RR	Lateral Flow	8
Northern Creek	Point Flow into Park	–	–	–	–
Drain 20 NR	Branch	Flow at Glasston (USGS 05092200)	Junction with RR	Lateral Flow	4
Pembina	Point Flow into RR	–	–	–	–
Tongue River	Point Flow into Pembina	–	–	–	–

Table 7-26. Data used to define the lateral weir at Thompson Bridge

Station (ft)	Distance (ft)	Sketch Graph and Remark
0.00	847.09	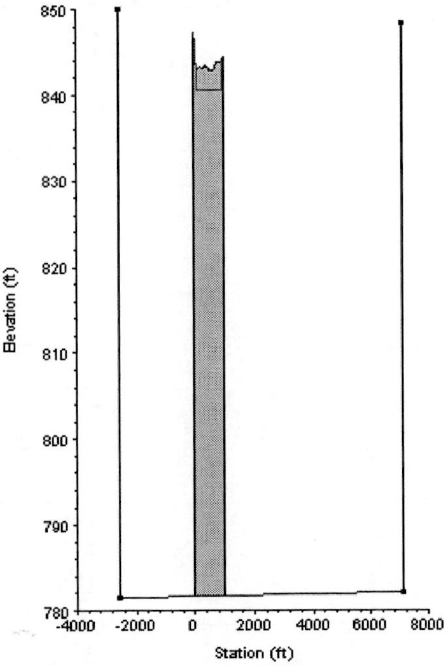 The lateral weir was modeled to have a breach bottom width of 800 ft with a side slope of 5 and a breach bottom elevation of 840.5 ft. The breach was assumed to be a result of overtopping that starts at a water surface elevation of 842.87 ft.
1.00	847.00	
12.50	846.00	
34.00	845.00	
55.30	844.00	
59.20	843.82	
170.52	843.03	
214.50	843.22	
272.30	843.00	
324.70	843.05	
345.60	843.05	
397.00	843.35	
405.30	843.23	
432.30	843.19	
464.30	843.19	
500.80	843.04	
507.50	842.88	
543.30	842.87	
663.05	842.87	
693.30	842.98	
700.90	843.02	
749.90	843.72	
793.90	843.78	
852.00	843.85	
923.70	843.98	
954.20	843.98	
957.00	844.19	
972.70	844.32	
988.64	844.40	

Table 7-27. Observed and EERC-M predicted peaks and volumes for the 1997 flood

Station	Observed Peak		Predicted Peak		Volume (from April 14 to May 10)		
	Magni-tude (cfs)	Timing	Magni-tude (cfs)	Timing	Obser-ved (ac-ft)	Predic-ted (ac-ft)	Error (%)
Halstad[1]	69,900	Apr. 19	69,900	Apr. 19	2,323,041	2,323,041	0.00
Grand Forks[2]	127,000	Apr. 18	102,420	Apr. 22	3,613,091	3,381,224	-6.42
Drayton	124,000	Apr. 24	121,859	Apr. 24	3,882,446	4,240,783	0.92
Pembina[3]	141,400	Apr. 26	140,430	Apr. 27	4,429,307	5,032,677	0.93
Emerson[4]	141,400	Apr. 26	140,488	Apr. 27	4,439,217	5,032,312	0.74

[1] As the model upper boundary, the predicted and observed values at this station are identical.

[2] The results are presented for information purposes only because the station was not used for model calibration.

[3] The observed flow hydrograph was derived by Dr. Xixi Wang, P.E., a research scientist at the University of North Dakota, using the data on observed stages and the rating curve provided by Mr. Steven Robinson from the U.S. Geological Survey (USGS).

[4] The observed flow hydrograph was provided by Mr. Alf Warkentin from Manitoba Water Stewardship. This corrected hydrograph considered the overflows occurred at the west bank of the Red River of the North in the vicinity of Emerson. In contrast, the USGS data did not consider the overflows.

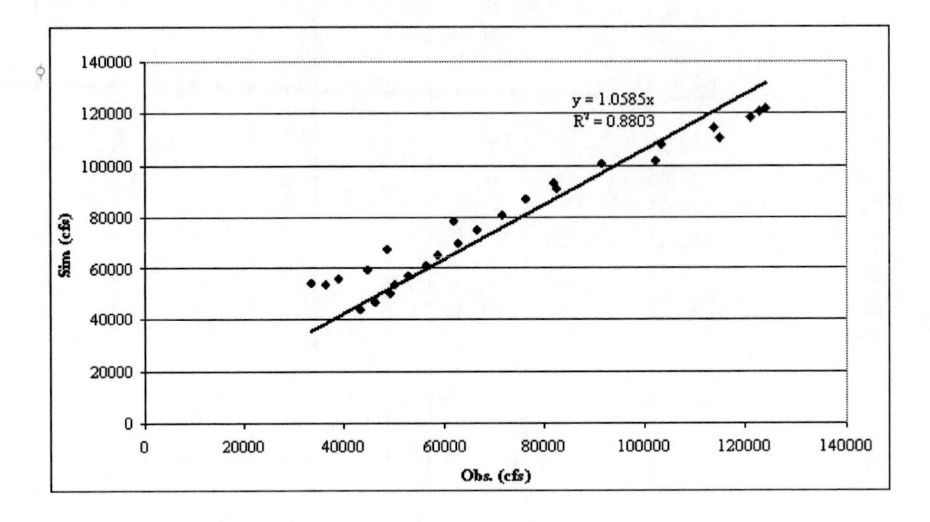

Figure 7-36. Plot showing the simulated versus observed daily discharges at Drayton.

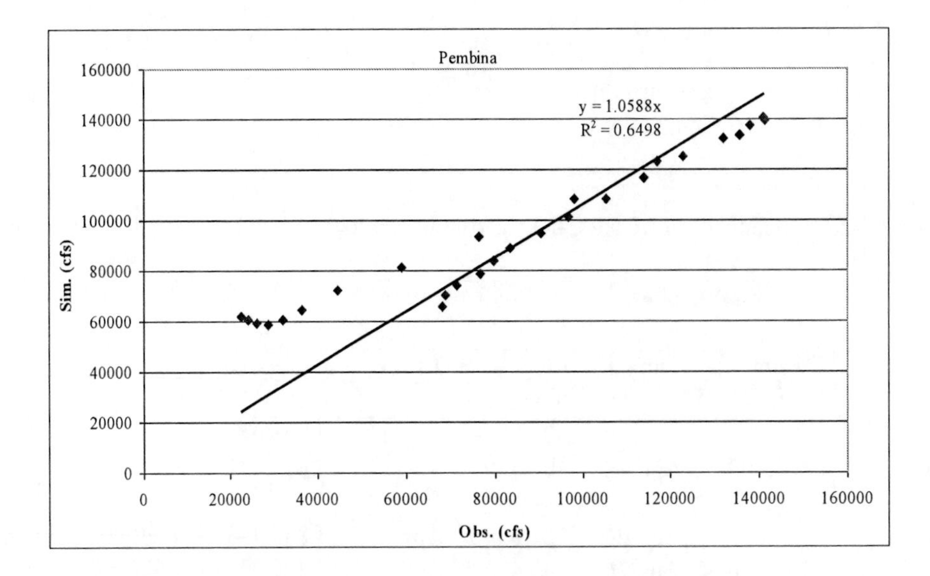

Figure 7-37. Plot showing the simulated versus observed daily discharges at Pembina.

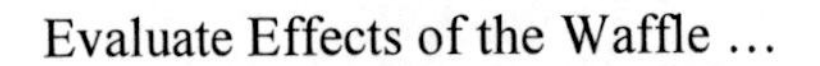

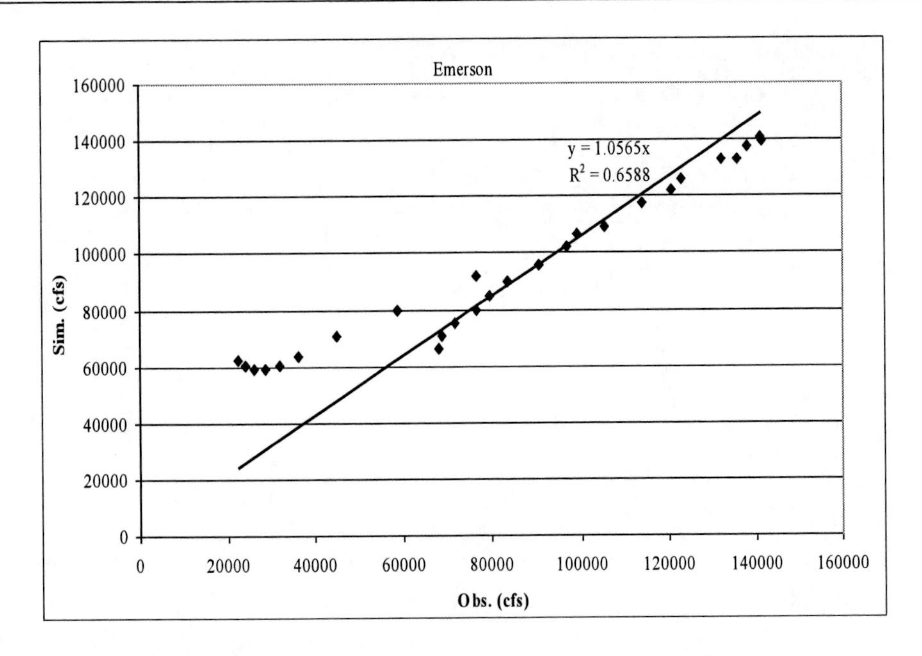

Figure 7-38. Plot showing the simulated versus observed daily discharges at Emerson.

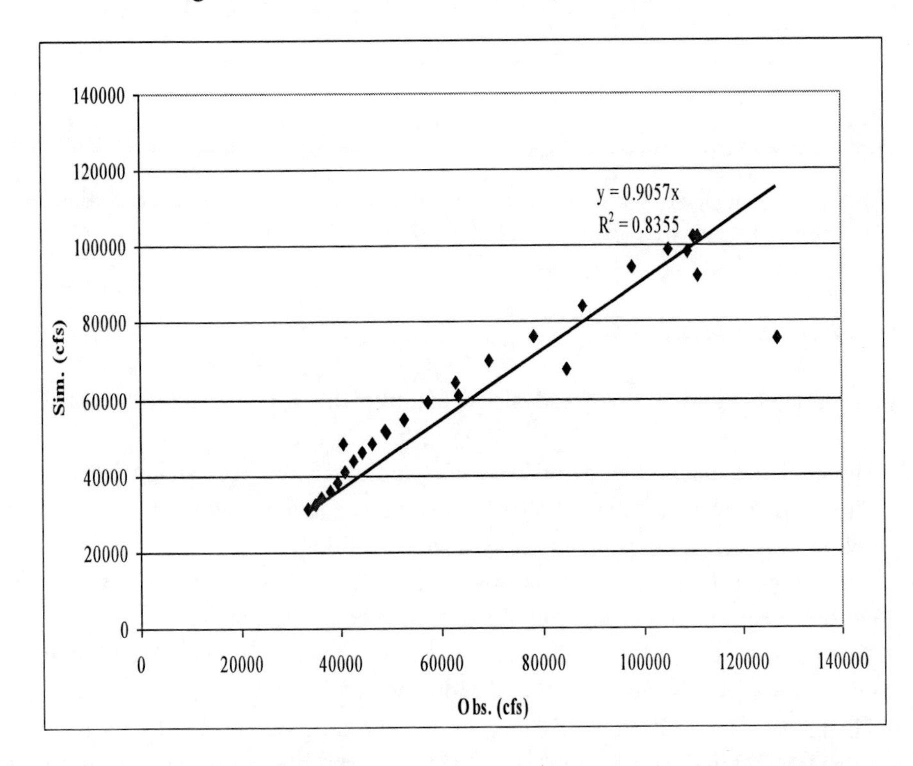

Figure 7-39. Plot showing the simulated versus observed daily discharges at Grand Forks. Note that this station was not used for model calibration. The results are shown for information purposes only.

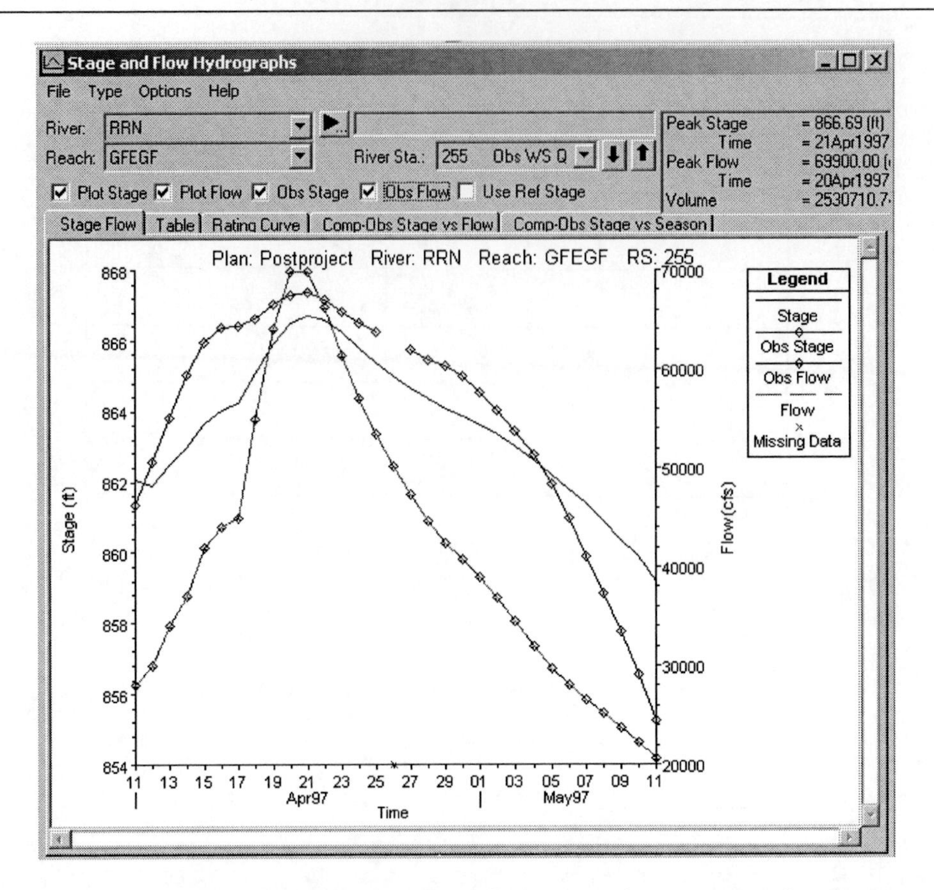

Figure 7-40. Plot showing the observed and predicted discharges and water surface elevations at Halstad. The observed stages were obtained from USACE.

Application Considerations

To appropriately apply ACE-M and EERC-M, following aspects should be considered:

- The models included the major tributaries and overflows occurred in 1997.
- The models were sufficiently calibrated in accordance with the 1997 flood, but neither was validated using any other historical flood.
- The models used flows simulated by the SWAT models, minimizing the uncertainty caused by speculated, inaccurate flows generated in the ungaged areas.
- The models did not consider ice jamming because no data were available for describing this specific hydraulic phenomenon.
- The models used the geometric data for the current ground truth. In particular, for the subreach within the Grand Forks/East Grand Forks city limits, the current topography and geomorphology have been dramatically changed from that in 1997.

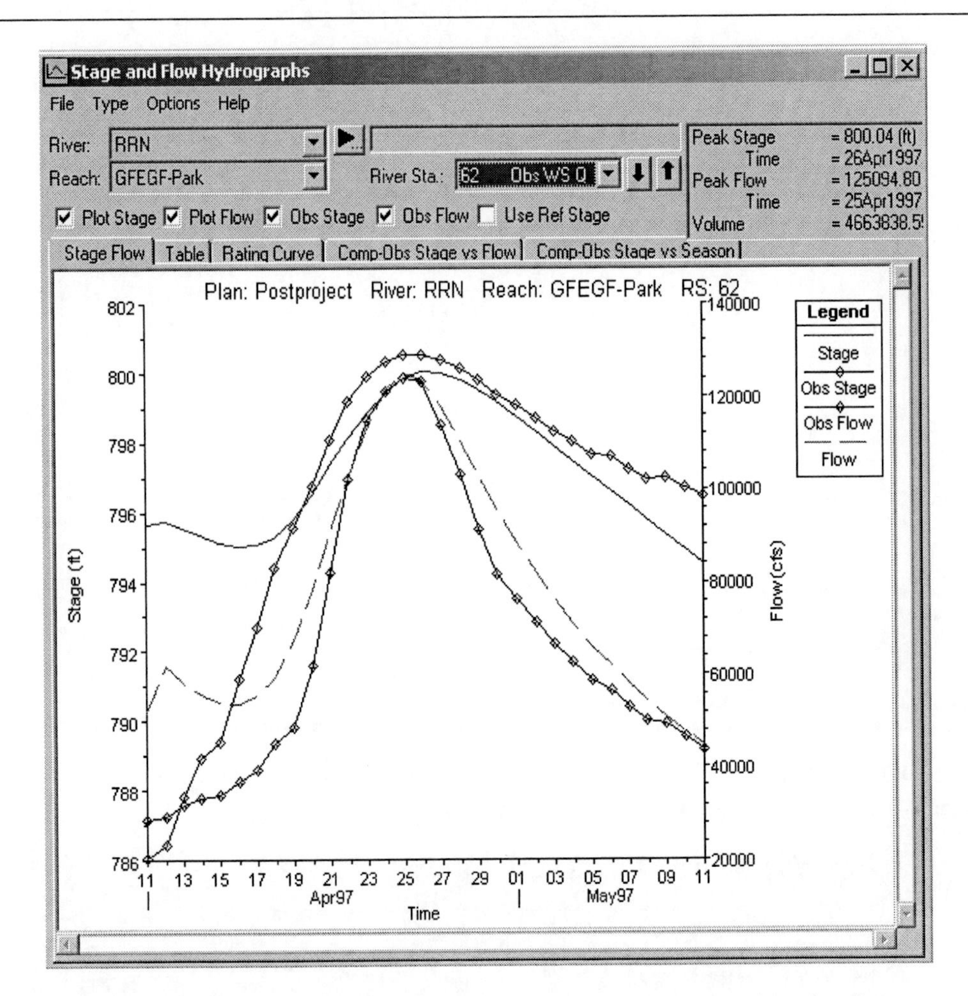

Figure 7-41. Plot showing the observed and predicted discharges and water surface elevations at Drayton. The observed stages were obtained from USACE.

- The models were set up to only simulate the spring flood. ACE-M has a simulation time window from March 25 to May 25, and EERC-M has a time window from April 10 to May 10. An earlier start date might leave more days for the models to converge. However, due to limited resources, the effort to use an earlier start date (e.g., March 10) was unsuccessful. With this regard, special tactics (e.g., pilot channels) have to be used to handle the frozen conditions of the northern streams, which tend to make the models divergent.
- When observed flows are available, the data can be used to substitute for the corresponding SWAT simulated values, which were used in these models. Also, the observed and SWAT simulated flows can be conjunctively used to conduct various scenario or "what-if" analyses.

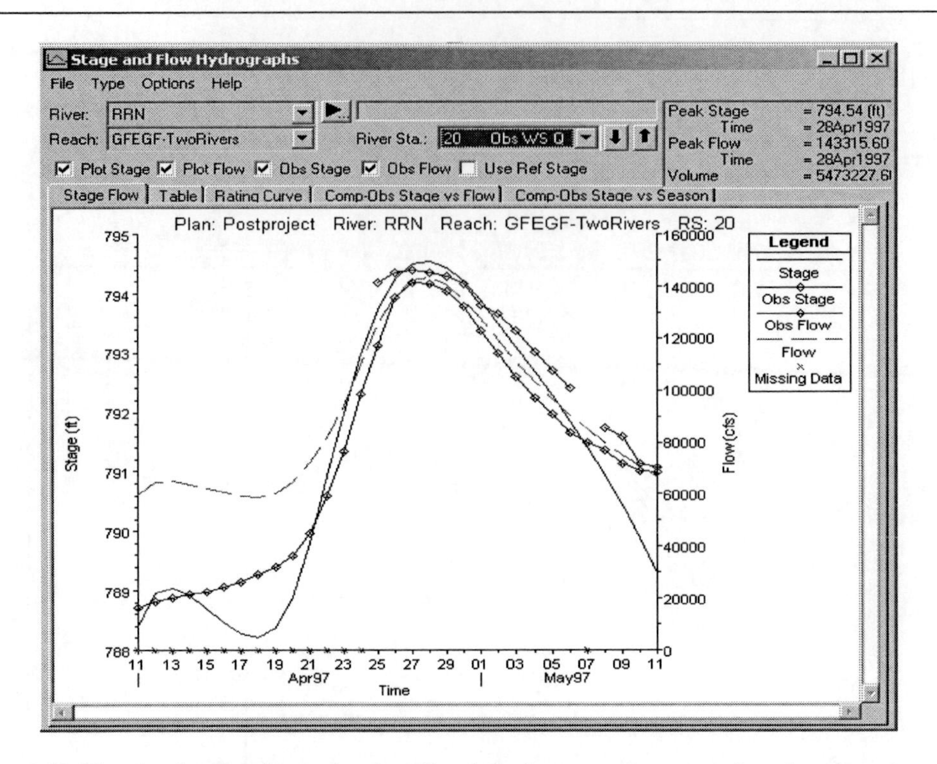

Figure 7-42. Plot showing the observed and predicted discharges and water surface elevations at Pembina. The observed stages were obtained from USGS.

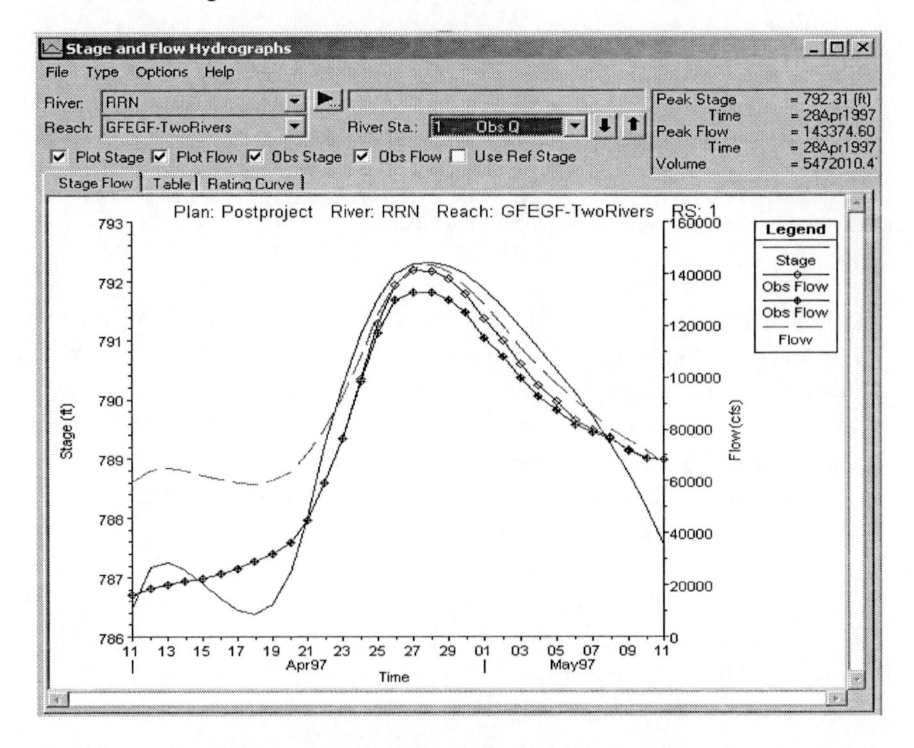

Figure 7-43. Plot showing the observed and predicted discharges and water surface elevations at Emerson. The observed stages were obtained from USACE.

MODELED FLOOD CREST REDUCTIONS ALONG THE MAINSTEM

Reductions at Control Locations

Along the mainstem, nine locations, namely Wahpeton, Hickson, Fargo, Halstad, Grand Forks, Olso, Drayton, Pembina, and Emerson, were selected to examine effects of Waffle® on reducing the 1997-type flood. These locations correspond to the USGS gauging stations (Figures 7-34 and 7-35). The observed daily stream flows were obtained from USGS and Manitoba Water Stewardship, whereas, the observed water surface elevations were obtained from USACE.

As a result of S-I, the 1997 flood crests would be lowered by 1.0 to 5.42 ft along the reach upstream of Pembina and by 0.85 ft at Emerson (Table 7-28). The crest at Wahpeton would be lowered by 5.42 ft and the crests at Fargo and Grand Forks would be reduced by 3.46 and 1.89 ft, respectively. Compared with that for S-I, the flood crests for S-II and S-III were predicted to be only 0.06 to 0.45 ft higher. This indicates that 50% of the identified Waffle® stoages would have a measurable effect on reducing the flood crests along the RR mainstem. S-III would reduce the flood crests at Wahpeton, Fargo, and Grand Forks by 5.12, 1.26, and 1.56 ft, respectively. At Emerson, the flood crest would be lowered by 0.72 ft. The predicted flow and water surface elevation hydrographs for the pre-Waffle® condition, S-I, S-II, and S-III at the nine locations are shown in Figures 7-46 to 7-52.

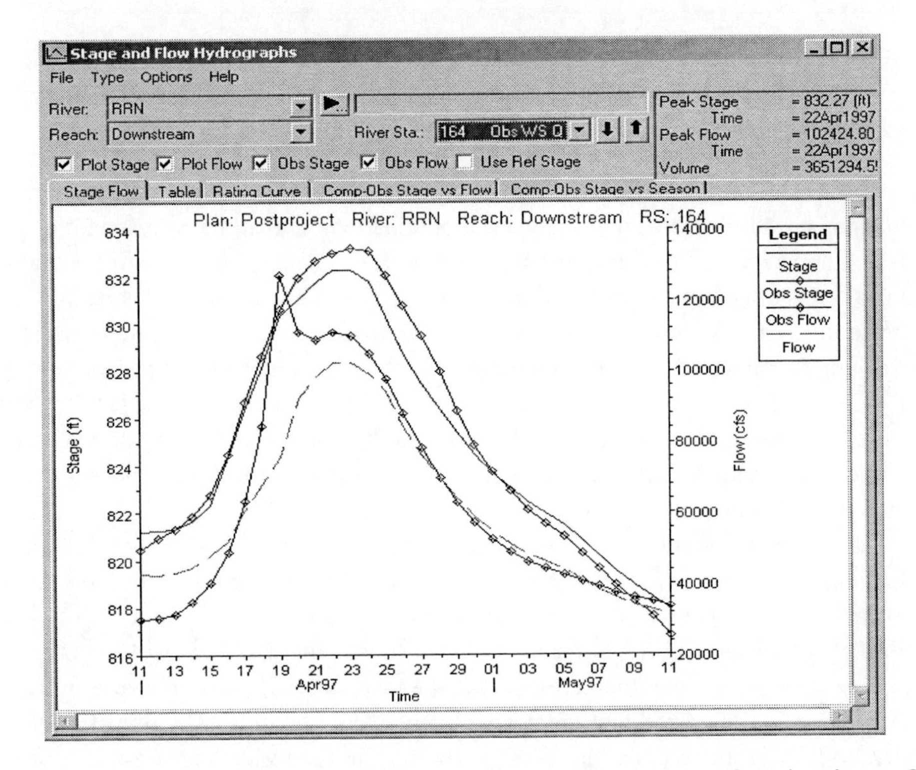

Figure 7-44. Plot showing the observed and predicted discharges and water surface elevations at Grand Forks. Note that this station was not used for model calibration. The results are shown for information purposes only. The observed stages were obtained from USACE.

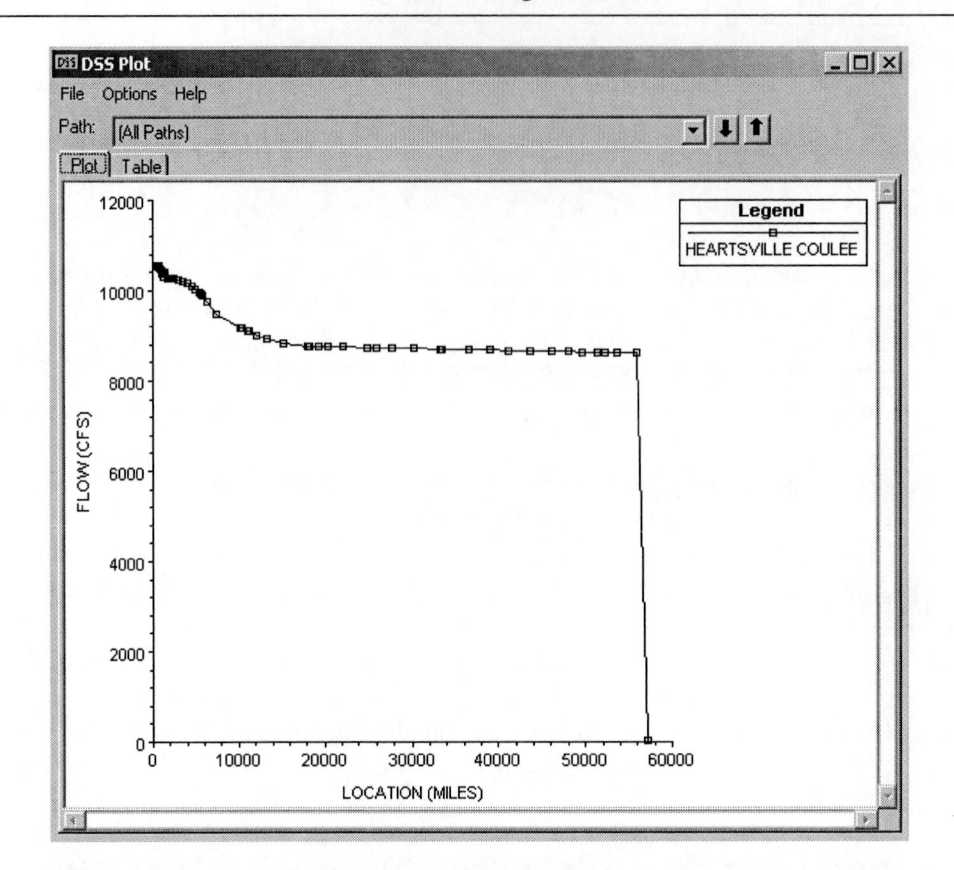

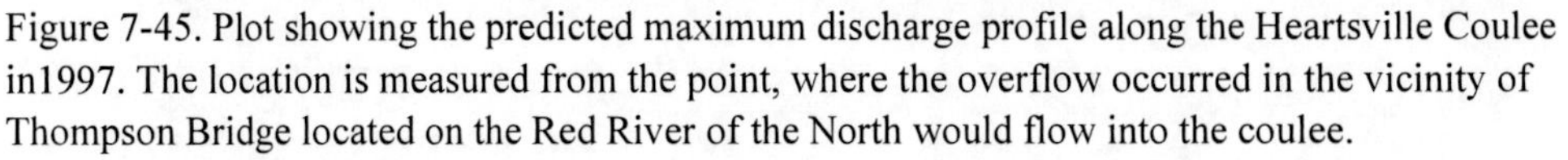
Figure 7-45. Plot showing the predicted maximum discharge profile along the Heartsville Coulee in1997. The location is measured from the point, where the overflow occurred in the vicinity of Thompson Bridge located on the Red River of the North would flow into the coulee.

To further investigate effects of the Waffle®, for each of the three scenarios, two combinations were formulated and analyzed. Combination I is that the corresponding scenario would be adopted for the watersheds upstream of Halstad but the downstream watersheds would use zero storage. In contrast, Combination II is that the watersheds upstream of Halstad would use zero storage but the downstream watersheds would adopt the corresponding scenario. For example, for S-I, Combination I is that the watersheds upstream of Halstad would use the corresponding identified Waffle® storages but the downstream watersheds would use zero storage, whereas, Combination II is that the watersheds upstream of Halstad would use zero storage but the downstream watersheds would use the corresponding identified Waffle® storages.

In Table 7-28, the results for Combination I are presented in the box brackets or [ ] and for Combination II in the parentheses or ( ). Overall, the predicted flood crests for Combination I were higher than the corresponding values for Combination II, implying that the Waffle® storages in the watersheds downstream of Halstad would have certain contributions to reducing the flood crests along the reach from Halstad to Emerson. However, a close examination indicated that the contributions would be only as high as 0.15 ft for S-I and 0.09 ft for S-III. In contrast, the Waffle® storages in the watersheds upstream of Halstad would be more important for reducing the flood crests along the entire RR mainstem.

Table 7-28. Predicted reductions of the 1997 flood crests along the Red River of the North mainstem

Station	Cross Section No.	Datum (ft)	Maximum Water Surface Elevation (ft)			
			Pre-Waffle®	Scenario I (S-I)	Scenario II (S-II)	Scenario III (S-III)
Wahpeton (USGS 05051500)	XS 548.595	942.97	962.07	961.79	961.84	961.84
Hickson (USGS 05051522)	XS 485.041	877.06	914.70	909.28	909.44	909.58
Fargo (USGS 05054000)	XS 452.92	861.80	901.36	897.90	898.06	898.18
Halstad (USGS 05064500)	XS 375.247	826.65	867.31	865.93	866.00	866.05
Grand Forks (USGS 05082500)	XS 163	779.00	831.99	830.10	830.25	830.43
				[830.70]	[830.76]	[830.81]
				(831.51)	(831.54)	(831.67)
Olso (USGS 05083500)	XS 107	772.65	810.95	809.92	810.45	810.53
				[810.25]	[810.65]	[810.67]
				(810.17)	(810.59)	(810.64)
Drayton (USGS 05092000)	XS 68	755.00	800.54	799.53	799.87	799.98
				[800.18]	[800.21]	[800.24]
				(800.10)	(800.15)	(800.21)
Pembina (USGS 05102490)	XS 16	739.45	794.39	793.29	793.35	793.44
				[793.97]	[793.99]	[794.01]
				(793.82)	(793.85)	(793.92)
Emerson (USGS 05102500)	XS 1	700.00	792.32	791.47	791.53	791.60
				[792.03]	[792.05]	[792.07]
				(791.91)	(791.94)	(791.99)

Note: The numbers in [ ] are for the combinations that the corresponding scenarios would be adopted for the watersheds upstream of Halstad but would not be adopted for the downstream watersheds. On the other hand, the numbers in ( ) are for the combinations that the corresponding scenarios would not be adopted for the watersheds upstream of Halstad but would be adopted for the downstream watersheds.

Water Surface Profiles

Figures 7-53 and 7-54 show the water surface profiles for the 1997-type flood along the RR mainstem reaches from Emerson to Halstad and from Halstad to White Rock Dam, respectively. In Figure 7-3, the profiles for the pre-Waffle® condition, S-I, and Combination II are drawn, and in Figure 7-54, the profiles for the pre-Waffle® condition, S-I, S-II, and S-

III are plotted. The profiles indicate that the Waffle® would have more effects on reducing the flood crests along the mainstem from just downsream of Grand Forks to Halstad and from approximately 18 mi downstram of Fargo to about 5 mi upstream of the Richard County Road 28 near Abercrombie. In addition, the Waffle® storages in the watersheds downstream of Halstad would have certain contributions to lowering the flood crests along the reach from Emerson to Halstad, but the contributions would be minor compared with that of the Waffle® storages in the upstream watersheds. Further, because of the small scale, the predicted differences among the three scenarios (i.e., S-I, S-II, and S-III) are hardly differentiable in Figure 7-54. For this reason, Figure 7-53 does not show the predicted profiles for S-II and S-III. The differences at nine control points, which are also USGS gauging stations that are located in the limits of the major cities or towns along the mainstem, can be identified in Table 7-28.

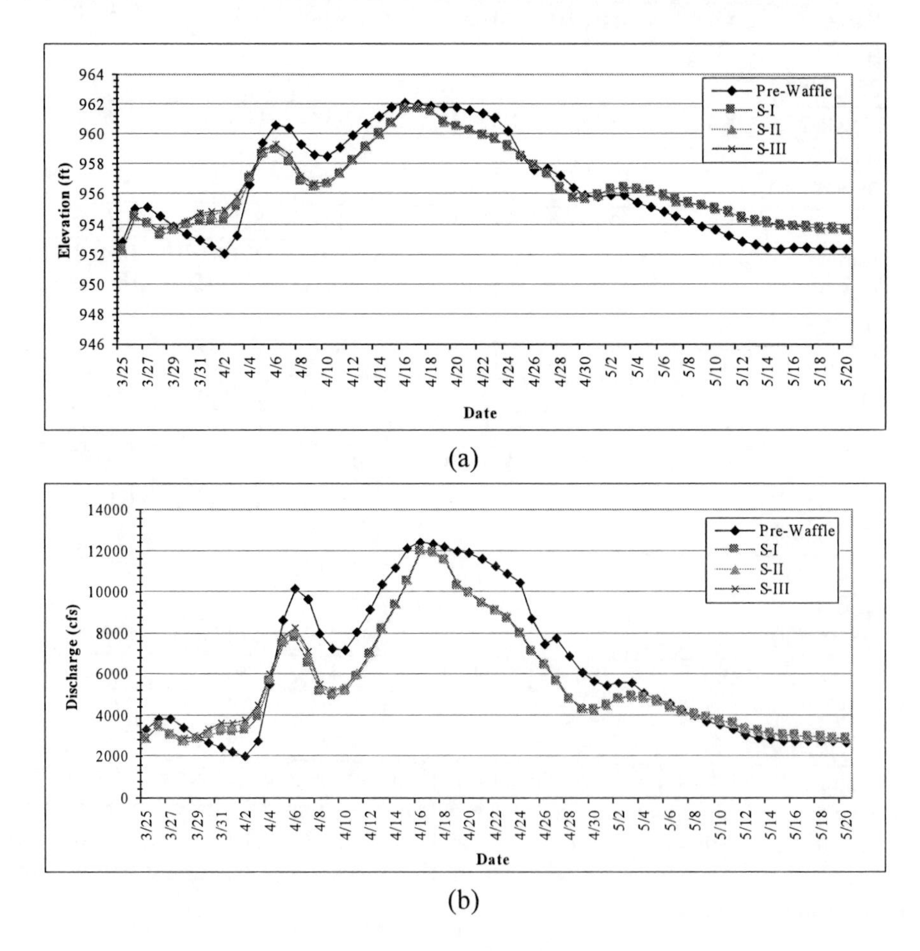

Figure 7-46. Predicted reductions of the 1997 flood (a) crest and (b) discaharge at Wahpeton.

CONCLUSIONS AND RECOMMENDATIONS

This study evaluated effects of the Waffle® on flood reduction in the Red River of the North Basin using coupled SWAT and HEC-RAS hydrodynamic models. The SWAT models were set up for 31 modeling domains, of which 17 are located in Minnesota and the other 14 in North Dakota. A modeling domain was defined in terms of the USGS 8-digit HUCs but redelineated to be completely located either in Minnesota or North Dakota, that is, a HUC that covers the two states was spitted into two modeling domains. The available data on observed daily stream flows for the 1997 flood were used to clibrate the SWAT models. When the data were unavailable, the models were verified based on scientific judgment and/or peak discharges obtained from various sources (e.g., consulting companies). In addition, the Minnesota SWAT models and one North Dakota model were validated using the other historical floods occurred in 1979, 1978, 1975, 1969, and 1966. The other 13 North Dakota SWAT models were not validated due to limited time and resources. The calibration and/or validation indicated that the SWAT models are accurate enough for evaluating effects of the Waffle®. Further, both ACE-M, the HEC-RAS model for the mainstem reach from White Rock Dam to Halstad, and EERC-M, the HEC-RAS model for the reach from Halstad to Emerson, were calibrated in accordance with the 1997 flood. However, neither of these models was validated using any other historical flood.

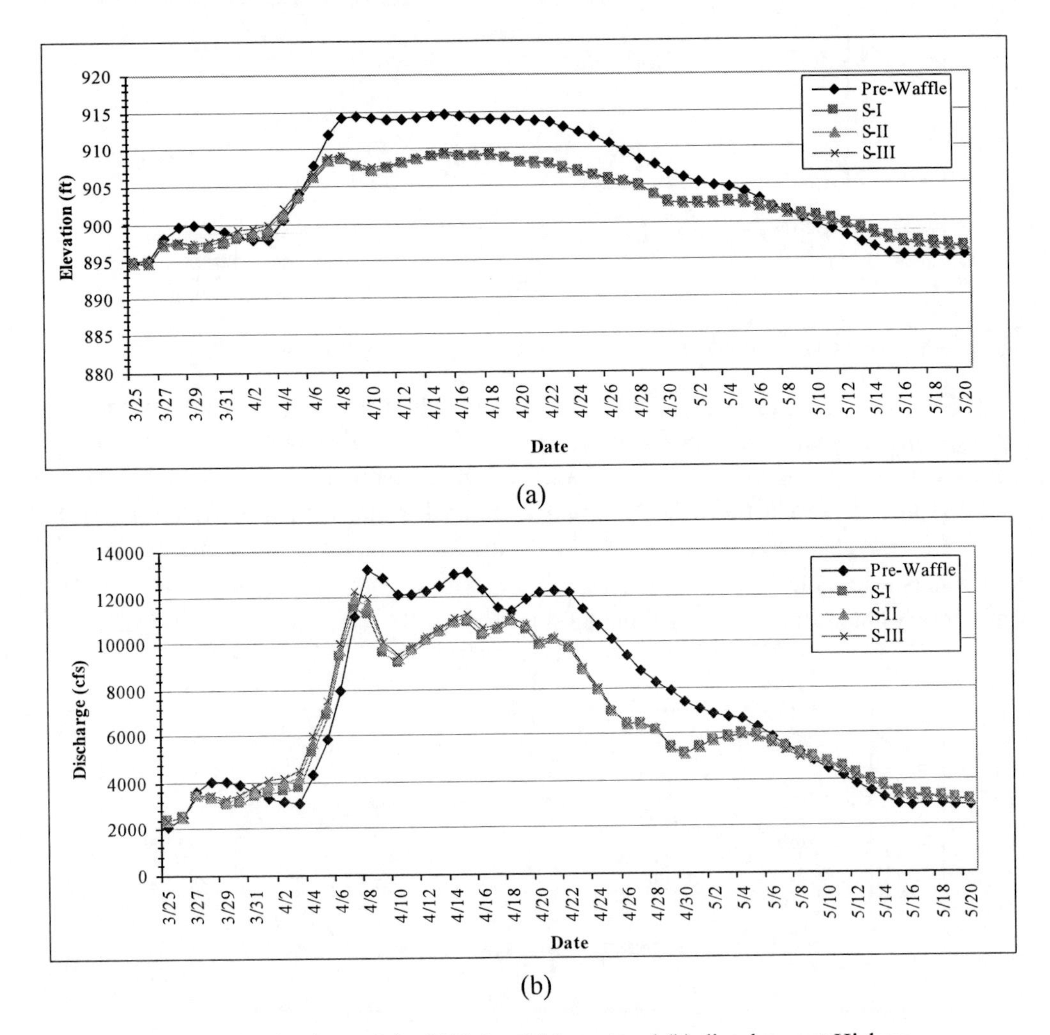

Figure 7-47. Predicted reductions of the 1997 flood (a) crest and (b) discacharge at Hickson.

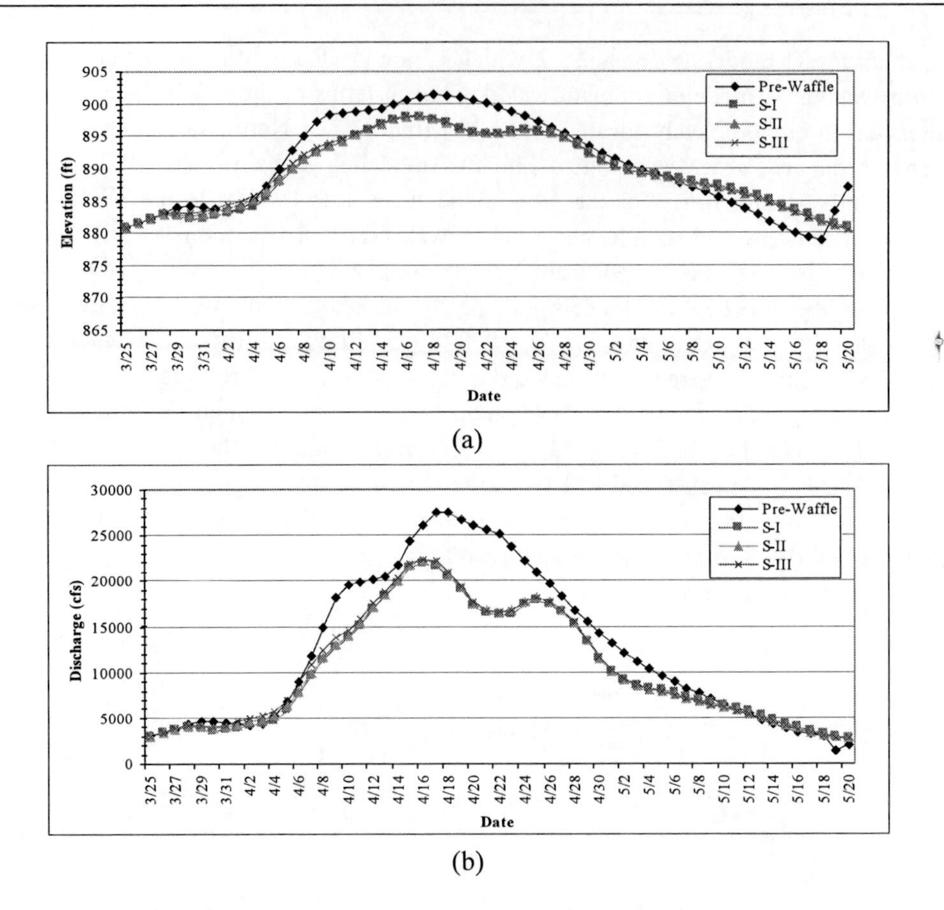

(a)

(b)

Figure 7-48. Predicted reductions of the 1997 flood (a) crest and (b) discaharge at Fargo.

The evaluation indicated that the Waffle® would reduce flooding within the watersheds as well as along the mainstem. For some watersheds, the Waffle® would reduce the 1997 peak discahres by as high as 59.2%, whereas, the percentage reductions for the other watersheds would be low. The reduction effects would be related to the factors of watershed size and shape and flood characteristics (i.e., magnitude and hydrograph shape). Moreover, the effects for a watershed would be also dependent on the Waffle® storage volume of, and the spatial distribution of the storage areas within, the watershed.

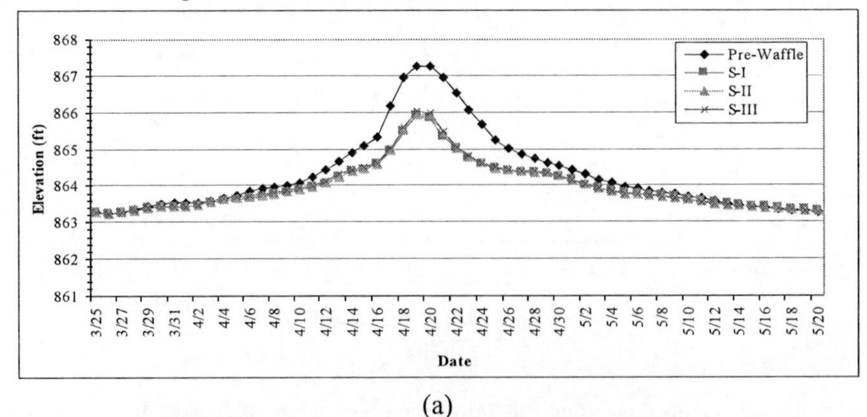

(a)

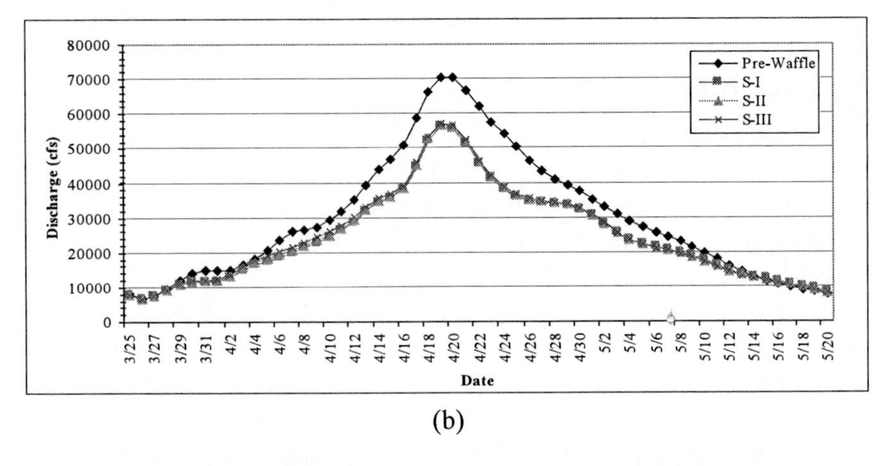

(b)

Figure 7-49. Predicted reductions of the 1997 flood (a) crest and (b) discaharge at Halstad.

The Waffle® would lower the 1997 flood crests by 1.0 to 5.42 ft along the mainstem reach upstream of Pembina and by 0.85 ft at Emerson. The Waffle® would have more effects on reducing the flood crests along the mainstem from just downstream of Grand Forks to Halstad and from approximately 18 mi downstream of Fargo to about 5 mi upstream of the Richard County Road 28 near Abercrombie. In addition, the Waffle® storages in the watersheds upstream of Halstad would be more important for reducing the flood crests along the entire mainstem than that in the downstream watersheds.

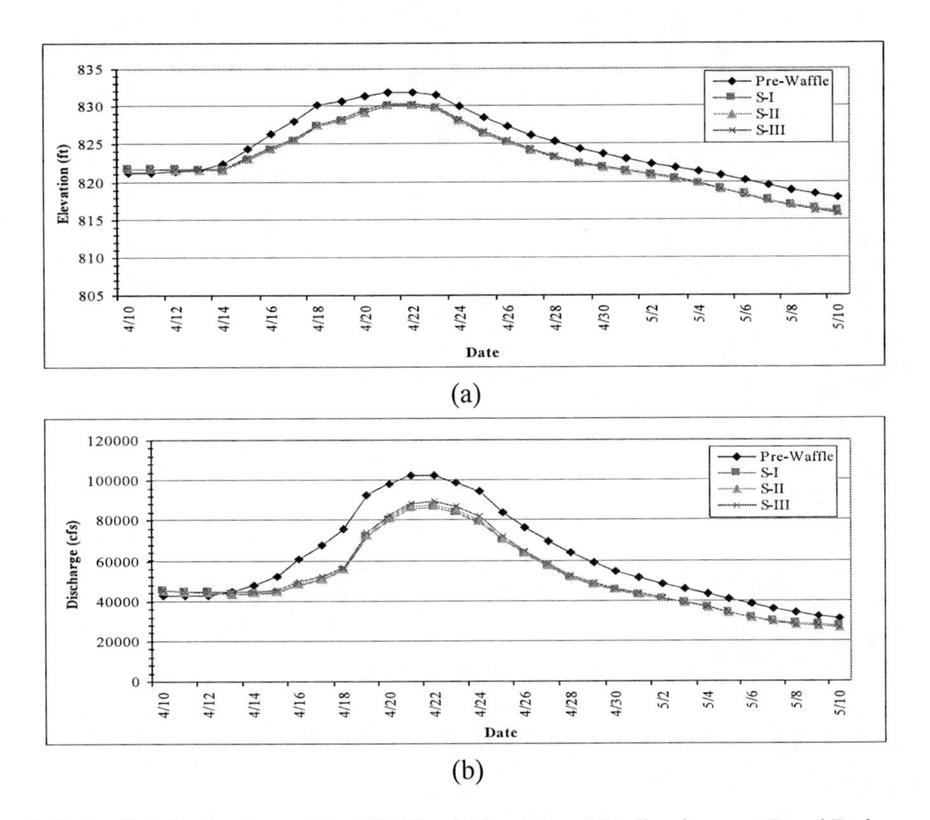

(b)

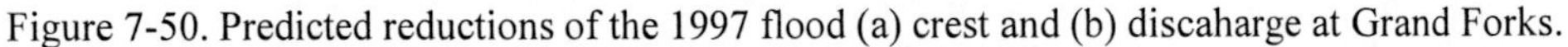

Figure 7-50. Predicted reductions of the 1997 flood (a) crest and (b) discaharge at Grand Forks.

Herein, we make the following recommendations for future research efforts:

- The North Dakota SWAT models should be validated using other historical flood events.
- The HEC-RAS models should be validated using other historical flood events. In addition, an earlier start date (e.g., March 10) may need to be used.
- An interface should be developed to automate the data transfer from the SWAT models to the HEC-RAS models.
- More Waffle® scenarios (e.g., 25% of the identified storages) and combinations should be analyzed to identify a set of optimal or cost-effective options that would use possibly few sections for storage but can still achieve required flood reductions.

The models should be expanded and enhanced for studies, as water quality and BMPs.

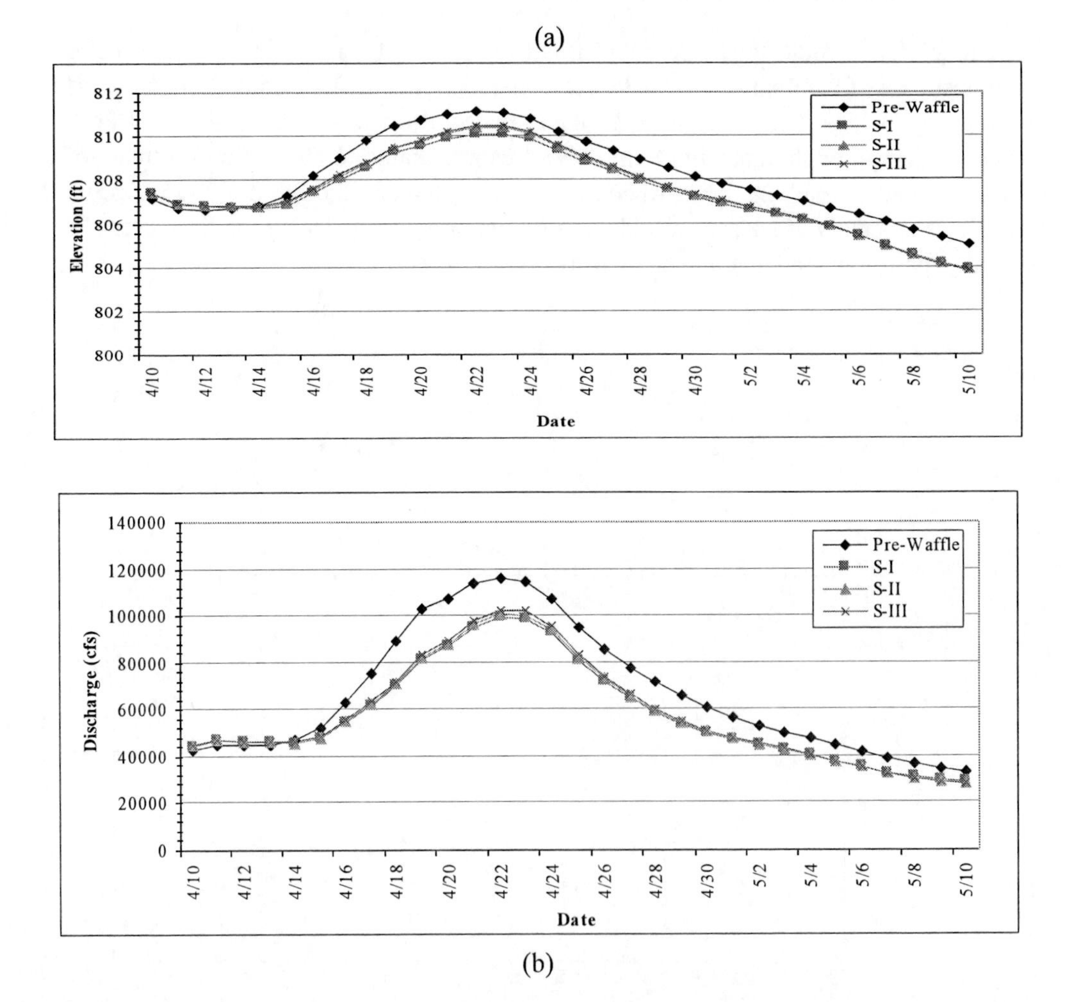

Figure 7-50. Predicted reductions of the 1997 flood (a) crest and (b) discaharge at Olso.

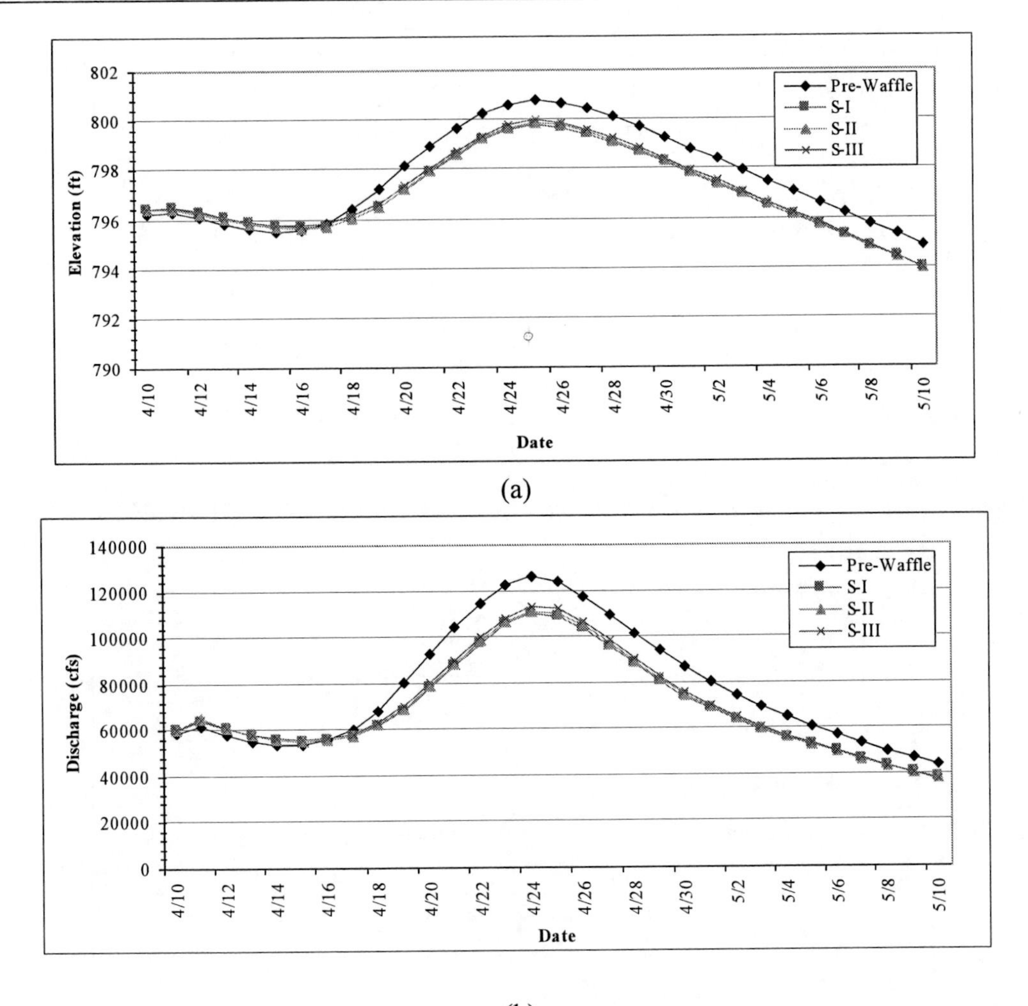

(a)

(b)

Figure 7-51. Predicted reductions of the 1997 flood (a) crest and (b) discaharge at Drayton.

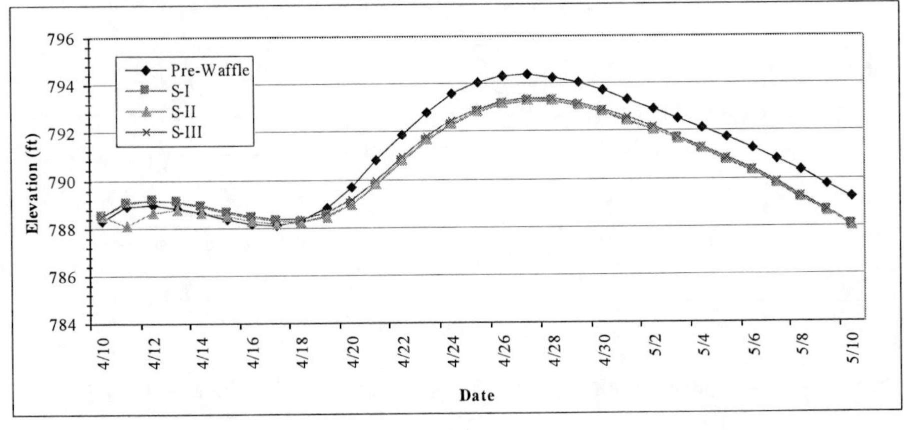

(a)

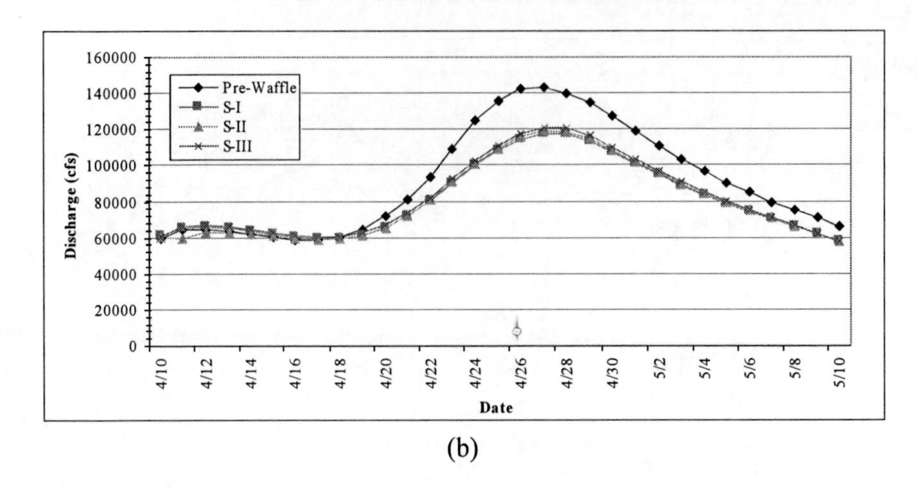

(b)

Figure 7-51. Predicted reductions of the 1997 flood (a) crest and (b) discaharge at Pembina.

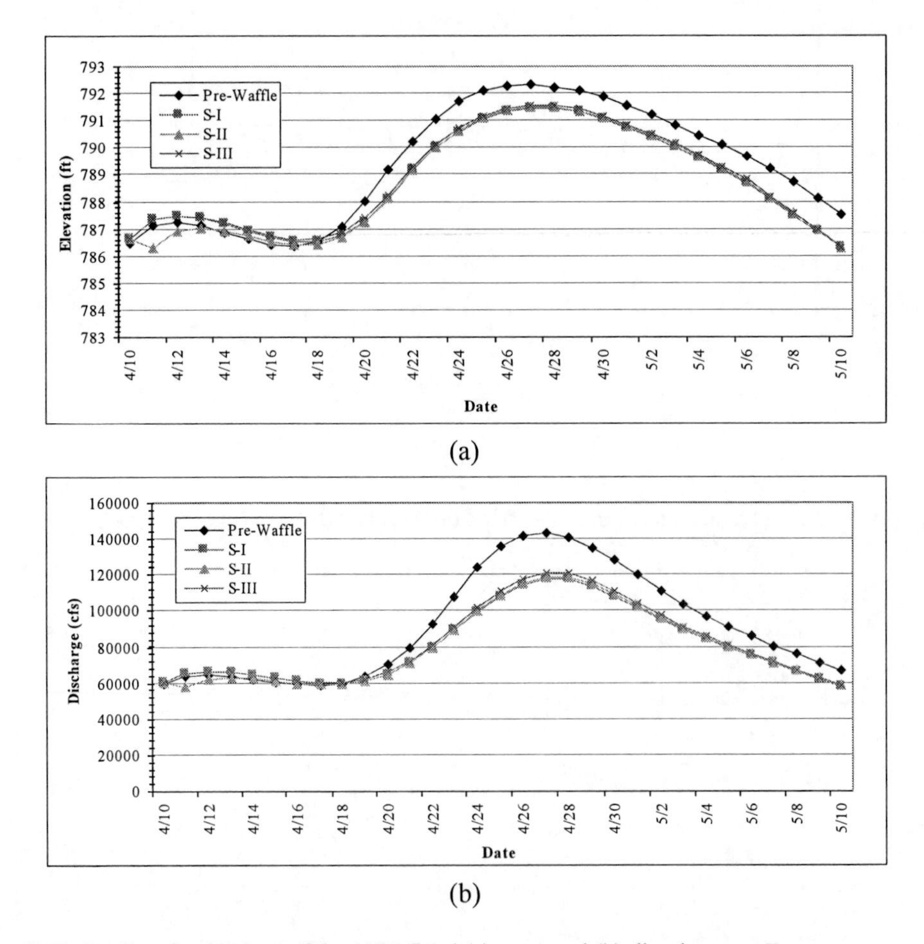

Figure 7-52. Predicted reductions of the 1997 flood (a) crest and (b) discaharge at Emerson.

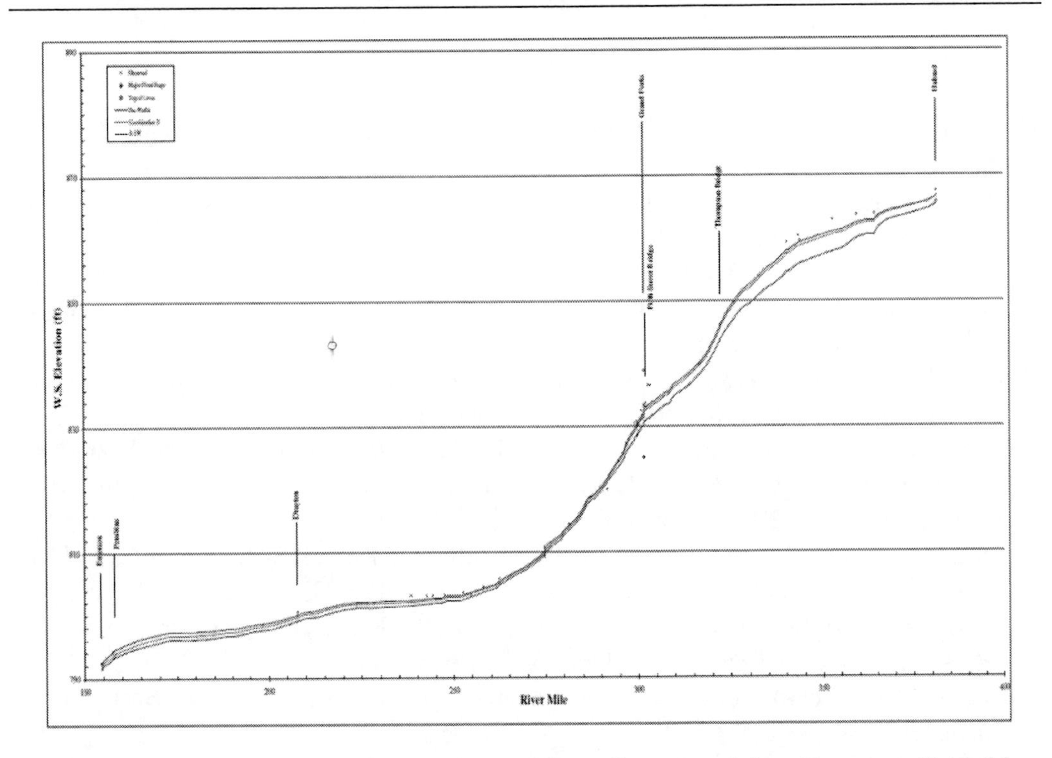

Figure 7-53. Predicted water surface elevations along the mainstem reach from Emerson to Halstad for the 1997-type flood. Combination II is that the watersheds upstream of Halstad would use zero storage but the downstream watersheds would adopt Scenario I (S-I), which corresponds to the 100% identified Waffle® storages.

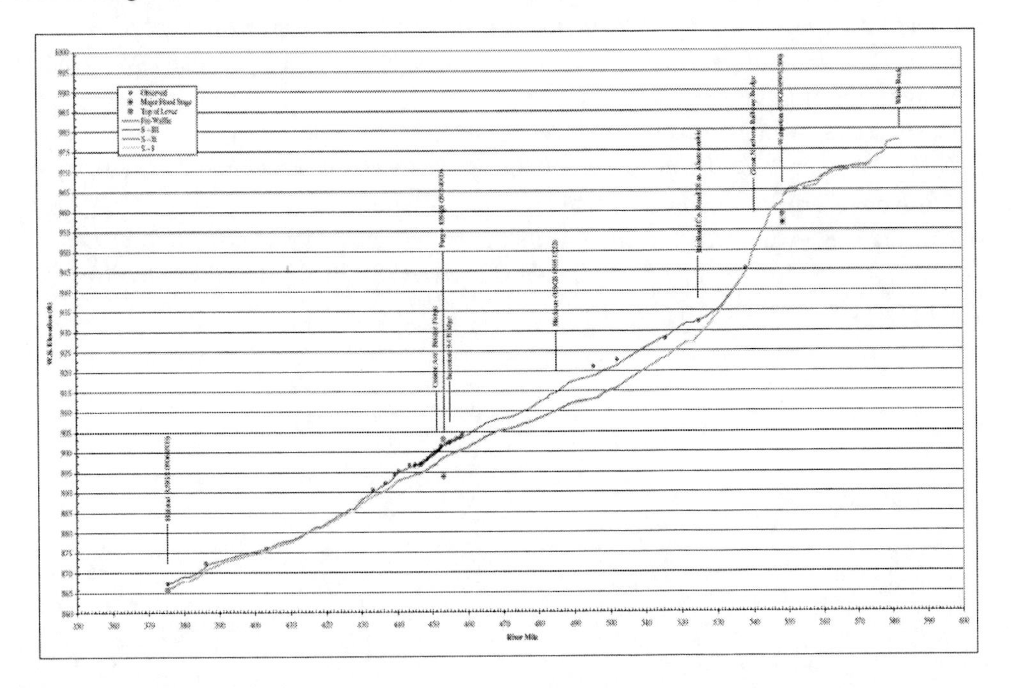

Figure 7-54. Predicted water surface elevations along the mainstem reach from Halstad to just downstream of White Rock Dam for the 1997-type flood.

BIBLIOGRAPHY

Dyhouse, G.R. 1995. Myths and Misconceptions of the 1993 Flood. *Available at* http://www.mvs.usace.army.mil/dinfo/pa/fl93info.htm, Accessed on November 1, 2009.

International Joint Commission. 1997. *Red River Flooding—Short-Term Measures*. Interim Report of the International Red River Basin Task Force to the International Joint Commission, Ottawa – Washington, 65 p.

International Joint Commission. 2000. *Living with the Red*. A Report to the Governments of Canada and the United States on Reducing Flood Impacts in the Red River Basin, 82 p.

Kingery, L.R.S., Frank, W.B., Luther, M. 1999. *Flood Peak Management Using Landscape Storage in the Pembina River Basin*. Unpublished paper.

LeFever, J.A., Bluemle, J.P., and Waldkirch, R.P. 1999. *Flooding in the Grand Forks–East Grand Forks – North Dakota and Minnesota Area*. Bismark, North Dakota: North Dakota Geological Survey, Educational Series No.25, 1999, 63 p.

Neitsch, S.L., Arnold, J.G., Kiniry, J.R., Williams, J.R., and King, K.W. 2002. *Soil and Water Assessment Tool: Theoretical Documentation* (Version 2000). Temple, TX: Grassland, Soil and Water Research Laboratory of Agricultural Research Service, and Blackland Ressearch Center of Texas Agricultural Experiment Station.

Red River Basin Board. 2000. *Drainage*. Moorhead, Minnesota: Red River Basin Board, Inventory Team Report.

Stoner, J.D., Lorenz, D.L., Wiche, G.J., and Goldstein, R.M. 1993. Red River of the North Basin, Minnesota, North Dakota, and South Dakota. *Water Resources Bulletin* 29 (4): 575–615.

USACE (the U.S. Army Corps of Engineers) and FEMA (the Federal Emergency Management Agency). 2003. *Regional Red River Flood Assessment Report: Wahpeton, North Dakota/Breckenridge,* Minnesota to Emerson, Manitoba. St. Paul, MN: the U.S. Army Corps of Engineers.

USACE (the U.S. Army Corps of Engineers). 2001. *Hydrologic Analyses: The Red River of the North Main Stem Wahpeton/Breckenridge to Emerson, Manitoba.* St. Paul, MN: the U.S. Army Corps of Engineers, Final Hydrology Report.

USACE (the U.S. Army Corps of Engineers). 2005. *Fargo-Moorhead Upstream Feasibility Study Phase I: Hydrology and Hysraulics Analysis.* St. Paul, MN: the U.S. Army Corps of Engineers, Final Report.

INDEX

A

B

C

D

E

F

G

H

Q

R

S